INTERMEDIATE ALGEBRA

INTERMEDIATE ALGEBRA

JOSEPH NEWMYER
Fullerton Junior College

GUS KLENTOS
Fullerton Junior College

CHARLES E. MERRILL PUBLISHING COMPANY
A Bell & Howell Company
Columbus, Ohio

Standard Book Number 0–675–09352–X

Library of Congress Catalog Card Number: 74–106504

2 3 4 5 6 7 8 9 10 — 74 73 21 71

PRINTED IN THE UNITED STATES OF AMERICA

to

Maria and Jean

To the Student;

1. The Audio Tutorial Approach

You are about to embark on a new adventure in learning intermediate algebra, namely the audio-tutorial method. This method utilizes a book designed to correlate with audio tapes. You will read through a unit and at the same time listen to an explanation of the problems and theory. Also, throughout each unit will be study exercises for you to work, in order to periodically check your progress.

This technique for the teaching of intermediate algebra was developed, tested, and used at Fullerton Junior College, Fullerton, California. It has been received with widespread student approval since its beginning, two years ago. In general, students have indicated that this technique greatly improved study habits, resulting in a better understanding of mathematics with a commensurate higher grade.

In the study of mathematics, most students are accustomed to the traditional lecture-textbook method where they read a certain section in the text, go to class to hear a lecture on the material, then are left to try to work the exercises. The frustrations of this type of situation are many. First, most students have difficulty reading a mathematics textbook; second, once the classroom lecture is over, the chalk board is erased and the explanatory lecture is lost forever; and third, many students have great difficulty working the homework assignment five or six hours after the lecture.

The audio-tutorial method of studying mathematics is an attempt to remedy the defects of the traditional lecture-textbook method. In the audio-tutorial approach, the lecture and other explanations are put on audio tape. The chalkboard illustrations are put in the text. Therefore, each student has a permanent record of the material generally presented in the class. The advantages of this system are numerous. The student may go through the lecture at his own rate and any part of the lecture may be reviewed as often as desired by simply reversing the recorder and turning back a few pages in the text.

2. Format of the Text

There are 29 units contained in the text. Each unit has been divided into the following parts:

(a) Frames
(b) Study Exercises
(c) Review Exercises
(d) Solutions to Review Exercises
(e) Solutions to Study Exercises
(f) Supplementary Problems

(a) Frames. Each unit consists of 30 to 50 frames. Each frame has a circled "frame number" located at the lower right hand corner. The frames are to be studied as you listen to the commentary on the audio tape. Remember to watch the circled "frame numbers", as this is how to keep the tape synchronized with the text.

(b) Study Exercises. Some of the frames are devoted to study exercises. When you come to a Study Exercise, turn off the recorder. The exercises should be worked and your answers checked before you turn on the recorder and proceed to the next frame. If you do poorly on a study exercise, you should repeat that part of the unit.

(c) Review Exercises. When you come to the end of the frames, turn off the recorder and do the review exercises. Check your answers with the ones immediately following. The review exercises are extremely important, in that they will measure your understanding of the entire unit. If you do poorly on the review exercises, you should go back and review the entire unit.

(d) Solutions to Review Exercises. The solutions to the review exercises immediately follow the review exercises. They are, for the most part, detailed solutions. Be sure to compare your work carefully with the details of these solutions. Many of your mistakes can be understood and corrected in this section.

(e) Solutions to Study Exercises. Immediately following the solutions to review exercises are the detailed solutions to the study exercises which were located in the frames. Before you begin each unit with the recorder, you will want to locate and mark this section for ready reference. Remember, as you proceed through the unit that you will be asked to work study exercises and to check your results with this section.

(f) Supplementary Problems. At the end of each unit you will find a set of supplementary problems. There are no answers furnished for these problems in the text. Your instructor may wish to use these for homework problems or for quizzes. However, you are encouraged to try as many of these as you wish. Remember, in mathematics, practice is the key to success.

3. Study Techniques

(a) The recorder may be stopped at any time to give more time for analysis of a frame.
(b) The recorder may be reversed to review a particular portion of a unit.
(c) Before an examination, you may go back and review all of the units involved and carefully examine the Review Exercises.
(d) Before the final examination, you should do the set of Comprehensive Review Exercises given on page 363–366.
(e) Mathematics pyramids. That is, it builds on itself. Don't get behind. In fact, it is a good idea to try and stay a unit ahead.
(f) Many students have indicated that they had tremendous success by doing a "constant review", that is, constantly going back and redoing earlier units.

The authors wish to express their appreciation to the students and the Mathematics Department of Fullerton Junior College. Without their cooperation, an undertaking of this magnitude would have been impossible.

We are particularly appreciative of the outstanding effort put forth by Jean Newmyer in typing the manuscript and her help in editing.

February, 1970

Gus Klentos
Joseph Newmyer

Table of Contents

Unit 1	INTRODUCTION TO SETS	1
	Solutions to Study Exercises	9
	Supplementary Problems	10
Unit 2	GRAPHS OF THE NUMBER LINE	12
	Solutions to Study Exercises	22
	Supplementary Problems	25
Unit 3	REVIEW OF FIELD PROPERTIES—PART I	27
	Solutions to Study Exercises	34
	Supplementary Problems	35
Unit 4	REVIEW OF FIELD PROPERTIES—PART II	37
	Solutions to Study Exercises	43
	Supplementary Problems	44
Unit 5	FACTORING	45
	Solutions to Study Exercises	51
	Supplementary Problems	53
Unit 6	FRACTIONS	54
	Solutions to Study Exercises	65
	Supplementary Problems	68
Unit 7	EXPONENTS	71
	Solutions to Study Exercises	82
	Supplementary Problems	84
Unit 8	RADICALS	85
	Solutions to Study Exercises	96
	Supplementary Problems	100
Unit 9	SOLUTION SETS OF LINEAR EQUATIONS	103
	Solutions to Study Exercises	111
	Supplementary Problems	113
Unit 10	SOLUTION SETS OF QUADRATIC EQUATIONS	115
	Solutions to Study Exercises	127
	Supplementary Problems	131
Unit 11	RELATIONS AND FUNCTIONS	133
	Solutions to Study Exercises	143
	Supplementary Problems	144

Unit 12	THE LINEAR FUNCTIONS	146
	Solutions to Study Exercises	158
	Supplementary Problems	162
Unit 13	THE QUADRATIC FUNCTION	164
	Solutions to Study Exercises	172
	Supplementary Problems	175
Unit 14	CONIC SECTIONS	177
	Solutions to Study Exercises	188
	Supplementary Problems	191
Unit 15	VARIATION	192
	Solutions to Study Exercises	195
	Supplementary Problems	197
Unit 16	INVERSE FUNCTIONS	199
	Solutions to Study Exercises	206
	Supplementary Problems	207
Unit 17	EXPONENTIAL FUNCTIONS	209
	Solutions to Study Exercises	215
	Supplementary Problems	216
Unit 18	PROPERTIES OF LOGARITHMS	217
	Solutions to Study Exercises	224
	Supplementary Problems	226
Unit 19	COMPUTATION WITH LOGARITHMS—PART I	227
	Solutions to Study Exercises	237
	Supplementary Problems	239
Unit 20	COMPUTATION WITH LOGARITHMS—PART II	240
	Solutions to Study Exercises	244
	Supplementary Problems	246
Unit 21	MORE ON SOLUTION SETS OF EQUATIONS AND INEQUALITIES	247
	Solutions to Study Exercises	262
	Supplementary Problems	267
Unit 22	DETERMINANTS AND SYSTEMS OF LINEAR EQUATIONS	270
	Solutions to Study Exercises	282
	Supplementary Problems	285
Unit 23	SYSTEMS OF EQUATIONS	287
	Solutions to Study Exercises	295
	Supplementary Problems	299
Unit 24	THE BINOMIAL THEOREM	301
	Solutions to Study Exercises	309
	Supplementary Problems	310
Unit 25	SEQUENCES AND SERIES	311
	Solutions to Study Exercises	321
	Supplementary Problems	324

Unit 26 THE COMPLEX NUMBER SYSTEM 325
Solutions to Study Exercises 334
Supplementary Problems 335

Unit 27 PERMUTATIONS 337
Solutions to Study Exercises 345
Supplementary Problems 346

Unit 28 COMBINATIONS 347
Solutions to Study Exercises 353
Supplementary Problems 355

Unit 29 INTRODUCTION TO PROBABILITY 356
Solutions to Study Exercises 361
Supplementary Problems 362

Comprehensive Review 365
Solutions to Comprehensive Review 367

unit

Introduction to Sets

1

Objectives

1. Know the various subsets of the real number system.
2. Know the basic operations on sets such as union and intersection.

①

Terms

set	empty set	subset	element
variable	constant	intersection	union
disjoint	one-to-one correspondence	finite	infinite

②

Set: A well defined collection of objects.
Elements: The individual components of a set.

An element of a set is also referred to as a *member* of a set.

③

Study Exercise One

For each of the following state whether there is or is not determined a well defined set.

1. All whole numbers between 5 and 10.
2. All states of the United States that are islands.
3. The ten best makes of cars.
4. The first three presidents of the United States.
5. The five most important people in Los Angeles.

④

How Sets Are Indicated

A set is referred to by listing the names of its elements within braces or by describing it.

Symbol For A Set

$\{\quad\}$

The symbol, $\{\quad\}$, is read, "the set of."

⑤

Example of Sets

Express in set notation:

(1) The set of all the days in the week whose names begin with "T."

Answer: {Tuesday, Thursday}

(2) The set of all months which have exactly 30 days.

Answer: {September, April, June, November}

(Frame 6, contd.)

(3) The set of all odd numbers between 1 and 20 which are exactly divisible by 3.

Answer: $\{3, 9, 15\}$

(6)

Further Examples of Sets

Express in set notation:

(1) The set of all men weighing 5 tons.

Answer: $\{\quad\}$; this set has no elements

(2) The set of all women presidents of the United States.

Answer: $\{\quad\}$

(3) The set of all living persons over 800 years old.

Answer: $\{\quad\}$

(7)

Empty Set

The set with no elements is called the *empty set* or *null set*.

The notation for the empty set is either $\{\quad\}$ or $\varnothing$.

It is incorrect to write $\{\varnothing\}$ for the empty set.

(8)

Uppercase letters will be used to represent or name the set.
Lowercase letters will be used to represent elements of a set.

Example: $T = \{a, b, x, y\}$

The set is T; the elements of the set are a, b, x, and y.

(9)

The Greek letter epsilon, $\in$, will be used to represent membership in a set.

$\in$ is read, belongs to, is a member of, or is an element of.
$\notin$ is read, does not belong to, or is not an element of.

Example: $T = \{a, b, x\}$
Set T has 3 elements:
$a \in T, b \in T, x \in T$
$d \notin T, \{a\} \notin T, o \notin T$

(10)

A **variable** is a symbol representing an unspecified element of a set containing more than one member. For example, if $A = \{2, 3, 4, 5\}$ and $x \in A$, then x is a variable.

If $B = \{6\}$ and $y \in B$, then y is not a variable since y is always 6 and y is said to be a *constant*.

(11)

Equal Sets

Two sets are equal if and only if they have the same elements.

If $A = \{1, 3\}$ and $B = \{1, 3\}$ then $A = B$.

If $C = \{1, 2, 3\}$ and $D = \{3, 1, 2\}$ then $C = D$.

If two sets E and F are not equal, we write $E \neq F$.

(12)

A set does not change if its elements are rearranged in a different order.

A set does not change if some or all of its elements are repeated.

$$\{7, 2, 0\} = \{2, 0, 7\}; \quad \{3, 1, 1, 1, 5, 5\} = \{3, 1, 5\}$$

(13)

Study Exercise Two

A. Determine if, $A = B$ or $A \neq B$

1. $A = \{1, 2, 3\}$, $B = \{2, 3, 1\}$
2. $A = \{1, 2, 4\}$, $B = \{4, 2, 1, 4\}$
3. $A = \{0\}$, $B = \varnothing$

B. True or False

4. $3 = \{3\}$
5. $\varnothing = 0$
6. $\{1\} \in \{1, 2, 5\}$
7. $\{0\} = \{\{0\}\}$
8. $5 \in \{5\}$

(14)

Subsets

Set A is a *subset* of set B if and only if every element of set A is an element of set B. The notation indicating subsets is $A \subseteq B$.

If $A = \{1, 5, 7\}$, $B = \{1, 2, 5, 7\}$ then $A \subseteq B$.
If $C = \{6, 8\}$, $D = \{6, 8\}$ then $C \subseteq D$.

(15)

Proper Subsets

Set A is a *proper subset* of set B if and only if set A is first a subset of set B and set B contains at least one element that is not in set A. The notation for this is $A \subset B$.

If $A = \{3, 5\}$, $B = \{1, 3, 5, 7\}$ then $A \subset B$.

(16)

Study Exercise Three

True or False

1. $\{1\} \subseteq \{1\}$
2. $\{1\} \subset \{1\}$
3. $\{3,5,6\} \subseteq \{3,5,6,7,8,9\}$
4. $\{8, 9\} \subset \{8, 9, 0\}$
5. $\{5, 6\} \subset \{5, 6, 7\}$
6. $\{5, 6\} \subseteq \{5, 6, 7\}$

(17)

Theorem: If A is a set, then $\varnothing \subseteq A$

Proof: To prove $\varnothing \subseteq A$, it must be shown that there are no elements of $\varnothing$ that are not elements of A. Since $\varnothing$ contains no elements, it follows that $\varnothing \subseteq A$.

(18)

The Subsets of $\{1, 2, 3\}$

The set $\{1, 2, 3\}$ has the following subsets:

$\{1\}$; $\{2\}$; $\{3\}$; $\{1, 2\}$; $\{1, 3\}$; $\{2, 3\}$; $\{1, 2, 3\}$; $\varnothing$.

(19)

Study Exercise Four

True or False

1. $\{2, 3\} \subset \{1, 2, 3\}$
2. $\varnothing \subseteq \{1, 3\}$
3. $\varnothing \subset \{1, 3\}$
4. $\{1, 2, 3\} \subseteq \{1, 2, 3\}$
5. $\{1\} \subset \{1\}$
6. $\varnothing \subset \{0\}$

(20)

Specifying Sets

Two methods by which a set may be specified are the *roster method* and the *set builder method.*

The roster method simply lists each element.

The set builder method gives a condition that enables us to determine whether a particular object is a member of the set.

(21)

The Set Builder Notation

$\{\square \mid \square$ meets the condition that...$\}$
The vertical line is read "such that."

Examples:
(1) $\{x \mid x \cdot x = 4\}$
(2) $\{y \mid y$ is an instructor at UCLA$\}$
(3) $\{* \mid *$ is a whole number between 1 and 2$\}$
(4) $\{x \mid x = 2 \cdot n$ and $n = 1, 2, 3\} = \{2, 4, 6\}$

(22)

Special Sets of Numbers

Natural Numbers: The set of Natural Numbers, denoted by N, is $N = \{1, 2, 3, 4, \ldots\}$.

Integers: The set of Integers, denoted by I, consists of the Natural Numbers, zero, and the negatives of the Natural Numbers.

$$I = \{0, 1\ -1, 2, -2, 3, -3, \ldots\}.$$

(23)

Primes: The set of primes, denoted by P, consists of all positive integers n, not equal to one, so that each n has only two exact positive integral divisors, itself and one.

3 is a prime since its only divisors are 3 and 1.
6 is not prime since it has more than two divisors.

(24)

Rationals: The set of Rationals, denoted by F, is the set of all numbers that may be expressed as a fraction in the form $\frac{a}{b}$ where $a, b \in I$ and $b \neq 0$. $F = \left\{\frac{a}{b} \mid a, b \in I, b \neq 0\right\}$

Irrationals: The set of Irrationals is the set of numbers with decimal representations that are nonterminating, nonrepeating numerals. For example, $\sqrt{2}$, $\sqrt{3}$, and π are irrational.

Reals: The set of Reals, denoted by R is the set of numbers made up of all rationals and irrationals.

(25)

The Real Numbers

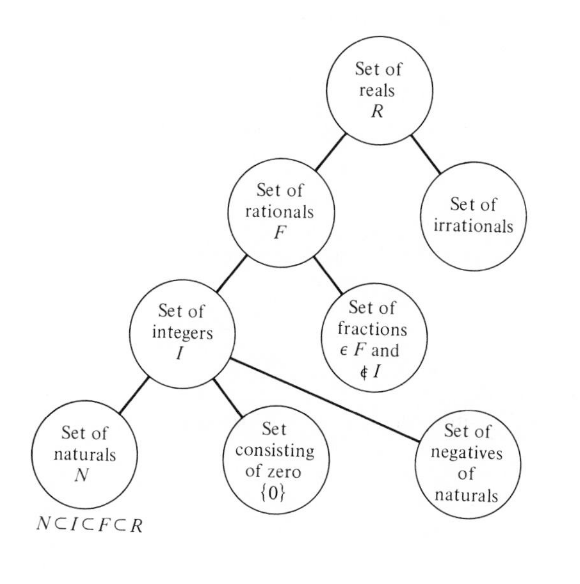

(26)

Examples:

Classify:	**Natural**	**Integer**	**Rational**	**Irrational**
5	X	X	X	
-3		X	X	
$\sqrt{5}$				X
0		X	X	
12.3			X	
$\frac{10}{2}$	X	X	X	
1/3			X	
$1 - \frac{11}{8}$			X	
.014			X	
$1 + \sqrt{2}$				X

(27)

Decimals

Any rational number may be written as a decimal by applying long division.

$$\frac{1}{4} = 0.25 \qquad \frac{1}{3} = 0.33\overline{3}$$

$$\frac{5}{11} = 0.45\overline{45} \qquad 3\frac{1}{5} = \frac{16}{5} = 3.2$$

These decimals either terminate or repeat. Repeating decimal means that either one digit or a block of digits repeat.

(28)

Examples:

Number:	Decimal form	Terminating?	Repeating?
$\frac{5}{9}$	$0.55\overline{5}$	No	Yes, 5 repeats
$\frac{69}{11}$	$6.27\overline{27}$	No	Yes, 27 repeats
$\frac{4}{5}$	0.8	Yes	No
$\frac{1}{7}$	$0.142857\overline{142857}$	No	Yes, 142857 repeats

(29)

The set of Rational Numbers is the set of numbers which can be represented as either terminating or repeating decimals.

Every terminating decimal represents a rational number.

Every repeating decimal represents a rational number.

(30)

Non terminating, non repeating decimals are irrational numbers. The decimal 0.101001000100001000001... represents an irrational number.

The number π is also irrational since $\pi = 3.1415926536\ldots$

Note that $3^1/_7$, 3.14, 3.1416 are close approximations of π and that $\pi \neq 3^1/_7$, $\pi \neq 3.14$, and $\pi \neq 3.1416$.

(31)

$$\sqrt{2} = 1.414214\ldots$$
$$\sqrt{3} = 1.732051\ldots$$
$$\sqrt{5} = 2.236068\ldots$$

$\sqrt{2}$, $\sqrt{3}$, $\sqrt{5}$, are also irrational.

Numbers expressible as decimals will either terminate, repeat, or will not repeat. If they terminate or repeat, they are rational. If they do not terminate and do not repeat, they are irrational.

(32)

Even and Odd Integers

Even Integers: The set of Even Integers, denoted by E, is the set of integers which are exactly divisible by two.

Odd Integers: The set of Odd Integers, denoted by O, is the set of integers which are not even.

(33)

$$E = \{\ldots -6, -4, -2, 0, 2, 4, 6, \ldots\}$$
$$O = \{\ldots -5, -3, -1, 1, 3, 5, \ldots\}$$

Zero is an even integer since zero is exactly divisible by 2.

(34)

Study Exercise Five

True or False

1. $\frac{2}{35} \in F$
2. $F \subset I$
3. $27 \in P$
4. $\pi = 3.14$
5. $.13\overline{13} \in F$
6. $I \subset N$

(35)

Intersection of Sets

The *intersection* of two sets A and B, denoted by $A \cap B$, is the set of all elements which belong to both A and B (that is, those elements which A and B have in common).

$$A \cap B = \{x | x \in A \text{ and } x \in B\}$$

(36)

Disjoint Sets

Two sets are said to be *disjoint* if their intersection is the empty set.

If $C = \{1, 2, 3\}$ and $D = \{4, 5\}$, then $C \cap D = \varnothing$.

Therefore, C and D are disjoint sets.

(37)

Union of Sets

The *union* of two sets A and B, denoted by $A \cup B$, is the set of all elements that are in A, or in B, or in both A and B.

$$A \cup B = \{x | x \in A \quad \text{or} \quad x \in B\}$$

The or is used in the inclusive sense. It means this, or that, or both.

(38)

Examples of Union and Intersection

$A = \{1, 2, 3, 5, 7\}$, $B = \{1, 3, 5, 7\}$, $C = \{2, 4, 6, 8\}$

$A \cup B = \{1, 2, 3, 5, 7\}$; $\quad A \cap B = \{1, 3, 5, 7\} = B$;

$A \cup C = \{1, 2, 3, 4, 5, 6, 7, 8\}$; $\quad A \cap C = \{2\}$;

$B \cap C = \varnothing$

(39)

Study Exercise Six

Operate as indicated and write the answer as a single set.

1. $\{a, f\} \cup \{f, g\} =$ ______.
2. $\{1, 6\} \cap [\{2, 7\} \cup \{3, 8\}] =$ ______.
3. $N \cap I =$ ______.
4. $N \cup I =$ ______.

(Frame 40, contd.)

5. $\{x \mid x \in N \text{ and } x \text{ is less than } 4\} \cup \{1, 4\} =$ ______.

6. $F \cap \{\text{all irrational numbers}\} =$ ______.

(40)

One To One Correspondence

Definition.
Two sets A and B are said to be in *one to one correspondence* when there exists a pairing of the elements of A with the elements of B such that each element of A corresponds to one and only one element of B, and each element of B corresponds to one and only one element of A.

(41)

Example:
Establish a one to one correspondence between the two sets $\{1, 2, 3\}$, $\{a, b, c\}$.

Answer: One possible correspondence is

$$\begin{matrix} 1 & 2 & 3 \\ \updownarrow & \updownarrow & \updownarrow \\ a & b & c \end{matrix}$$

Another possibility is

$$\begin{matrix} 1 & 2 & 3 \\ \updownarrow & \updownarrow & \updownarrow \\ c & b & a \end{matrix}$$

(42)

Finite and Infinite

A nonempty set contains n elements if it can be put into a one to one correspondence with the set $\{1, 2, 3, \ldots n\}$ where $n \in N$.

A set is *finite* if it contains n elements or if it is the null set.

A set which is not finite is said to be *infinite*.

(43)

Examples of Finite Sets:

(1) $\{1, 3, 5, 6\}$ is finite since it contains four elements.

(2) $\{x \mid x \in N \text{ and } x \text{ is less than ten}\}$ is finite since it contains nine elements.

(3) $\{y \mid y \text{ is a positive even integer and prime}\}$ is finite since it contains one element.

Examples of Infinite Sets:

(1) $\{x \mid x \text{ is a real number greater than five}\}$.

(2) $\{y \mid y \text{ is an odd integer}\}$.

(3) $\{w \mid w \text{ is a natural number}\}$.

(44)

Study Exercise Seven

The set A is the set of all positive integers greater than two. The set B is the set of all positive integers greater than five.

1. Write A in set notation by establishing a pattern and using three dots.

2. Write B in a similar fashion.

3. Is $A \subseteq B$?

4. Is $B \subseteq A$?

5. Find $A \cap B$.

6. Find $A \cup B$.

(45)

REVIEW EXERCISES

A. True or False

1. $5 \in \{5\}$ **2.** $\{5\} \subset N$ **3.** $5 = \{5\}$

4. $\{1, 2\} = \{2, 1\}$ **5.** $F \subset R$ **6.** $I \subset R$

7. $N \subset I$ **8.** Zero is an integer

9. The set consisting of the grains of sand in the world is an infinite set.

10. The set of even integers and the set of odd integers are disjoint sets.

11. A one to one correspondence is not possible with $\{3, 5, 6\}$ and $\{-1, 3, 4\}$

12. $\pi \neq 3\frac{1}{7}$

B. Which of these sets, {a, t}, {2, 6} {h, a, t}, {all even natural numbers}, are subsets of the following sets?

13. $\{2, 4, 6, 8, 10\}$

14. $\{m, a, t, h\}$

15. $\{t, e, n, a\}$

16. N

C. Given $A = \{3, 5, 6\}$, $B = \{-1, 3, 4\}$

17. Find $A \cup B$

18. Find $A \cap B$

19. List all the subsets of B.

D. 20. Find the decimal form for 3/8

21. Refer to the previous problem. Is the decimal form for 3/8 terminating, repeating, or non repeating?

22. Is $0.11\overline{1}$ a rational number?

23. Is -1 an integer?, a rational?, a real?

(46)

SOLUTIONS TO REVIEW EXERCISES

A. 1. True **2.** True

3. False, since the right side is a set and the left side is not.

4. True **5.** True **6.** True **7.** True **8.** True **9.** False

10. True, since their intersection is empty **11.** False

12. True, since π is not rational

B. 13. $\{2, 6\}$ **14.** $\{a, t\}$, $\{h, a, t\}$ **15.** $\{a, t\}$

16. {2, 6}, {all even natural numbers}

C. 17. $\{-1, 3, 4, 5, 6\}$

18. $\{3\}$

19. $\{-1\}$, $\{3\}$, $\{4\}$, $\{-1, 3\}$, $\{-1, 4\}$, $\{3, 4\}$, $\{-1, 3, 4\}$, and $\varnothing$

D. 20. 0.375, (divide 8 into 3)

21. Terminating

22. Yes, since it is a repeating decimal

23. -1 is an integer, a rational, and a real number

(47)

SOLUTIONS TO STUDY EXERCISES

Study Exercise One (Frame 4)

1. Well defined. **2.** Well defined. **3.** Not well defined.

4. Well defined. **5.** Not well defined.

(4A)

SOLUTIONS TO STUDY EXERCISES, CONTD.

Study Exercise Two (Frame 14)

1. $A = B$ **2.** $A = B$ **3.** $A \neq B$
4. False **5.** False **6.** False
7. False **8.** True

14A

Study Exercise Three (Frame 17)

1. True **2.** False **3.** True
4. True **5.** True **6.** True

17A

Study Exercise Four (Frame 20)

1. True **2.** True **3.** True
4. True **5.** False **6.** True

20A

Study Exercise Five (Frame 35)

1. True **2.** False **3.** False
4. False **5.** True **6.** False

35A

Study Exercise Six (Frame 40)

1. $\{a, f, g\}$ **2.** $\{\quad\}$ or $\varnothing$ **3.** N
4. I **5.** $\{1, 2, 3, 4\}$ **6.** $\{\quad\}$ or $\varnothing$

40A

Study Exercise Seven (Frame 45)

1. $A = \{3, 4, 5, 6, \ldots\}$ **2.** $B = \{6, 7, 8, 9, \ldots\}$ **3.** No
4. Yes **5.** B **6.** A

45A

UNIT 1—SUPPLEMENTARY PROBLEMS

True or False

1. $5 \subseteq 4, 5, 6$ **2.** $\{5\} \in 4, 5, 6$ **3.** $0 \in I$
4. $F \subseteq I$ **5.** $\{3\} \subset \{2, 3\}$ **6.** $\{3\} \subseteq \{2, 3\}$
7. $F \subset R$ **8.** $I \subset R$ **9.** $I \subset F$
10. $N \subset F$ **11.** $N \subset I$

Specify by listing the members:

12. {The odd natural numbers between 4 and 10}
13. {The even natural numbers between 3 and 11}
14. {The positive primes less than 11}
15. Complete the following by one of : $\varnothing, N, I, F, R$

a) $F \cap I =$ **b)** $F \cup I =$
c) $N \cup F =$ **d)** $R \cap \varnothing =$

If $A = \{1, 3, 7\}$ and $B = \{2, 5, 7\}$, find:

16. $B \cup A$ **17.** $B \cup B$
18. $A \cap B$ **19.** $(A \cap B) \cup A$

UNIT 1—SUPPLEMENTARY PROBLEMS, CONTD.

20. List the subsets of $\{1, 5, 9\}$

21. Show that 0.6 is rational by using the definition.

22. What is the decimal representation of $1/32$?

23. Which of the terms — real, rational, irrational, integer — apply to:

a) $.003\overline{003}$ **b)** -1.6 **c)** $4 \cdot 2$

24. Give an example of two disjoint sets.

25. If $A = \{\text{odd integers}\}$, $B = \{\text{even integers}\}$, $C = \{1, 2, 3, 4, 5, 6\}$, and $D = \{-4, -2, 0, 2, 4\}$ find:

a) $A \cap B$ **b)** $(A \cap D) \cup B$ **c)** $C \cup D$

d) $(A \cap B) \cup C$ **e)** $(A \cup C) \cap (B \cup D)$

unit

2

Graphs On The Number Line

Objectives

1. Understand the meaning of the symbols $<$, $\nless$, $\leq$, $>$, $\ngtr$, $\geq$, $a < x < b$, $|x| < a$, and $|x| > a$.
2. Plot points on the real number line.
3. Plot intervals on the real number line.

①

Numbers and The Number Line

(1)

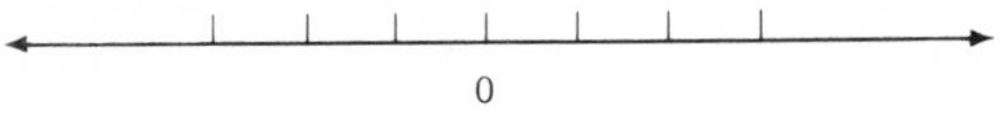

0

(2)

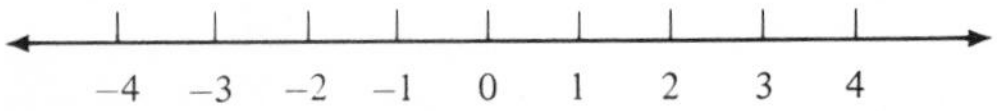

−4 −3 −2 −1 0 1 2 3 4

(3)

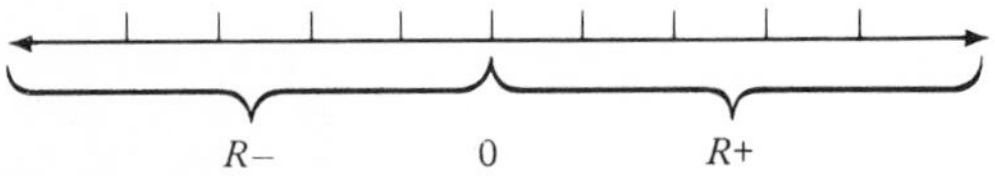

R− 0 R+

②

To each point on the number line, we associate a real number.

To each real number there corresponds a point on the number line.

Thus we have established a *one to one correspondence* between the points on the number line and the real numbers.

③

The Real Number Line

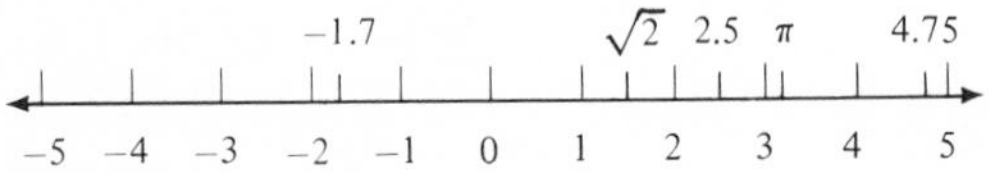

−1.7 $\sqrt{2}$ 2.5 π 4.75

−5 −4 −3 −2 −1 0 1 2 3 4 5

To the right of zero, numbers are positive. To the left of zero, numbers are negative.

④

Graphs of Number Sets

Example 1: Graph $\{1, 3, -2\}$.
Solution:

Example 2: Graph $\{x \mid x$ is less than two and $x \in N\}$.
Solution:

(5)

Example 3: Graph $\{y \mid y \in N\}$.
Solution:

Example 4: Graph $\{1, 3, 4\}, \cup \{-2, 1, 0\}$.
Solution:

(6)

Study Exercise One

Sketch the graph of the following sets on the real number line.

1. $\{-3, -2, 0, 1\}$

2. $\{1, 3, 4\} \cap \{x \mid x \in N\}$

3. $\{x \mid x$ is a positive integer less than eleven and x is a prime$\}$

(7)

Order

If $a, b \in R$ and the graph of a lies to the left of the graph of b, then a is *less than* b. In symbols write $a < b$.

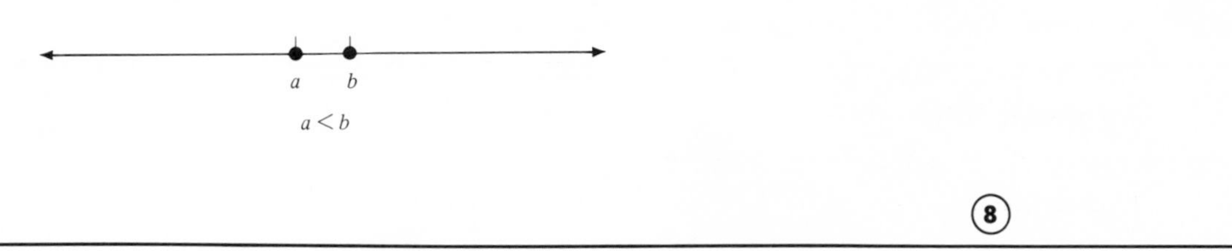

(8)

If $c, d \in R$ and the graph of c lies to the right of the graph of d, then c is *greater than d*. In symbols write $c > d$.

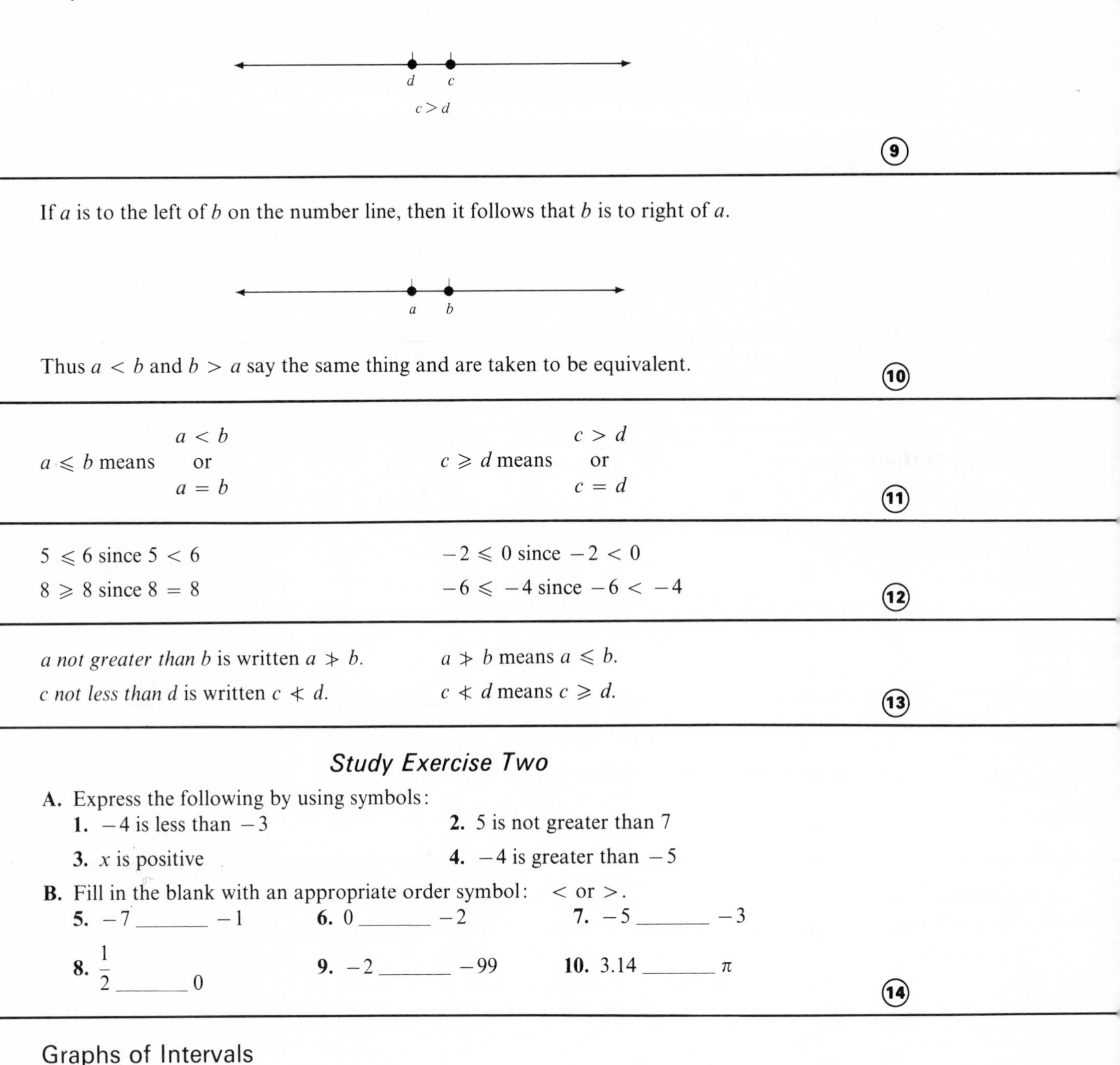

(9)

If a is to the left of b on the number line, then it follows that b is to right of a.

Thus $a < b$ and $b > a$ say the same thing and are taken to be equivalent.

(10)

$a \leqslant b$ means $a < b$ or $a = b$

$c \geqslant d$ means $c > d$ or $c = d$

(11)

$5 \leqslant 6$ since $5 < 6$

$8 \geqslant 8$ since $8 = 8$

$-2 \leqslant 0$ since $-2 < 0$

$-6 \leqslant -4$ since $-6 < -4$

(12)

---

a not greater than b is written $a \ngtr b$.

c not less than d is written $c \nless d$.

$a \ngtr b$ means $a \leqslant b$.

$c \nless d$ means $c \geqslant d$.

(13)

Study Exercise Two

A. Express the following by using symbols:

1. -4 is less than -3

2. 5 is not greater than 7

3. x is positive

4. -4 is greater than -5

B. Fill in the blank with an appropriate order symbol: $<$ or $>$.

5. -7 ______ -1

6. 0 ______ -2

7. -5 ______ -3

8. $\frac{1}{2}$ ______ 0

9. -2 ______ -99

10. 3.14 ______ π

(14)

Graphs of Intervals

Example 1: Graph $\{x \mid x > 2, x \in R\}$

Solution:

−5 −4 −3 −2 −1 0 1 2 3 4

(Frame 15, contd.)

Example 2: Graph $\{x \mid x \leqslant 1, x \in R\}$

Solution:

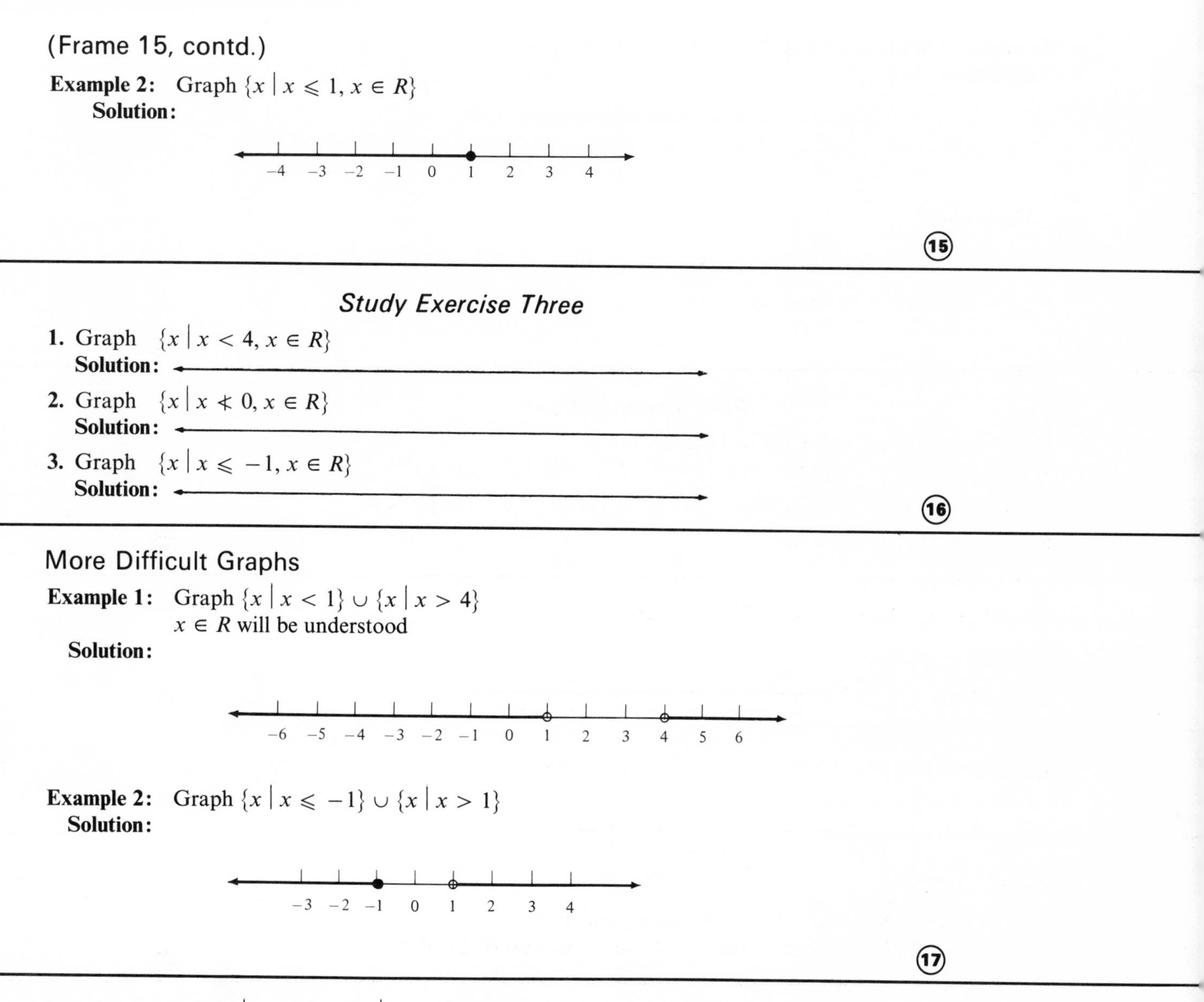

(15)

Study Exercise Three

1. Graph $\{x \mid x < 4, x \in R\}$
   Solution:
2. Graph $\{x \mid x \nless 0, x \in R\}$
   Solution:
3. Graph $\{x \mid x \leqslant -1, x \in R\}$
   Solution:

(16)

More Difficult Graphs

Example 1: Graph $\{x \mid x < 1\} \cup \{x \mid x > 4\}$
$x \in R$ will be understood

Solution:

Example 2: Graph $\{x \mid x \leqslant -1\} \cup \{x \mid x > 1\}$

Solution:

(17)

In order to graph $\{x \mid x > 1\} \cap \{x \mid x < 3\}$, first graph each set on the number line and then select the overlap for the final result.

It is better to show the final result on another number line.

(18)

Example: Graph $\{x \mid x \leqslant 4\} \cap \{x \mid x > 0\}$

Solution:

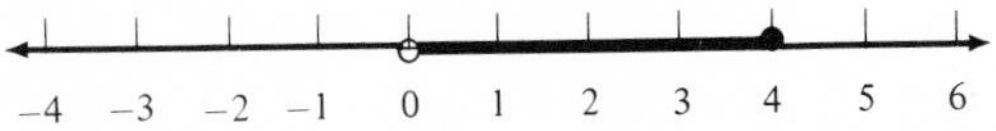

Final result:

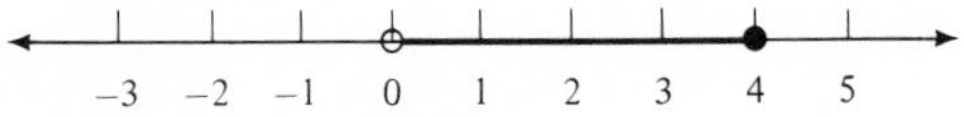

(19)

Study Exercise Four

Graph the following sets on the real number line.

1. $\{x \mid x < 3\} \cup \{x \mid x \geqslant 5\}$

2. $\{x \mid x > 2\} \cup \{x \mid x < -2\}$

3. $\{x \mid x < 3\} \cap \{x \mid x \geqslant 2\}$

4. $\{x \mid x < 0\} \cup \{x \mid x > 0\}$

5. $\{x \mid x \nless 0\} \cap \{x \mid x = 0\}$

(20)

The set $\{x \mid x < 4\} \cap \{x \mid x > 2\}$ has as a graph;

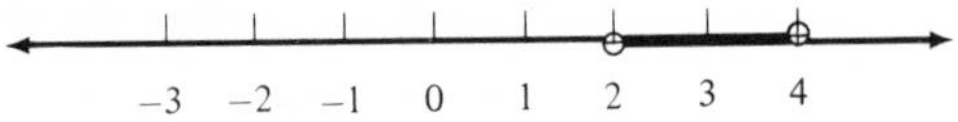

Solution:

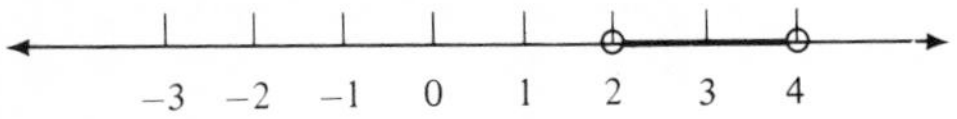

This set can be described as $\{x \mid x \text{ is between 2 and 4}\}$.

$2 < x < 4$ will mean $x > 2$ and $x < 4$ or x is between 2 and 4

(21)

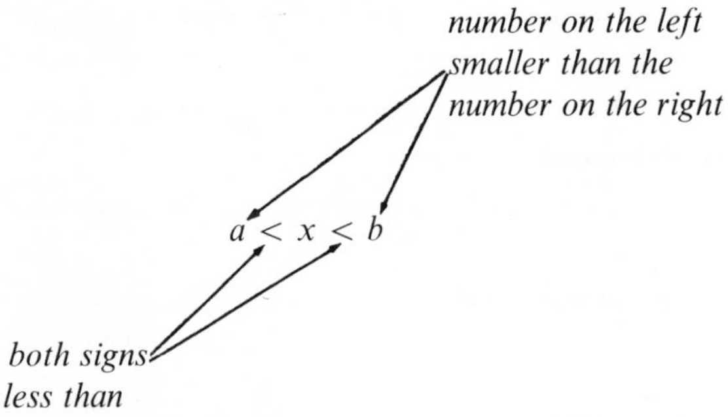

number on the left smaller than the number on the right

$a < x < b$

both signs less than

$a < x < b$ means x is between a and b.

$a < x < b$ means $x > a$ and $x < b$.

(22)

Example 1: $1 < x < 3$ means x is between 1 and 3 *or* $x > 1$ and $x < 3$.

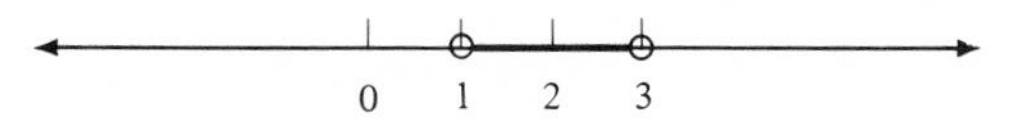

Example 2: $-2 \leqslant x \leqslant 0$ means x is between -2 and 0 inclusive, *or* $x \geqslant -2$ and $x \leqslant 0$.

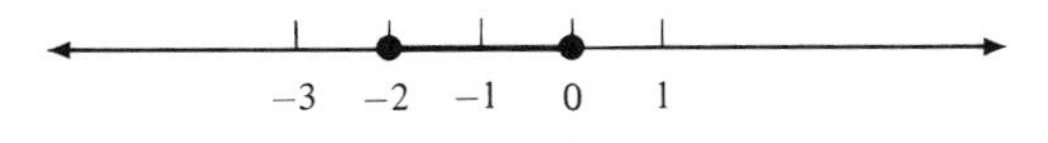

(23)

Study Exercise Five

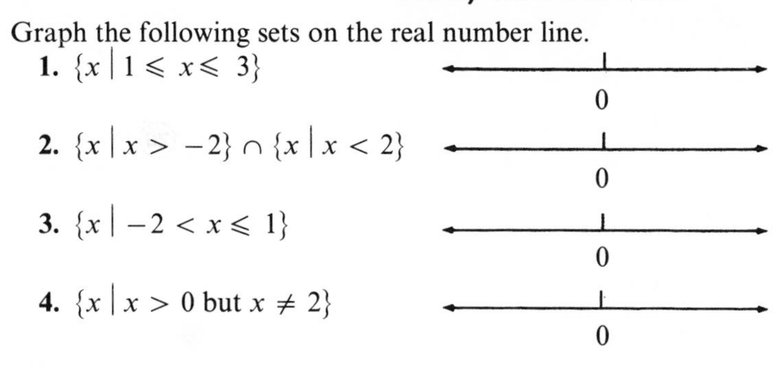

Graph the following sets on the real number line.

1. $\{x \mid 1 \leqslant x \leqslant 3\}$ — 0
2. $\{x \mid x > -2\} \cap \{x \mid x < 2\}$ — 0
3. $\{x \mid -2 < x \leqslant 1\}$ — 0
4. $\{x \mid x > 0 \text{ but } x \neq 2\}$ — 0

(24)

Absolute Value

The graphs of 4 and -4 lie the same distance, but on opposite sides of the origin.

−5 −4 −3 −2 −1 0 1 2 3 4 5

If we wish to refer only to the distance of the graph of a number from the origin and not to the side of the origin on which it is located, we will use the term *absolute value.*

(25)

The *absolute value* of a number refers to the distance that number is on either side of the origin and thus will never be negative.

The absolute value of 5 refers to the distance the number 5 is from the origin. The absolute value of 5 is 5.

The absolute value of -2 refers to the distance the number -2 is from the origin. The absolute value of -2 is 2.

(26)

The symbol $|5|$ is used to indicate the absolute value of 5.

Example 1: $|4| = 4$
Example 2: $|-2| = 2$
Example 3: $|-4| = 4$
Example 4: $|0| = 0$
Example 5: $|2^1/_2| = 2^1/_2$

(27)

Suppose the absolute value of some number(s) is 6. What numbers are there whose distance is 6 from the origin?

In symbols, let x stand for the possible numbers.

Then $|x| = 6$. The only numbers that are 6 units from the origin are 6 and -6.

(28)

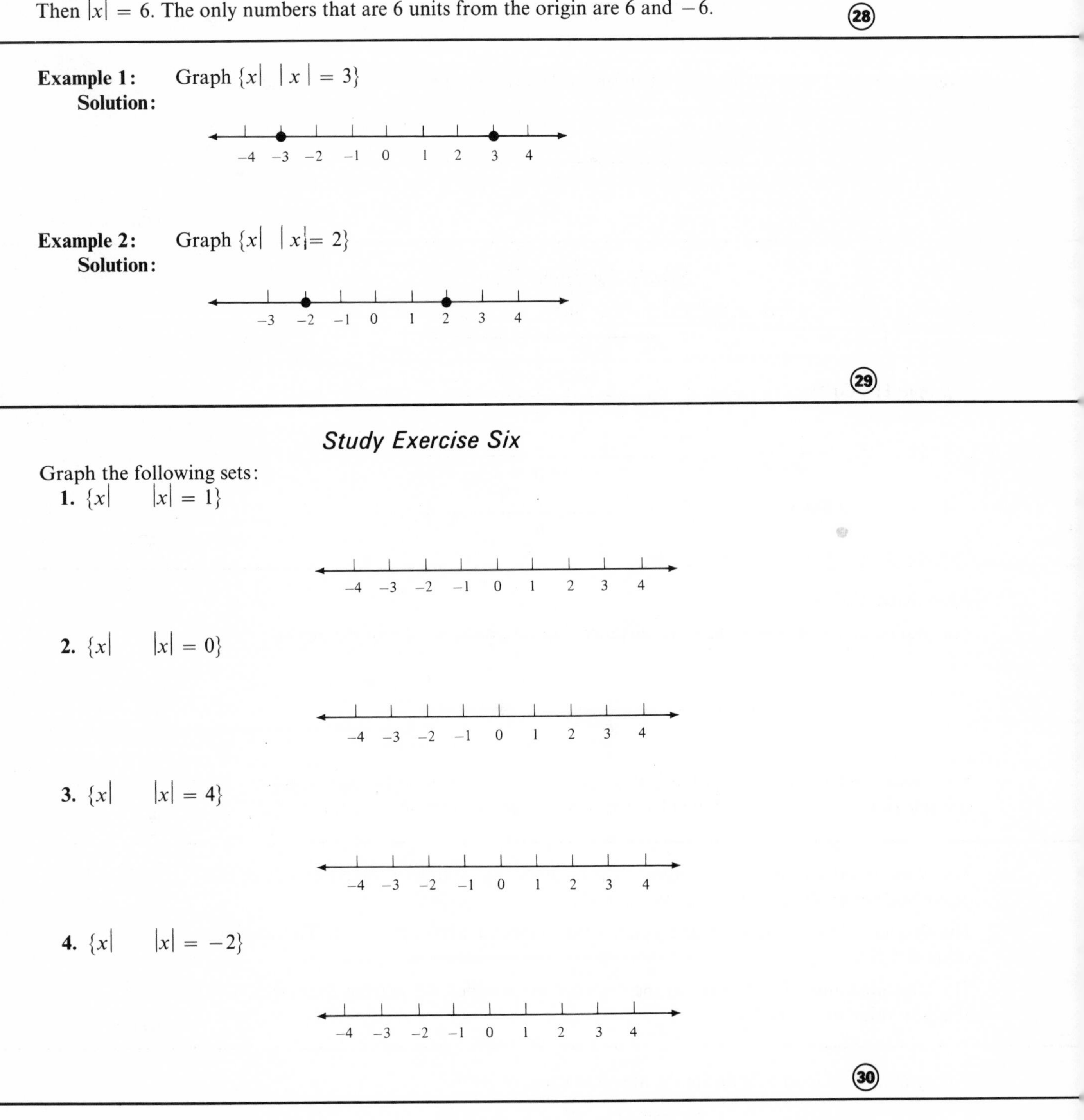

Example 1: Graph $\{x| \; |x| = 3\}$

Solution:

Example 2: Graph $\{x| \; |x| = 2\}$

Solution:

(29)

Study Exercise Six

Graph the following sets:

1. $\{x| \quad |x| = 1\}$

2. $\{x| \quad |x| = 0\}$

3. $\{x| \quad |x| = 4\}$

4. $\{x| \quad |x| = -2\}$

(30)

What does $\{x \mid |x| > 1\}$ contain?

The set contains all numbers x whose distance from the origin is more than one.

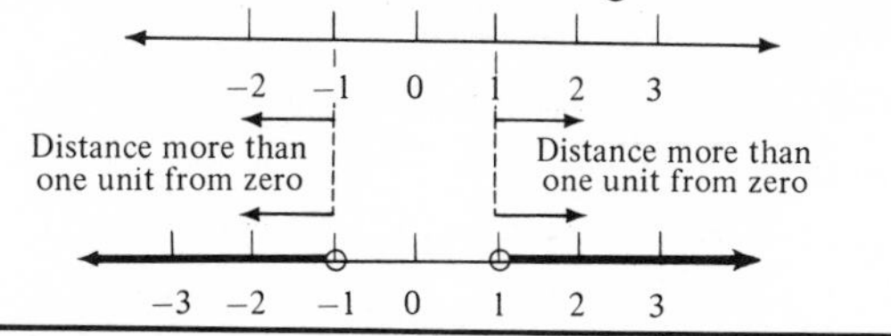

(31)

What does $\{x \mid |x| < 1\}$ contain?

The set contains all numbers whose distance from the origin is less than one unit.

$\{x \mid |x| < 1\}$ contains all real numbers between -1 and 1.

Example 1: Graph $\{x \mid |x| < 3\}$

Solution.

Example 2: Graph $\{x \mid |x| \geqslant 2\}$

Solution:

(33)

If $a > 0$, then

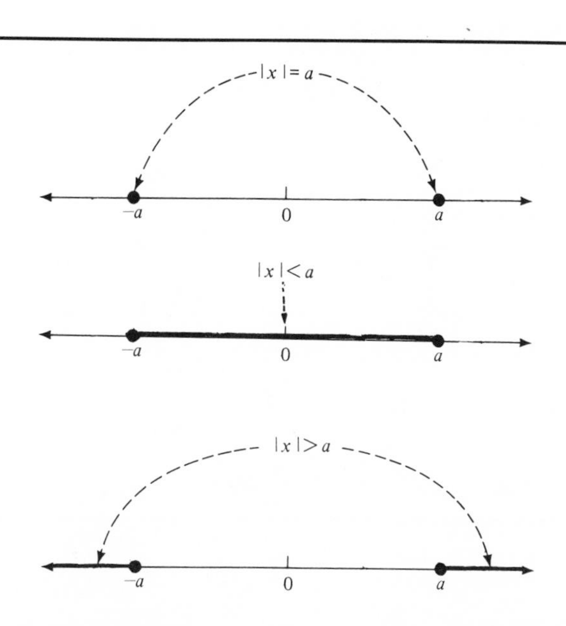

In exercises dealing with absolute values, common sense and a grasp of meaning take precedence over a flurry of mechanical rules.

(34)

Study Exercise Seven

Graph on the real number line:

1. $\{x \mid |x| \leq 2\}$

2. $\{x \mid |x| > 3\}$

3. $\{x \mid |x| < 3\}$

4. $\{x \mid |x| > 1\} \cup \{x \mid |x| = 1\}$

(35)

REVIEW EXERCISES

A. Fill in the blank with either $<$ or $>$ to make a true statement.

1. 3 ______ -3 **2.** -11 ______ -13 **3.** -15 ______ -8

4. $|-3|$ ______ -3 **5.** -3 ______ $|3|$

B. Graph on the real number line.

6. $\{x|x > 0 \text{ and } x < 3, x \in R\}$ **7.** $\{x|x \in I \text{ and } x \cdot x \leqslant 4\}$

8. $\{x|x > 2\} \cup \{x|x \leqslant 0\}$ **9.** $\{x| \ |x| \geqslant 2\}$

10. $\{x|-3 < x \leqslant 1\}$ **11.** $\{x|x > -2\} \cap \{x|x < 2\}$

(36)

SOLUTIONS TO REVIEW EXERCISES

A. **1.** $>$ **2.** $>$ **3.** $<$ **4.** $>$ **5.** $<$

B. **6.**

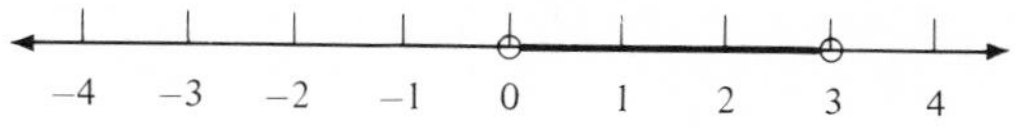

7.

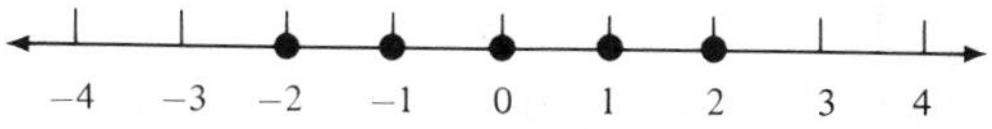

8.

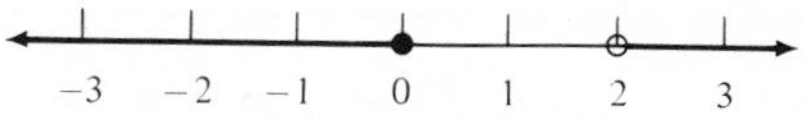

9.

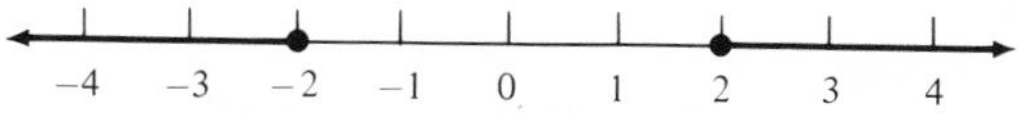

10.

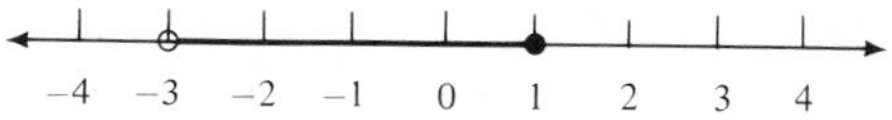

11.

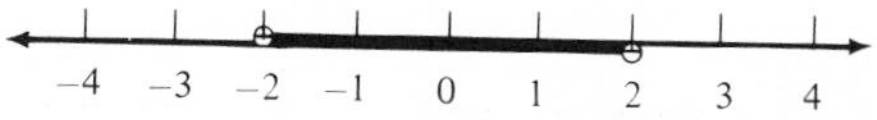

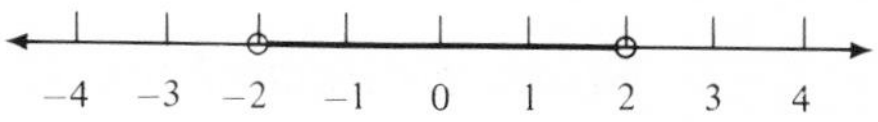

Final result

(37)

SOLUTIONS TO STUDY EXERCISES

Study Exercise One (Frame 7)

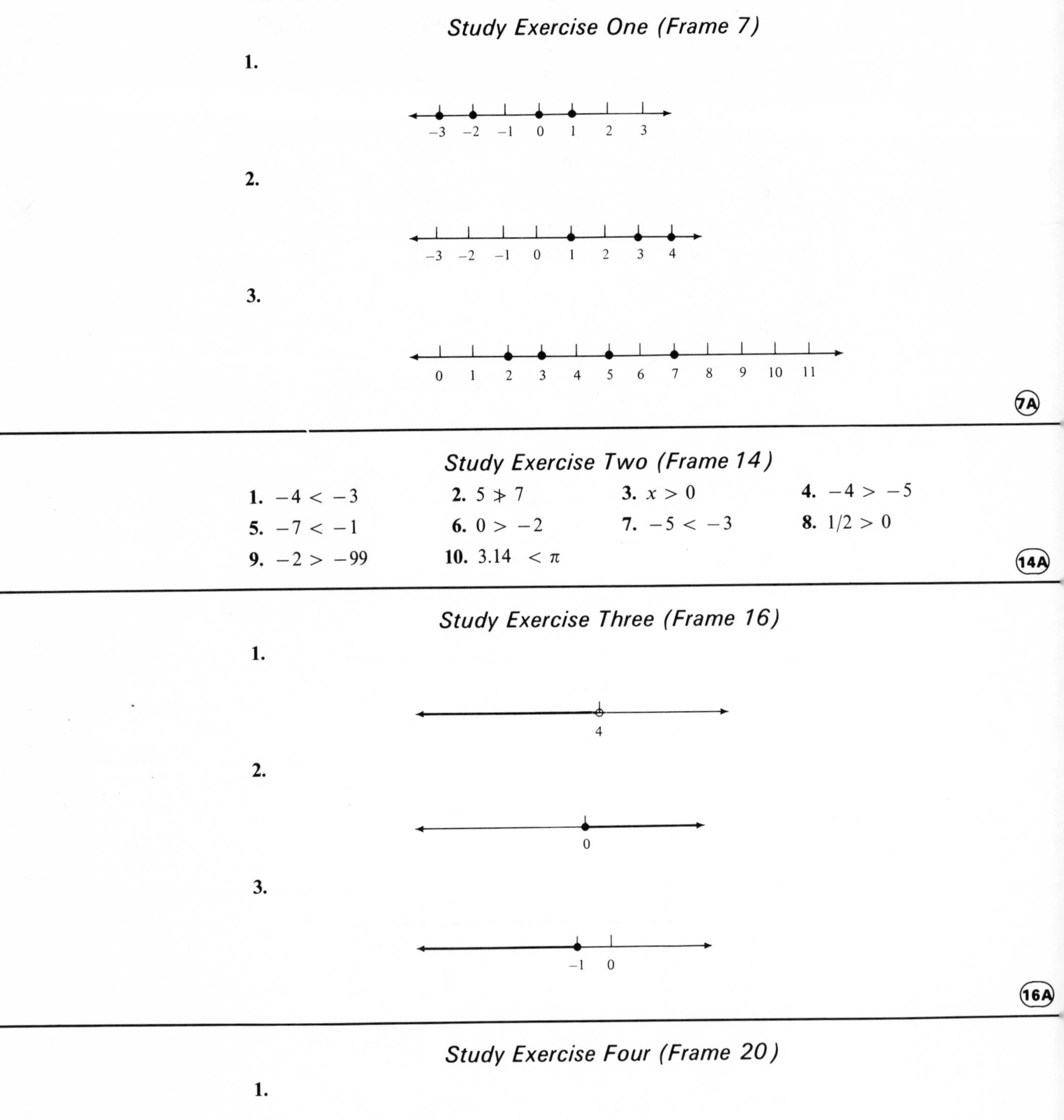

1.

2.

3.

7A

Study Exercise Two (Frame 14)

1. $-4 < -3$ **2.** $5 \ngtr 7$ **3.** $x > 0$ **4.** $-4 > -5$

5. $-7 < -1$ **6.** $0 > -2$ **7.** $-5 < -3$ **8.** $1/2 > 0$

9. $-2 > -99$ **10.** $3.14 < \pi$

14A

Study Exercise Three (Frame 16)

1.

2.

3.

16A

Study Exercise Four (Frame 20)

1.

SOLUTIONS TO STUDY EXERCISES, CONTD.

Study Exercise Four (Frame 20, contd.)

2.

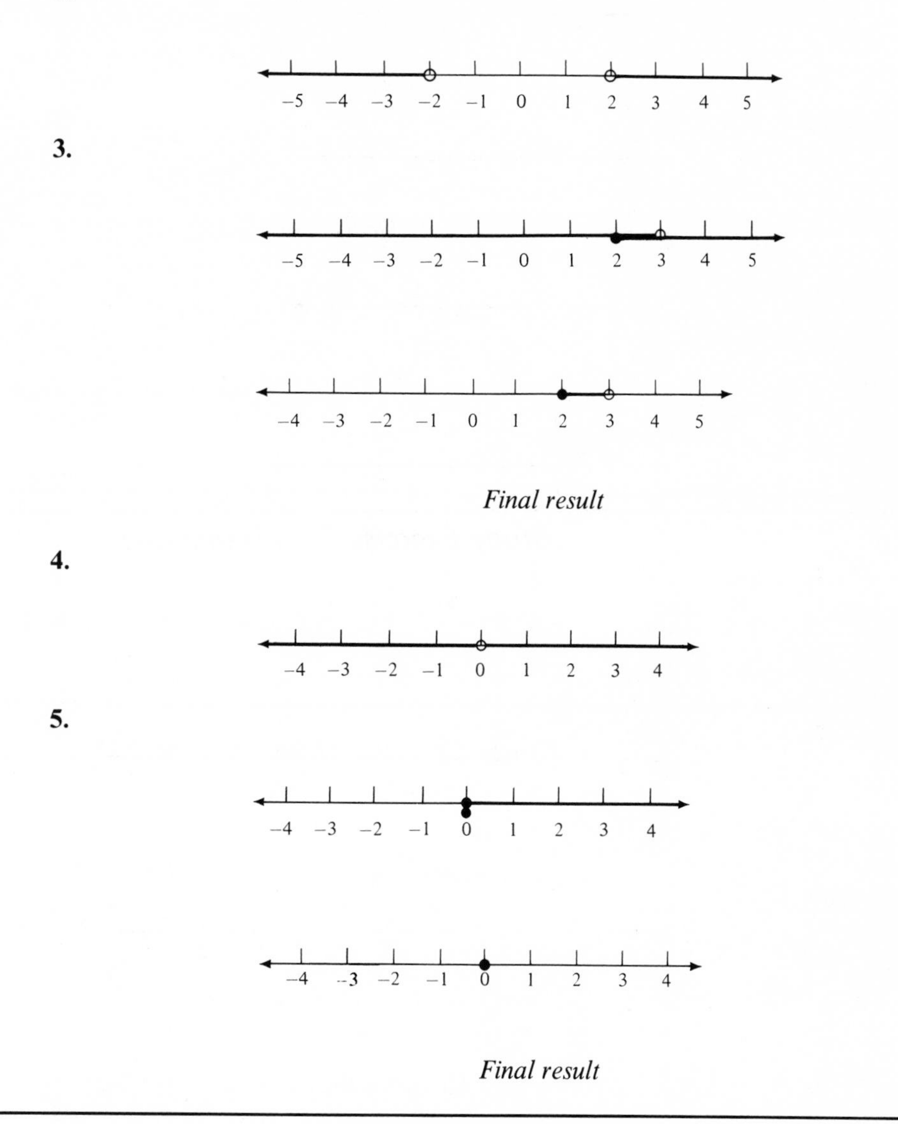

Final result

Final result

20A

Study Exercise Five (Frame 24)

1.

1 3

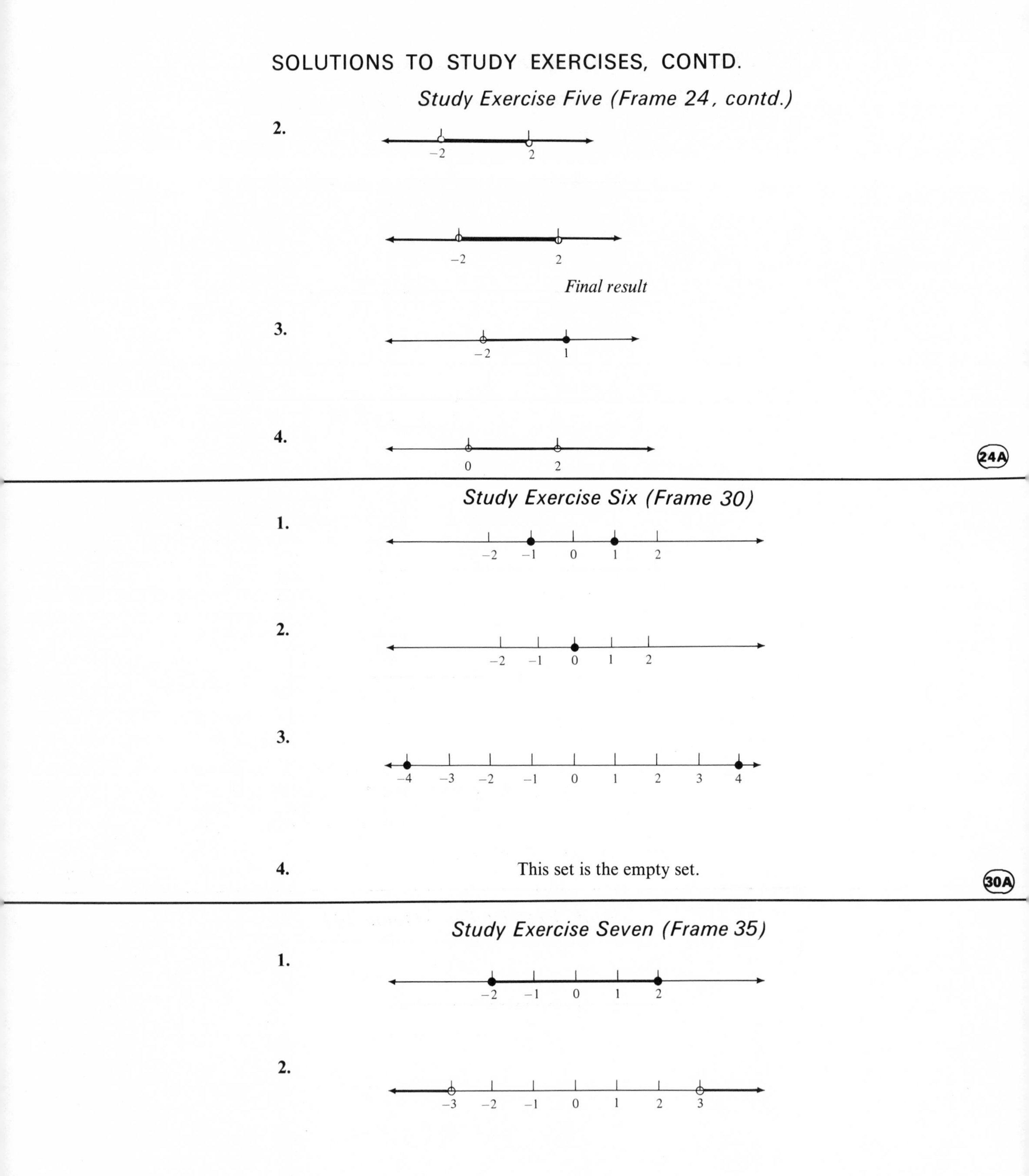
SOLUTIONS TO STUDY EXERCISES, CONTD.
Study Exercise Five (Frame 24, contd.)
2.
−2
2
−2
2
Final result
3.
−2
1
4.
0
2
24A
Study Exercise Six (Frame 30)
1.
−2 −1 0 1 2
2.
−2 −1 0 1 2
3.
−4 −3 −2 −1 0 1 2 3 4
4.
This set is the empty set.
30A
Study Exercise Seven (Frame 35)
1.
−2 −1 0 1 2
2.
−3 −2 −1 0 1 2 3

SOLUTIONS TO STUDY EXERCISES, CONTD.

Study Exercise Seven (Frame 35, contd.)

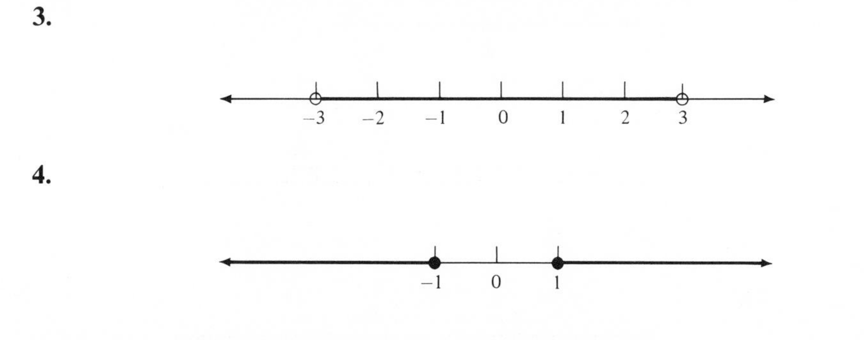

35A

UNIT 2—SUPPLEMENTARY PROBLEMS

A. In each blank insert the proper sign from among $<, =, >$.
Refer to the graph for a, b, x, and y.

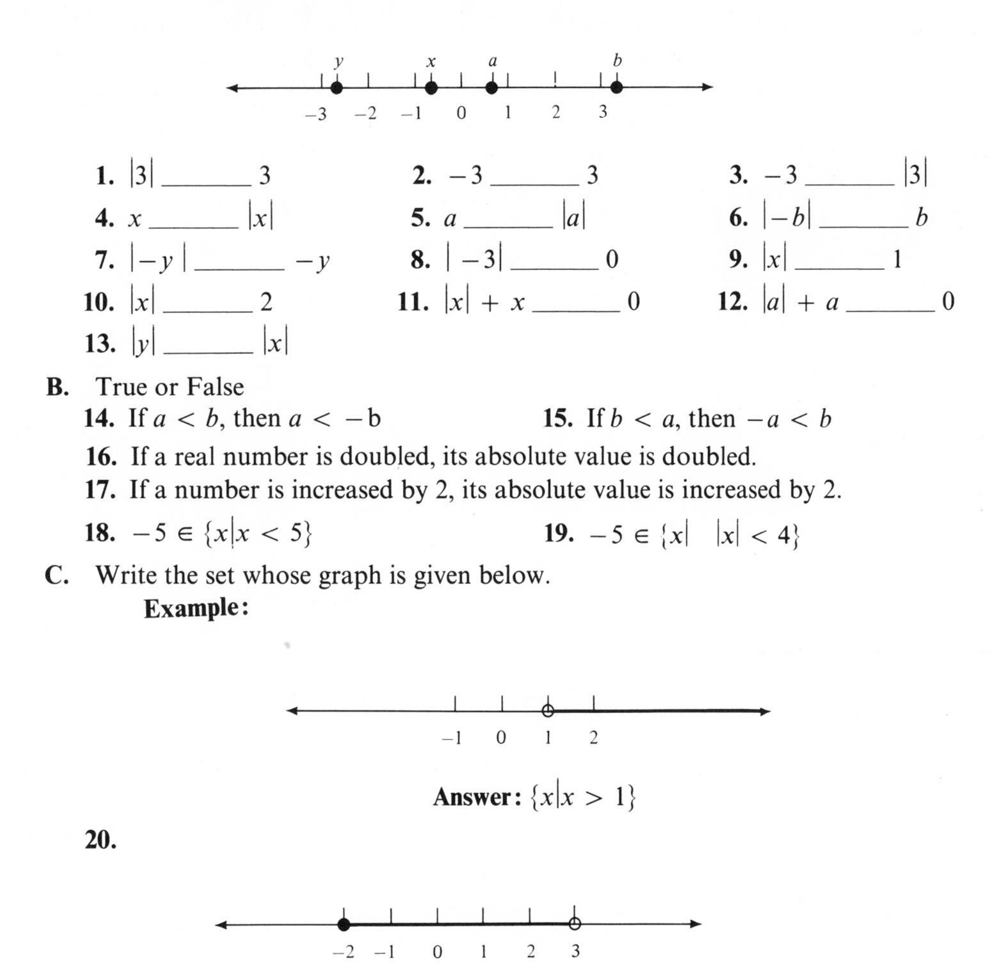

1. $|3|$ ______ 3 **2.** -3 ______ 3 **3.** -3 ______ $|3|$

4. x ______ $|x|$ **5.** a ______ $|a|$ **6.** $|-b|$ ______ b

7. $|-y|$ ______ $-y$ **8.** $|-3|$ ______ 0 **9.** $|x|$ ______ 1

10. $|x|$ ______ 2 **11.** $|x| + x$ ______ 0 **12.** $|a| + a$ ______ 0

13. $|y|$ ______ $|x|$

B. True or False

14. If $a < b$, then $a < -b$ **15.** If $b < a$, then $-a < b$

16. If a real number is doubled, its absolute value is doubled.

17. If a number is increased by 2, its absolute value is increased by 2.

18. $-5 \in \{x|x < 5\}$ **19.** $-5 \in \{x| \; |x| < 4\}$

C. Write the set whose graph is given below.

Example:

−1 0 1 2

Answer: $\{x|x > 1\}$

20.

−2 −1 0 1 2 3

UNIT 2—SUPPLEMENTARY PROBLEMS, CONTD.

21.

22.

23.

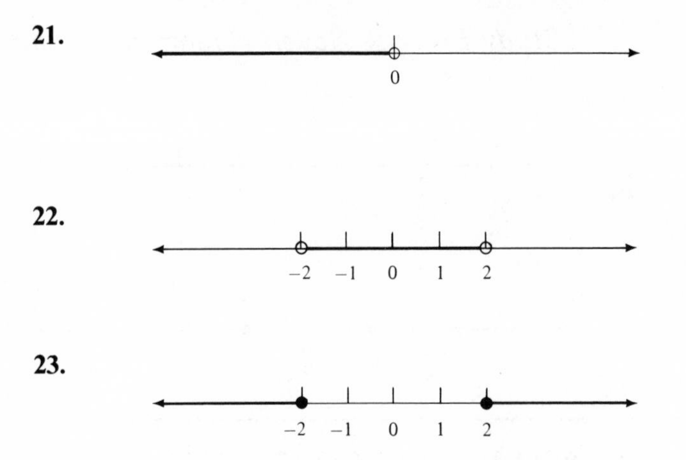

D. Graph on the real line:

24. $\{x|x \geqslant 2\} \cup \{x|x < -2\}$

25. $\{x|x > -1\} \cap \{x|x \leqslant 4\}$

26. $\{x| \ |x| \leqslant 3\}$

27. $\{x| \ |x| > 4\}$

28. $\{x|-2 < x \leqslant 0\} \cup \{x|x > 2\}$

29. $\{x|x$ is non-negative$\}$

unit

3

Review of Field Properties-I

Objectives

1. Know the field axioms, axioms of equality, and basic theorems.
2. Know the order of operations convention.
3. Be able to operate with signed numbers.

Terms

axiom binary operation reciprocal theorem factor

①

The study of algebra is largely a study of certain properties of the real numbers. In mathematics we make formal assumptions about numbers and their properties. These assumptions are called *axioms* or *postulates*.

Axioms will be formal statements about properties that we propose to assume as always valid.

②

Axioms of Equality

We assume the "is equal to" $(=)$ relationship has the following properties for $a, b, c, \in R$:

Reflexive Law: $a = a$
Symmetric Law: If $a = b$, then $b = a$
Transitive Law: If $a = b$ and $b = c$, then $a = c$
Substitution Law: If $a = b$, then b may be replaced by a or vice versa in any statement without altering the truth or falsity of the statement.

③

Study Exercise One

Each of the following is an application of one of the axioms of equality. Justify each by quoting the appropriate axiom.

1. If $x = 2$, then $2 = x$

2. If $-3 = y$, then $y = -3$

3. If $x = y$ and $2x = 4$, then $2y = 4$

4. If $3 = a$ and $a = b$, then $3 = b$.

5. $x = x$

④

Binary Operation

A binary operation in a set A is a rule that assigns to any pair of elements of A, taken in a definite order, another element of A.

In our study of algebra, the set A will be the real numbers. We will use two binary operations, addition $(+)$ and multiplication $(\cdot)$.

⑤

In regard to the operations of addition and multiplication, we postulate eleven basic laws of real numbers.
Parenthesis used in some of these laws indicate that symbols within parentheses are to be viewed as representing a single entity.

Any set with a pair of binary operations satisfying the eleven laws is called a *Field* under these operations.

(6)

Properties of Real Numbers

Given: Set R, the set of real numbers, $a, b, c \in R$, and two binary operations, addition and multiplication.

1. *Closure Property of Addition (clpa)*
$$(a + b) \in R$$
2. *Closure Property of Multiplication (clpm)*
$$(a \cdot b) \in R$$
3. *Commutative Property of Addition (cpa)*
$$a + b = b + a$$
4. *Commutative Property of Multiplication (cpm)*
$$a \cdot b = b \cdot a$$

(7)

---

5. *Associative Property of Addition (apa)*
$$(a + b) + c = a + (b + c)$$
6. *Associative Property of Multiplication (apm)*
$$(a \cdot b) \cdot c = a \cdot (b \cdot c)$$
7. *Identity For Addition (add iden)*
There exists a unique number $0 \in R$ such that
$$a + 0 = 0 + a = a$$
8. *Identity Element For Multiplication (mult iden)*
There exists a unique number $1 \in R$ such that
$$a \cdot 1 = 1 \cdot a = a$$

(8)

---

9. *Inverse For Addition (add Inv)*
For each a there exists a unique number $-a \in R$ such that
$$(a) + (-a) = (-a) + (a) = 0$$
10. *Inverse For Multiplication (mult Inv)*
For each $a \neq 0$, there exists a unique number $1/a \in R$ such that
$$a \cdot (1/a) = (1/a) \cdot a = 1$$
11. *Distributive Property Multiplication Over Addition (dpma)*
$$a \cdot (b + c) = a \cdot b + a \cdot c$$

(9)

Multiplicative Inverse

For each $a \neq 0$, there exists a unique number $1/a \in R$, so that
$$a \cdot (1/a) = (1/a) \cdot a = 1.$$

Comments:

1. Every number except zero has a multiplicative inverse.
2. Sometimes the multiplicative inverse of a number is called the *reciprocal* of the number.

(10)

Examples: Justify each statement by giving the abbreviation of one of the *Field Properties:*

1. $6 + 0 = 0 + 6$ — **1. cpa**
2. $6 + 0 = 6$ — **2. add iden.**
3. $1 \cdot 6 = 6 \cdot 1$ — **3. cpm**
4. $(4 + 2) + 1 = 4 + (2 + 1)$ — **4. apa**
5. $2 \cdot 3 + 2 \cdot 4 = 2 \cdot (3 + 4)$ — **5. dpma**
6. $(4 + 2) + 1 = (2 + 4) + 1$ — **6. cpa**

(11)

Study Exercise Two

Justify each statement by quoting a Field Property, Equality Property, or Definition.

1. $5 + (-2) = (-2) + 5$
2. $a \cdot b + a \cdot c = b \cdot a + a \cdot c$
3. $(a + b) + c = c + (a + b)$
4. $1 \cdot 1/1 = 1$
5. $0 + 0 = 0$
6. $x + y = x \cdot 1 + y \cdot 1$
7. $1/-4 \cdot (-4) = 1$
8. $(-5) + [- \ (-5)] = 0$
9. $1 \cdot (x + y) = 1 \cdot x + 1 \cdot y$
10. $5 = 5$
11. If $b = x$, then $x = b$
12. $|3| = 3$

(12)

Theorems

The Field Properties together with the Properties of Equality imply other properties of the Real Numbers. These implications are generally stated as *theorems*. A theorem is an assertion of a fact that follows logically from previous axioms, definitions, and theorems.

(13)

Theorem: Addition Law of Equality Theorems

If $a, b, c \in R$ and $a = b$, then $a + c = b + c$.

Proof:

1. $a, b, c \in R, a = b$	**1 Given**
2. $(a + c) \in R$	**2. clpa**
3. $a + c = a + c$	**3. reflexive law of equality**
4. $a + c = b + c$	**4. substitution**

(14)

Theorem: Multiplication Law of Equality

If $a, b, c \in R$ and $a = b$, then $ac = bc$.

Theorem: Cancellation Law For Addition

If $a, b, c \in R$ and $a + c = b + c$, then $a = b$.

Theorem: Cancellation Law For Multiplication

If $a, b, c \in R$, $c \neq o$, and $ac = bc$, then $a = b$.

(15)

Study Exercise Three

Below is the proof for the Cancellation Law for Addition. Justify each statement with the appropriate reason.

Theorem. If $a, b, c \in R$ and $a + c = b + c$, then $a = b$.

Proof:

1. $a, b, c \in R, a + c = b + c$	1. ____________
2. $(a + c) + (-c) = (b + c) + (-c)$	2. ____________
3. $a + [c + (-c)] = b + [c + (-c)]$	3. ____________

(Frame 16, contd.)

4. $a + 0 = b + 0$ 4. ____________

5. $a = b$ 5. ____________

(16)

Theorem. Multiplication By Zero Law: For every $a \in R$, $a \cdot 0 = 0$

Proof:

1. $a \in R$	1. **Given**
2. $0 + 0 = 0$	2. **add iden**
3. $a \cdot (0 + 0) = a \cdot 0$	3. **multiplication law for equality**
4. $a \cdot (0 + 0) = 0 + a \cdot 0$	4. **add iden**
5. $a \cdot 0 + a \cdot 0 = 0 + a \cdot 0$	5. **dpma**
6. $a \cdot 0 = 0$	6. **cancellation law for addition**

(17)

Study Exercise Four

Justify each statement by quoting an axiom, definition, or theorem.

1. If $a, b \in R$ and $a + b = c$, then $c \in R$.
2. If $x = 5$, then $x + 3 = 5 + 3$.
3. $3 \cdot 0 = 0 \cdot 3$
4. $3 \cdot 0 = 0$
5. If $2 \cdot r = 2 \cdot t$, then $r = t$.
6. If $(a + b) + 3 = z + 3$, then $(a + b) = z$
7. $(a + b) + 2 = (b + a) + 2$

(18)

Theorem. Double Negative Law: If $a \in R$, then $-(-a) = a$.

Proof:

1. $a \in R$	1. ____________
2. $(-a) + -(-a) = 0$	2. ____________
3. $a + (-a) = 0$	3. ____________
4. $-a + -(-a) = a + (-a)$	4. ____________
5. $-(-a) + (-a) = a + (-a)$	5. ____________
6. $-(-a) = a$	6. ____________

(19)

The Negative of a Number

The negative of the number 3 is -3.
The negative of (-2) is $-(-2)$ or 2.
The negative of 5 is -5.
The negative of $(-1\frac{1}{2})$ is $-(-1\frac{1}{2})$ or $1\frac{1}{2}$.

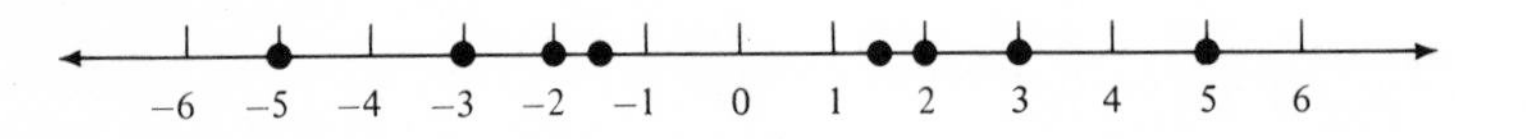

The negative of a number lies an equal distance on the opposite side of zero.

(20)

Opposites

The negative of a number is sometimes called the *opposite* of a number.

The opposite of a positive number is a negative number.
The opposite of a negative number is a positive number.

(21)

If a is positive, what is $(-a)$?

If $(-a)$ is positive, what is a?

What kind of a number is $(-b)$?

Is $-(-a)$ positive?

(22)

Axiom. If $a, b \in R, a > 0, b > 0$, then $(a + b) > 0$.
The sum of two positive numbers is a positive number.

Axiom. If $a, b \in R, a > 0, b > 0$, then $(a \cdot b) > 0$.
The product of two positive numbers is a positive number.

(23)

Theorem. If $a, b \in R$, then $-(a + b) = (-a) + (-b)$.
Proof:

1. $a + (-a) = 0$
2. $b + (-b) = 0$
3. $0 + 0 = 0$
4. $[a + (-a)] + [b + (-b)] = 0$
5. $[a + b] + [(-a) + (-b)] = 0$
6. By the uniqueness of the add inv, $-(a + b) = (-a) + (-b)$

(24)

We have established the theorem $(-a) + (-b) = -(a + b)$. Let $a > 0$ and $b > 0$. Then $(-a)$ is negative and $(-b)$ is negative.

$(-a)$	$+$	$(-b)$	$=$	$-$	$(a + b)$
↑		↑			↑
negative number		negative number			*Sum of two positive numbers is positive*

(negative) + (negative) is the opposite of (positive)

(25)

A negative number plus a negative number yields a negative number. *The sum of two negative numbers is the negative of the sum of their absolute values.*

Example 1: $(-2) + (-3) = -(2 + 3)$
$= -5$

Example 2: $(-1) + (-7) = -(1 + 7)$
$= -8$

Example 3: $(-4) + (-5) = -(4 + 5)$
$= -9$

(26)

Subtraction

$a - b$ means the number c such that $c + b = a$.

Example 1: $4 - 2 = 2$ since $2 + 2 = 4$
Example 2: $8 - 3 = 5$ since $5 + 3 = 8$
Example 3: $10 - 1 = 9$ since $9 + 1 = 10$
Example 4: $6 - 0 = 6$ since $6 + 0 = 6$

(27)

Let us attempt to add (8) and (-2).

Line (1) $8 + (-2) = (6 + 2) + (-2)$
Line (2) $= 6 + [(2) + (-2)]$
Line (3) $= 6 + [0]$
Line (4) $= 6$

(28)

Let us attempt to add (-5) and 2.

Line (*1*) $(-5) + (2) = [(-3) + (-2)] + (2)$
Line (*2*) $= (-3) + [(-2) + (2)]$
Line (*3*) $= (-3) + [0]$
Line (*4*) $= -3$

(29)

Thus, to add two numbers with unlike signs we take the difference of the absolute values of the numbers (smaller absolute value from larger absolute value) and the sign of the result is the same as the sign of the number with the greater absolute value. If absolute values are equal, the sum is zero.

(30)

$$(8) + (-3) = +(8 - 3) = 5$$

positive sign used since 8 is greater in absolute — *absolute value of 8* — *absolute value of 3*

(31)

Example (1): $(-5) + (+1) = -(5 - 1)$
$= -4$

Example (2): $(+7) + (-4) = +(7 - 4)$
$= +3$

Example (3): $(-8) + (5) = -(8 - 5)$
$= -3$

Example (4): $(5) + (-11) = -(11 - 5)$
$= -6$

Example (5): $(-4) + (+4) = 0$

(32)

Study Exercise Five

Perform the indicated operation.

1. $3 + (-5)$ **2.** $(-5) + (-1)$ **3.** $(-18) + (+18)$

4. $(5) + (+3) + (-2)$ **5.** $(-2) + (-3) + (-4)$

(33)

The definition of subtraction says $a - b$ is the number c such that $c + b = a$.

(1) $c + b = a$	
(2) $(c + b) + (-b) = a + (-b)$	**addition law of equality**
(3) $c + [b + (-b)] = a + (-b)$	**apa**
(4) $c + 0 = a + (-b)$	**add inv**
(5) $c = a + (-b)$	**add iden**
(6) therefore, $a - b = a + (-b)$	**substitution**

(34)

Since the difference $a - b$ is given by $a + (-b)$, we may consider $a - b$ as representing:

(1) the difference of a and b *or* **(2)** the sum of a and $(-b)$

(35)

Example 1: $5 - 3 = (5) + (-3)$
$= 2$

Example 2: $(-6) - (2) = (-6) + (-2)$
$= -8$

(Frame 36, contd.)

Example 3: $8 - (-3) = 8 + -(-3)$
$= 8 + (3)$
$= 11$

(36)

Uses of Plus and Minus

Minus signs have been used in three ways:

1. To denote negative numbers.
2. As a sign of operation to denote subtraction.
3. To denote an opposite.

Plus signs have been used in three ways:

1. To denote positive numbers.
2. As a sign of operation to denote sum.
3. To denote the same as.

(37)

Uses of The Symbol Minus

Example 1:

−3 −2 −1 0 1 2 3

Minus signs for negative numbers.

Example 2: $(5) - (3)$
Minus sign to denote subtraction.

Example 3: $(7) + (-x)$
Minus sign to denote the opposite.

(38)

Uses of The Symbol Plus

Example 1:

−3 −2 −1 0 +1 +2 +3

Plus signs for positive numbers.

Example 2: $(5) + (3)$; Plus sign to denote *sum.*

Example 3: $+x = x$; Plus sign to denote *same as.*

(39)

Sometimes the phrase subtract a from b is used.
To subtract a from b means "*to b add the opposite of a.*"
To subtract a from b means $b+(-a)$.

(40)

Example 1: Subtract 7 from 8
Solution: $8 + (-7) = 1$

Example 2: Subtract -3 from -9
Solution: $(-9) + -(-3) = (-9) + 3$
$= -6$

(41)

REVIEW EXERCISES

A. Supply the real number that should appear in the blank to make the statement true. Also state the axiom that is used.

1. $5 \cdot (4 + \_\_\_\_) = 5 \cdot 4 + 5 \cdot 7$ **2.** $3 \cdot (\_\_\_\_ + 4) = 3 \cdot 7 + \_\_\_\_ \cdot 4$

3. $3 + 5 = \_\_\_\_ + 3$ **4.** $0 \cdot (1 + 2) = (1 + 2) \cdot \_\_\_\_$

B. Supply reasons in the following proof that $(x + 2) + a = 2 + (a + x)$:

5. $(x + 2) + a = (2 + x) + a$
6. $= 2 + (x + a)$
7. $= 2 + (a + x)$

C. Perform the indicated operations

8. $14 + (-4) + 18 - 3$ **9.** $-(-2) + (+3) - (6)$

10. $|(4 - 7)| + 3$ **11.** $5 + (6 - 8) \cdot 0$

12. $-6 + (+6) + 1 - 1 + 0$

D. Justify each statement:

13. If $2 = x$, then $x = 2$ **14.** $0(1 + 2) = 0$ **15.** If $5 \cdot x = 5 \cdot y$, then $x = y$.

(42)

SOLUTIONS TO REVIEW EXERCISES

A. **1.** 7; dpma **2.** 7, 3; dpma **3.** 5; cpa **4.** 0; cpm

B. **5.** cpa **6.** apa **7.** cpa

C. **8.** 25 **9.** -1 **10.** 6 **11.** 5 **12.** 0

D. **13.** symmetric law of equality **14.** zero multiplication law

15. cancellation law for multiplication.

(43)

SOLUTIONS TO STUDY EXERCISES

Study Exercise One (Frame 4)

1. Symmetric law of equality
2. Symmetric law of equality
3. Substitution law
4. Transitive law or substitution law
5. Reflexive law of equality

(4A)

Study Exercise Two (Frame 12)

1. cpa
2. cpm
3. cpa
4. mult inv
5. add iden
6. mult iden
7. mult inv
8. add inv
9. dpma
10. reflexive law of equality
11. symmetric law of equality
12. definition of absolute value

(12A)

Study Exercise Three (Frame 16)

1. given **2.** addition law for equality **3.** apa **4.** add inv **5.** add iden

(16A)

SOLUTIONS TO STUDY EXERCISES, CONTD.

Study Exercise Four (Frame 18)

1. clpa
2. addition law for equality
3. cpm
4. multiplication by zero law
5. cancellation law for multiplication
6. cancellation law for addition
7. cpa

18A

Double Negative Law (Frame 19)

1. given
2. add inv.
3. add inv.
4. substitution
5. cpa
6. cancellation for addition

19A

Study Exercise Five (Frame 33)

1. -2 **2.** -6 **3.** 0 **4.** 6 **5.** -9

33A

UNIT 3—SUPPLEMENTARY PROBLEMS

1. Simplify to a basic numeral:

$$[3 - (-2) + 4] - [2 + (8 - 4)]$$

2. Simplify to a basic numeral:

$$-5 + [3 + (-17)] - 18 + \{-5 - [1 + (-7 + 4)]\}$$

3. Name the property that justifies each statement:

a) $6 + 0 = 6$ ____________________
b) $5 + (-5) = 0$ ____________________
c) $-(3 + 8) \cdot 1 = -(3 + 8)$ ____________________
d) $(4 + 2) \cdot 0 = 0$ ____________________
e) $0 \cdot 1 = 0$ ____________________
f) $0 + 0 = 0$ ____________________
g) if $x + 3 = 0$, then $(x + 3) + (-3) = 0 + (-3)$ ____________________

4. Simplify: $[3 - (+2) - 4] - [2 - (8 + 4)]$

5. How is each sign used in $-(-6) - (+3)$?

6. Subtract the second expression from the first:

$$a - b + c - d, \qquad c - a + d - b.$$

7. Supply reasons in the following proof:

a) $a + (-1) \cdot a = a + (-1) \cdot a$ ____________________
b) $= a \cdot 1 + (-1) \cdot a$ ____________________
c) $= a \cdot 1 + a(-1)$ ____________________
d) $= a[1 + (-1)]$ ____________________
e) $= a \cdot 0$ ____________________
f) $= 0$ ____________________

But

g) $a + (-a) = 0$ ____________________
h) $a + (-a) = a + (-1)a$ ____________________
i) $-a = (-1) \cdot a$ ____________________

8. *Prove*: If $a, b, c \in R$ and $a + b = c$, then $b = (-a) + c$

9. Give an example which shows that subtraction is not commutative.

10. What is the additive inverse of?

a) -4
b) $(a + b)$
c) $x - 1$
d) 1

11. Supply reasons for the following statements:

a) $3x + 5x = (3 + 5)x$
b) $5by + (a + c) = (a + c) + 5by$

SUPPLEMENTARY PROBLEMS, CONTD.

12. *Prove*: If $a = b$, then $-a = -b$

1. $a = b$	**1.** ____________________
2. $a + (-a) = b + (-a)$	**2.** ____________________
3. $0 = b + (-a)$	**3.** ____________________
4. $0 = (-a) + b$	**4.** ____________________
5. $0 + (-b) = [(-a) + b] + (-b)$	**5.** ____________________
6. $0 + (-b) = (-a) + [b + (-b)]$	**6.** ____________________
7. $0 + (-b) = (-a) + 0$	**7.** ____________________
8. $-b = -a$	**8.** ____________________
9. $-a = -b$	**9.** ____________________

unit

4

Review of Field Properties—Part 2

Terms

1. quotient **2.** factor **3.** exponent
4. algebraic expression **5.** term **6.** monomial
7. binomial **8.** trinomial **9.** polynomial
10. degree of a polynomial **11.** coefficient **12.** like terms.

①

Theorem. If $a, b \in R$, then $-(ab) = a(-b) = (-a) \cdot b$

Proof:

Line (1) $b + (-b) = 0$
Line (2) $a(b + [-b]) = a \cdot 0$
Line (3) $ab + a(-b) = a \cdot 0$
Line (4) $ab + a(-b) = 0$
Line (5) $a(-b)$ must equal the opposite of (ab)
Line (6) $a(-b) = -(ab)$

②

Similarly, if we follow the same pattern and begin with $a + (-a) = 0$, we can show that $b(-a) = -(ab)$:

Thus; $-(ab) = a(-b) = (-a) \cdot b$

③

Remember; $-(ab) = a(-b)$.

Let $a > 0, b > 0$. Then $-(ab) = a \; (-b)$

$-(ab)$	$=$	a	$(-b)$
↑		↑	↑
opposite of a positive number		*positive number*	*negative number*

[(positive number) · (negative number)] is a (negative number).

④

Now consider,

Line (1) $b + (-b) = 0$
Line (2) $-a[b + (-b)] = (-a) \cdot 0$
Line (3) $-a(b) + (-a) \cdot (-b) = (-a) \cdot 0$
Line (4) $-a(b) + (-a)(-b) = 0$
Line (5) $-(ab) + (-a)(-b) = 0$
Line (6) $(-a)(-b)$ must be the opposite of $-(ab)$
Line (7) $(-a)(-b) = ab$

⑤

Remember, $(-a)(-b) = ab$. Let $a > 0, b > 0$.

Then $\quad (-a) \quad (-b) \quad = \quad ab$

$\uparrow$ negative number $\quad \uparrow$ negative number $\quad \uparrow$ positive number

The product of two negative numbers is a positive number.

(6)

Example 1: $(-2)(5) = -(2 \cdot 5) = -10$

Example 2: $-(8[-4]) = -(-32) = 32$

Example 3: $-([-2][-3]) = -(6)$
$= -6$

Example 4: $(-1)(7) + (-8)(-2) = -7 + 16$
$= 9$

Example 5: $0(-8) + (-7)(-2)(-3) = 0 + [(-7)(-2)](-3)$
$= 0 + (14)(-3)$
$= 0 + (-42)$
$= -42$

(7)

Study Exercise One

Write each product as a basic numeral.

1. $(-3)(3)(-3)$

2. $-2[\{(3)(-2)\} + (-2)]$

3. $(-23)(0)(-5)$

4. $(-6)(-4) - [(-8)(-9)]$

5. $2(-3)(6) - [(-4)(+5)(+7)]$

(8)

Division

The *quotient* of two numbers a and b is defined as the number q such that $b \cdot q = a$ where $b \neq 0$.

The quotient of a and b will be written a/b; sometimes the form $a \div b$ is used. Thus $a/b = q$ where $b \cdot q = a, b \neq 0$

(9)

Consider a, b, and q all positive. To maintain consistency with the laws of multiplication, we define

$+a/+b = +q \qquad -a/-b = +q \qquad +a/-b = -q \qquad -a/+b = -q$

(10)

Another Form For The Quotient of a and b

Recall that the quotient of a and b is the number q such that $b \cdot q = a$.

Now $b(a[1/b]) = (a[1/b]) \cdot b$
$= a([1/b] \cdot b)$
$= a \cdot 1$
$= a$

Therefore, $a/b = a(1/b)$ since $b \cdot [a(1/b)] = a$
$a/b = a(1/b), b \neq 0$

(11)

Example 1: $6/6 = 6 \cdot 1/6$
$= 1$

Example 2: $8/2 = 8 \cdot 1/2$
$= (4 \cdot 2) \cdot 1/2$
$= 4(2 \cdot 1/2)$
$= 4 \cdot 1$
$= 4$

Example 3: $3/0 =$ does not exist

(12)

Order of Operations

What does $4 + 8 \div 2$ equal?
Is the answer 8?
Is the answer 6?
Does $4 + 8 \div 2$ mean
$4 + [8 \div 2]$ or $[4 + 8] \div 2$?

(13)

If an expression contains addition, subtraction, multiplication, and division operations without grouping symbols to indicate order, the multiplication and division is to be done before the addition and subtraction, moving from left to right.

According to this agreement, $4 + 8 \div 2$ is taken to mean that $4 + (8 \div 2)$ and $4 + (8 \div 2) = 4 + 4 = 8$.

(14)

If an expression contains nothing but multiplication and division without grouping symbols, perform the operations moving from left to right.

According to this agreement, $20 \div 5 \cdot 2$ is taken to mean $(20 \div 5) \cdot 2$ and $(20 \div 5) \cdot 2 = 4 \cdot 2 = 8$.

(15)

If an expression contains additions and subtractions and no grouping symbols, perform the operations moving from left to right.

Thus, $a + b + c = (a + b) + c$ and $a - b - c = (a - b) - c$

(16)

Factors

Each of the numbers multiplied in an indicated product is called a *factor* of that product.

1. $2 \cdot 7 \cdot 5$ has factors 2, 7, and 5
2. $3xy$ has factors 3, x, and y
3. $5(x + y)$ has factors 5 and $(x + y)$
4. $1 \cdot 2 \cdot 2 \cdot 2$ has factors 1 and 2 (2 as a factor three times).

(17)

Exponents

An *exponent* is a natural number written to the right and a little above another number called the *base* that tells how many times the base is used as a factor.

exponent

$$2^3 = \underbrace{2 \cdot 2 \cdot 2}$$

base — *base used as a factor three times*

(18)

$4^3 = (4)(4)(4)$; The base is 4, the exponent is 3.

$2x^5 = 2 \cdot x \cdot x \cdot x \cdot x \cdot x$; The base is x, the exponent is 5.

$(-4)^2 = (-4)(-4)$; The base is (-4), the exponent is 2.

(19)

What is the base and exponent for -3^2?

Is the base 3 or -3? Does -3^2 equal 9 or -9?

(20)

---

1. $-3^2 = -[3]^2$
   $= -9$
2. $-2^3 = -[2 \cdot 2 \cdot 2]$
   $= -8$
3. $-x^4 = -[x \cdot x \cdot x \cdot x]$
   $= -x^4$
4. $(-x)^4 = (-x)(-x)(-x)(-x)$
   $= x^4$

(21)

Study Exercise Two

Simplify completely:

1. $(-5)(+4) - (-8)(-9)$
2. $-2 \cdot [4 + (-4)] - 3 \cdot (-5) - 10$
3. $(-10) \div (-2) \cdot 5 - (-7)(+8)$
4. $(-9) - (5) \cdot (-2) \div (-2) + (-7)(+6)$
5. $-10 + (-12) \cdot (+4) \div (-2)(+3)$
6. $-2^4 - (-3)^2 \cdot (-1)^4$
7. $-(-2)^2 + (-1)^{1971}$

(22)

More on D p m a

By **dpma** we know that $a(b + c) = ab + ac$. But what does $(b + c) \cdot a$ equal?

$$\begin{aligned} (b + c) \cdot a &= a(b + c) && \textbf{cpm} \\ &= ab + ac && \textbf{dpma} \\ &= ba + ca && \textbf{cpm} \end{aligned}$$

(23)

$$(b + c) \cdot a = ba + ca$$

$a(b + c) = ab + ac$ left hand distributive law. $(b + c) \cdot a = ba + ca$ right hand distributive law.

ldpma ⟷ **rdpma** ⟷ **dpma**

(⟷ means "is equivalent to")

(24)

What about $a(b - c)$?

Line (1)	$a(b - c) = a(b + [-c])$
Line (2)	$= ab + a[-c]$
Line (3)	$= ab + [-(ac)]$
Line (4)	$= ab - ac$

Thus, we state the distributive law of multiplication over subtraction:

$$a(b - c) = ab - ac$$

(25)

The distributive law can also be used to change the basis product $(a + b)(c + d)$ into a basis sum.

$$\begin{aligned} (a + b)(c + d) &= \boxed{(a + b)}\,(c + d) \\ &= (a + b) \cdot c + (a + b) \cdot d \\ &= ac + bc + ad + bd \end{aligned}$$

(26)

Study Exercise Three

Using the distributive property, express as basic sums.

1. $5(x - 3)$ **2.** $(x + 2)(y + 1)$

3. $(a + b)(c + 2)$ **4.** $(a + b + c)(x + 1)$

(27)

Let us now apply the distributive law on $(a - b)(c - d)$.

$$\begin{aligned}(a - b)(c - d) &= (a - b)c - (a - b)d \\ &= ac - bc - (ad - bd) \\ &= ac - bc - ad + bd\end{aligned}$$

(28)

Study Exercise Four

Use the distributive property and write as basic sums.

1. $(3a + 1)(x - 2)$ **2.** $(a + b + c)(x - y + 1)$ **3.** $(x - 1)(a - b - c)$

(29)

Any grouping of constants and variables generated by applying a finite number of the basic operations is called an *algebraic expression*.

In any algebraic expression of the form $A + B + C + D + \ldots$, where $A, B, C, D, \ldots$ are algebraic expressions, $A, B, C, D, \ldots$ are called *terms of the expression*.

(30)

In $3x + 2y$ there are two terms, $3x$ and $2y$.
In $x - y + 7$ there are three terms, x, $-y$, and 7.

In $2x^2y$ there is only one term.
In $3x - 2(y - 1) + 5y^2$ there are three terms, namely $3x$, $-2(y - 1)$, and $5y^2$.

(31)

An algebraic expression of one term is called a *monomial*.

An algebraic expression of two terms is called a *binomial*.

An algebraic expression of three terms is called a *trinomial*.

(32)

3, $2x^2y$, $5x$, $2/x$ are examples of monomials.

$x + y$, $x^2 - 2y^3$, $\frac{1}{x} + \frac{1}{y}$ are binomials.

$1 + x + y$, $x^2 + 2x + 1$ are trinomials.

(33)

Any factor or group of factors in a term is said to be the *coefficient* of the remaining factors in the term.

Consider the term $5xy$.
The coefficient of 5 is xy.
The coefficient of xy is 5.
The coefficient of $5y$ is x.
The coefficient of $5xy$ is 1.
The numerical coefficient is 5.

(34)

Terms that differ only in their numerical coefficients are called *like terms* or *similar terms*.

Like terms are combined by use of the distributive property.

(35)

Example 1: Combine $5x + 3x$

Solution: $5x + 3x = (5 + 3)x$
$= 8x$

Example 2: Combine $7x^2y - 2x^2y$

Solution: $7x^2y - 2x^2y = (7 - 2)x^2y$
$= 5x^2y$

Example 3: Combine $4x^3 - x^3 + 2x^3$

Solution: $4x^3 - x^3 + 2x^3 = (4 - 1 + 2)x^3$
$= 5x^3$

(36)

An algebraic expression is in *simplest form* when no variable is included in a grouping symbol, all like terms are combined, and the variable does not appear in any term more than once.

Example: Write in simplest form: $2 - [3x - (x + 1) + 2]$

Solution: $2 - [3x - x - 1 + 2] = 2 - [2x + 1]$
$= 2 - 2x - 1$
$= 1 - 2x$

(37)

Study Exercise Five

Write in simplest form:

1. $(2x + 3) - (-x + 2)$

2. $(x + 2y) - (2x - 3y)$

3. $4 - [3x + (x - 3) - 1]$

4. $x - \{3y - [x - (x - y)] + x\}$

5. $-(x - 1) + [2x - (3 + x) - 2]$

(38)

A *polynomial* is an expression of one term or sum of terms of the form ax^n where a is a real number and where n is a natural or each term is of the form c where c is a constant.

Examples of polynomials include: $3x^3 + 2x - 5$, $5x$, 7, and $x^2 - 4$.

(39)

If a polynomial does not contain a variable it is called a *constant polynomial.*

$3x + 2$ is a polynomial
5 is a constant polynomial
$x^2 + \frac{5}{x} + 2$ is not a polynomial since the term $\frac{5}{x}$ is not of the form $a \cdot x^n$.

(40)

A polynomial in one variable can always be written in decreasing powers of that variable. When the polynomial is written in this form, the *degree* of the polynomial is the highest exponent on the variable in the leading term.

The constant polynomial has degree zero if the constant is not zero. We will agree that the zero polynomial has no degree.

(41)

Examples:

1. $x^2 + 2x + 1$ is of degree 2.

2. $5x - 7 + 4x^3$ is of degree 3.

3. $4 - x$ is of degree 1.

4. $5x$ is of degree 1.

5. 7 is of degree 0.

6. 0 has no degree.

(42)

REVIEW EXERCISES

A. Evaluate:

1. $(-3)(-2) - (-3) + (-1)$ **2.** $[4 + (2 - 5)] - [6 - (5 - 2)]$

3. $(-2)[3(4 - 2) + 6] + 4 \div 2 \cdot 2$ **4.** $-4^2 - 1^2 + 2(-3)^2$

B. Express as basic sums and write in simplest form:

5. $(x + 1)^2$ **6.** $(x + 1)(x + 3)$

7. $(x - y + 1)(x + 1)$

C. Miscellaneous

8. Simplify $7a - (-4) \cdot b + (-5) \cdot a - (+6) \cdot b$

9. If $15 = 3 \cdot 5$, then 3 and 5 are __________ of 15.

10. The numerical coefficient of $3x^2y$ *is* __________.

11. The expression $5x^2y^3z$ consists of __________ term(s).

12. The base of the exponent 3 in the expression -2^3 is __________.

13. $10/0 =$ __________.

(43)

SOLUTIONS TO REVIEW EXERCISES

A. **1.** $(6) + 3 + (-1) = 8$ **2.** $[4 + (-3)] - [6 - 3] = 1 - (3) = -2$

3. $-2[3 \cdot 2 + 6] + 2 \cdot 2 = -2(12) + 4 = -20$ **4.** $-16 - 1 + 18 = 1$

B. **5.** $x^2 + 2x + 1$ **6.** $x^2 + 4x + 3$

7. $(x - y + 1)x + (x - y + 1) \cdot 1 = x^2 - xy + x + x - y + 1 = x^2 - xy + 2x - y + 1$

C. **8.** $7a + 4b - 5a - 6b = 2a - 2b$ **9.** factors. **10.** 3

11. one **12.** 2 **13.** does not exist

(44)

SOLUTIONS TO STUDY EXERCISES

Study Exercise One (Frame 8)

1. 27 **2.** $-2[(-6) + (-2)] = -2(-8) = 16$ **3.** 0

4. $24 - [72] = -48$ **5.** $(-36) - [-140] = -36 + 140 = 104$

(8A)

Study Exercise Two (Frame 22)

1. $(-20) - (72) = -92$ **2.** $-2[0] + 15 - 10 = 0 + 5 = 5$

3. $5 \cdot 5 - (-56) = 25 + 56 = 81$

4. $-9 - [-10 \div (-2)] + (-42) = -9 - 5 + (-42) = -56$

5. $-10 + [(-48) \div (-2)](3) = -10 + 24(3) = 62$

6. $-16 - (9)(1) = -25$ **7.** $-4 + (-1) = -5$

(22A)

SOLUTIONS TO STUDY EXERCISES, CONTD.

Study Exercise Three (Frame 27)

1. $5x - 15$
2. $(x + 2) \cdot y + (x + 2) \cdot 1 = xy + 2y + x + 2$
3. $(a + b) \cdot c + (a + b) \cdot 2 = ac + bc + 2a + 2b$
4. $(a + b + c) \cdot x + (a + b + c) \cdot 1 = ax + bx + cx + a + b + c$

27A

Study Exercise Four (Frame 29)

1. $(3a + 1)\,x - (3a + 1)\,(2) = 3ax + x - 6a - 2$
2. $(a + b + c)\,x - (a + b + c)\,y + (a + b + c)\,1 =$
   $ax + bx + cx - ay - by - cy + a + b + c$
3. $(x - 1)\,a - (x - 1)\,b - (x - 1)\,c = ax - a - bx + b - cx + c$

29A

Study Exercise Five (Frame 38)

1. $3x + 1$ **2.** $-x + 5y$ **3.** $4 - [4x - 4] = -4x + 8$
4. $x - \{3y - x + x - y + x\} = x - \{2y + x\}$
   $= -2y$
5. $-x + 1 + [2x - 3 - x - 2] = -4$

38A

UNIT 4—SUPPLEMENTARY PROBLEMS

1. Find a basic numeral for each of the following:
(a) $-3\,[(2 + (-3)) + 3]$ **(b)** $(-2 + 1)\,(-3 + [-3])$
(c) $[-5 + (-2)]\,[-2 + 1]\,(0)$ **(d)** $-(-12)\,(-2) + (-1)\,(-3)$
(e) $(-7 + 8)\,[-15 + (-7)]$

2. Write a multiplication statement equivalent to:
(a) $55 \div 5 = 11$ **(b)** $5/1 = 5$

3. Write a division statement equivalent to:
(a) $9(3x) = 27x$ **(b)** $7(x + 1) = 7x + 7$

4. If $x = -1$, $y = 3$, $z = 2$, evaluate:
(a) $-y^2$ **(b)** $4x^3y^2 - 3xz^2$
(c) $(z + 2x)y^3$ **(d)** $x^2 - xz + x$

5. Simplify and collect like terms:
(a) $3 - \{2x - [1 - (x + y)] + [x - 2y]\}$ **(b)** $4 - 2x\,\{x + [x + 2z] - 1\}$

6. Apply the distributive law and collect like terms:
(a) $(x - 1)\,(x^2 + x + 1)$ **(b)** $(y + 4)\,(y^2 - 4y + 16)$
(c) $(x + 3)\,(x^2 - x + 2)$

7. What is the degree of:
(a) $x + 1$ **(b)** $1 + x - x^2$

8. Express as a product of primes:
(a) 36 **(b)** 50

9. Explain why the statement $x/x = 1$ is not true for all x.

10. Find
(a) -2^2 **(b)** $0/x$ **(c)** $x/0$ **(d)** $4/0$ **(e)** $0/4$

11. If $x < 0$, is $-x$ positive, negative, or zero.?

Factoring

unit **5**

Objectives

1. Know the first two laws of exponents.
2. Know the basic factoring types.

(1)

First Law of Exponents

If $m, n \in N$, then x^m means x is used as a factor m times and x^n means x is used as a factor n times.

$$\begin{aligned} x^m \cdot x^n &= \underbrace{(x \cdot x \cdot x \cdot \ldots x)}_{m\ x\text{'s}} \underbrace{(x \cdot x \cdot x \cdot \ldots x)}_{n\ x\text{'s}} \\ &= \underbrace{(x \cdot x \cdot x \cdot \ldots x)}_{(m+n)\ x\text{'s}} \\ x^m \cdot x^n &= x^{m+n},\ m, n \in N \qquad \textbf{First Law of Exponents} \end{aligned}$$

(2)

Example 1: $$\begin{aligned} (x^5)(2x^3) &= (x^5)(x^3 \cdot 2) \\ &= (x^5 \cdot x^3) \cdot 2 \\ &= x^{5+3} \cdot 2 \\ &= x^8 \cdot 2 \\ &= 2x^8 \end{aligned}$$

Example 2: $$\begin{aligned} x^n(3x + x^2) &= x^n(3x) + x^n \cdot x^2 \\ &= 3x^{n+1} + x^{n+2},\ n \in N \end{aligned}$$

(3)

Study Exercise One

Write each in simplest form.

1. $2(-5a)(a^2)$
2. $-2x(xy)(y^3)$
3. $ab(a^3 - ab + b)$
4. $x^n \cdot x^n,\ n \in N$
5. $(x^{2n} - 2)(x^{2n} + 2),\ n \in N$

(4)

To simplify x^6/x^2 we proceed as follows:

$$\begin{aligned} \frac{x^6}{x^2} &= \frac{(x^4 \cdot x^2)}{x^2} \\ &= (x^4 \cdot x^2) \cdot 1/x^2 \\ &= x^4(x^2 \cdot 1/x^2) \\ &= x^4 \cdot 1 \\ &= x^4,\ x \neq 0 \end{aligned}$$

(5)

Second Law of Exponents

In simplifying $\frac{x^m}{x^n}$, $m, n \in N, m > n$, we get

$$\begin{aligned}\frac{x^m}{x^n} &= \frac{x^{m-n} \cdot x^n}{x^n} \\ &= (x^{m-n} \cdot x^n) \cdot \frac{1}{x^n} \\ &= x^{m-n}\left(x^n \cdot \frac{1}{x^n}\right) \\ &= x^{m-n} \cdot 1 \\ &= x^{m-n}, x \neq 0\end{aligned}$$

(6)

$\frac{x^m}{x^n} = x^{m-n}$, $m, n \in N$ and $m > n$, $x \neq 0$, is called the *Second Law of Exponents.*

(7)

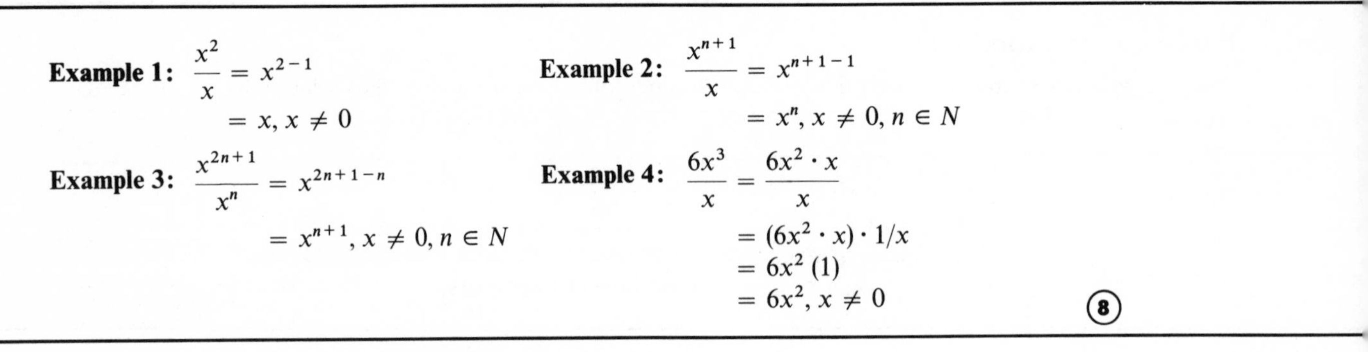

Example 1: $\frac{x^2}{x} = x^{2-1} = x, x \neq 0$

Example 2: $\frac{x^{n+1}}{x} = x^{n+1-1} = x^n, x \neq 0, n \in N$

Example 3: $\frac{x^{2n+1}}{x^n} = x^{2n+1-n} = x^{n+1}, x \neq 0, n \in N$

Example 4: $\frac{6x^3}{x} = \frac{6x^2 \cdot x}{x} = (6x^2 \cdot x) \cdot 1/x = 6x^2\,(1) = 6x^2, x \neq 0$

(8)

Study Exercise Two

Simplify by using the laws of exponents:

1. $\frac{x^8}{x^2}$
2. $\frac{x^{3n+1}}{x^{2n}}, n \in N$
3. $\frac{12ab^2}{ab}$
4. $\frac{-a^3b}{ab}$
5. $\frac{y^{2n}}{y^{n+1}}, n \in N, n > 1$
6. $\frac{x^{10}}{x^2} \cdot \frac{x^5}{x^3}$
7. $\frac{25x^2}{5x}$

(9)

The *distributive property* allows a basic sum to be written as a basic product. $ab + ac = a(b + c)$

Example 1: $ax + 2x = x(a + 2)$

Example 2: $2x^2 + 4x = 2x(x + 2)$ *or* $2(x^2 + 2x)$

(10)

Writing a basic sum into a basic product is sometimes called *undistributing.*

Example 1: $-2x + 2 = -2(\ ?\ )$
Solution: $-2x + 2 = -2(x - 1)$

Example 2: $x^{n+1} + x = x\,(?)$
Solution: $x^{n+1} + x = x(x^n + 1)$

(11)

Study Exercise Three

Supply the missing factors:

1. $3x^2 - 3xy + x = x(\_\_\_\_)$

2. $4x^2 + 8x^3 = 4x^2(\_\_\_\_)$

3. $x^{a+2} - 2x^2 = x^2(\_\_\_\_), a \in N$

4. $-6x - 9 = -3(\_\_\_\_)$

5. $-x^{5n} + x^{2n} = \_\_\_\_(x^{3n} - 1), n \in N$

6. $-2x + 2 = -2(\_\_\_\_)$

(12)

By using the distributive property:

$(x + a)(x + b) = (x + a)x + (x + a)b = x^2 + ax + bx + ab$

$$\overset{ax}{\frown}$$

$(x + a)(x + b) = x^2 + ax + bx + ab$ is a shortcut method.

$$\underset{bx}{\smile}$$

(13)

$$(x + 4)(x + 5) = x^2 + 4x + 5x + 20 \quad (\text{inner: } 4x, \text{ outer: } 5x)$$
$$= x^2 + 9x + 20$$

$$(3x - 1)(2x + 3) = 6x^2 - 2x + 9x - 3 \quad (\text{inner: } -2x, \text{ outer: } 9x)$$
$$= 6x^2 + 7x - 3$$

(14)

Other Products

Consider $(x + a)(x - a)$. This could be called the product of the sum and difference of the same two quantities.

$$(x + a)(x - a) = x^2 + ax - ax - a^2 \quad (\text{inner: } ax, \text{ outer: } -ax)$$
$$= x^2 - a^2$$

The result is called the *difference of two squares.*

(15)

Two other products occur quite often.

$$\begin{aligned}(x + a)^2 &= (x + a)(x + a)\\ &= x^2 + ax + ax + a^2\\ &= x^2 + 2ax + a^2\\ (ax + by)(cx + dy) &= axcx + bycx + axdy + bydy\\ &= acx^2 + bcxy + adxy + bdy^2\end{aligned}$$

(16)

Study Exercise Four

Write as basic sums.

1. $(x + 5)(x - 3)$

2. $(3x + 2)(x + 4)$

3. $(2x + 4y)(x - y)$

4. $(x + 4)(x - 4)$

5. $(x + 3y)(x + 3y)$

(17)

Other Special Products

$$\begin{aligned}(x - a)(x^2 + ax + a^2) &= (x - a)x^2 + (x - a)ax + (x - a)a^2 \\ &= x^3 - ax^2 + ax^2 - a^2x + a^2x - a^3 \\ &= x^3 - a^3\end{aligned}$$

Difference of 2 cubes: $x^3 - a^3 = (x - a)(x^2 + ax + a^2)$

(18)

$$\begin{aligned}(x + a)(x^2 - ax + a^2) &= (x + a)x^2 - (x + a)ax + (x + a)a^2 \\ &= x^3 + ax^2 - ax^2 - a^2x + a^2x + a^3 \\ &= x^3 + a^3\end{aligned}$$

Sum of 2 cubes: $x^3 + a^3 = (x + a)(x^2 - ax + a^2)$

(19)

Factoring

Factoring is essentially a process of finding two or more factors whose product is the given expression. Factoring converts basic sums into basic products. That is, the factored form must consist of one term.

(20)

So that factoring will result in a unique answer, we will insist to factor a polynomial with integral coefficients that;

(1) *it is written as a product of polynomials*
(2) *no polynomial, other than a monomial contains a polynomial factor with integral coefficients.*

(21)

Thus to factor $4x^2 + 8x^3$, we could write:

$$\begin{aligned}&x(4x + 8x^2) \\ \text{or } &x^2(4 + 8x) \\ \text{or } &4(x^2 + 2x^3) \\ \text{or } &4x(x + 2x^2) \\ \text{or } &4x^2(1 + 2x)\end{aligned}$$

But, only $4x^2(1 + 2x)$ is correct.

(22)

Factoring Types

Two Terms

1. Remove the monomial factor
2. Difference of two squares
3. Difference of two cubes
4. Sum of two cubes

(23)

The difference of two squares: $x^2 - y^2 = (x + y)(x - y)$

The difference of two cubes: $x^3 - y^3 = (x - y)(x^2 + xy + y^2)$

The sum of two cubes: $x^3 + y^3 = (x + y)(x^2 - xy + y^2)$

(24)

Factoring over the integers means factor using integral coefficients.

Example 1: Factor $8x + 16$
Solution: $8(x + 2)$

Example 2: Factor $x^2 - 49$
Solution: $(x)^2 - (7)^2 = (x + 7)(x - 7)$

Example 3: Factor $2a^2b^2 - 8$
Solution: $2(a^2b^2 - 4) = 2[(ab)^2 - (2)^2]$
$= 2(ab + 2)(ab - 2)$

(25)

More Factoring Problems

Example 1: Factor $x^4 - 1$
Solution: $(x^2)^2 - (1)^2 = (x^2 + 1)(x^2 - 1)$
$= (x^2 + 1)(x + 1)(x - 1)$

Example 2: Factor $8x^3 - y^3$
Solution: $(2x)^3 - (y)^3 = (2x - y)(4x^2 + 2xy + y^2)$

Example 3: Factor $x^3 + 27$
Solution: $(x)^3 + (3)^3 = (x + 3)(x^2 - 3x + 9)$

(26)

Study Exercise Five

Factor completely over the integers.

1. $5ax + 15ay$
2. $x^3y - xy^3$
3. $9a^2 - 4b^2$
4. $9x^2 - 36$
5. $a^3 + b^3c^3$
6. $8x^3 - 27y^3$
7. $(a^2b)^2 - 4$
8. $2x^4 - 18$

(27)

Factoring Types

Three terms:

1. Remove the monomial factors.
2. Perfect square trinomials.
3. Trinomials that are not perfect squares.

(28)

Perfect Square Trinomials

$$x^2 + 2xy + y^2 = (x + y)^2$$
$$x^2 - 2xy + y^2 = (x - y)^2$$

(29)

Trinomials that are not Perfect Squares

Example: $6x^2 + 13xy - 5y^2$

6 is the product of 6 and 1 or 3 and 2.
5 is the product of 5 and 1.
Possibilities include $(3x + 5y)(2x - y)$, $(3x - 5y)(2x + y)$, $(3x + y)(2x - 5y)$, and $(3x - y)(2x + 5y)$

The correct factorization is: $(3x - y)(2x + 5y)$.

(30)

Example 1: Factor $5x^3 + 10x + 20$
Solution: $5(x^3 + 2x + 4)$

Example 2: Factor $4x^2 + 12x + 9$
Solution: $(2x + 3)^2$

Example 3: Factor $x^2 + 5x + 6$
Solution: $(x + 3)(x + 2)$

(31)

Study Exercise Six

Factor completely over the integers:

1. $x^2 + x - 2$ **2.** $6x^2 - x - 1$ **3.** $2x^2 - 11x + 5$

4. $y^2 - 8y + 16$ **5.** $36x^2 - 12x + 1$ **6.** $10x^2 - 3x - 18$

7. $x^4 - 6x^2 - 27$ **8.** $x^4 + 3x^2 - 4$

(32)

More Difficult Factoring

Example 1: Factor $x^2(a + 2b) - 4(a + 2b)$

undistribute $(a + 2b)$ $(a + 2b)(x^2 - 4)$

difference of 2 squares $(a + 2b)(x + 2)(x - 2)$

Example 2: Factor $x^4 - (y^2 + 1)^2$

difference of 2 squares $(x^2)^2 - (y^2 + 1)^2$

$[x^2 + (y^2 + 1)][x^2 - (y^2 + 1)]$

$(x^2 + y^2 + 1)(x^2 - y^2 - 1)$

(33)

Example: Factor $(x + y)^3 + 27$

$(x + y)^3 + (3)^3$

$[(x + y) + 3][(x + y)^2 - 3(x + y) + 9]$

$[x + y + 3][x^2 + 2xy + y^2 - 3x - 3y + 9]$

(34)

Study Exercise Seven

Factor completely over the integers:

1. $(x + 1)^2 - y^2$ **2.** $1 - (x - y)^3$

3. $(x - y - 1)^2 - (2x + y + 1)^2$ **4.** $x^2(x + 2) - 4(x + 2)$

(35)

Factoring 4 Or More Terms

Group in some manner and consider as two or three terms. Then apply the techniques for two or three terms.

Example: Factor $x^3 + 2x^2 - 7x - 14$

Group the first two and last two terms: $(x^3 + 2x^2) - (7x + 14)$

Factor each grouping separately: $x^2(x + 2) - 7(x + 2)$

Undistribute the common factor: $(x + 2)(x^2 - 7)$

(36)

Example: Factor $x^3 + x^2 - 4x - 4$

Solution: $(x^3 + x^2) - (4x + 4)$

$x^2(x + 1) - 4(x + 1)$

$(x + 1)(x^2 - 4)$

Note that $x^2 - 4$ is still factorable.

$(x + 1)(x + 2)(x - 2)$

(37)

Study Exercise Eight

Factor completely:

1. $y^3 - 2y^2 + 5y - 10$ **2.** $x^3 + 3x - 7x^2 - 21$

3. $2x - 6 - xy^2 + 3y^2$ **4.** $4x^3 - 12x^2 - x + 3$

(38)

REVIEW EXERCISES

A. Simplify:

1. $\frac{x^{10}}{x^2}$

2. $3x^2(x^5y^2)$

3. $\frac{x^{2n+1}}{x^{n-1}}, n \in N, n \geqslant 2$

4. $x^8 \cdot x^4$

5. $\frac{4x^4y^2}{xy}$

B. Factor completely over the integers:

6. $x^4 - 1$

7. $2x^2 - 9x - 5$

8. $16x^8 - 7x^6 + 6x^7$

9. $3x^3 + 3$

10. $3x^3 + 2x^2 - 12x - 8$

11. $9x^2 - 36$

12. $xy + xz$

13. $4y^4 - (x - y)^2$

14. $27x^6 - 8$

(39)

SOLUTIONS TO REVIEW EXERCISES

A. **1.** $x^{10-2} = x^8, \quad x \neq 0$

2. $3x^{2+5} \cdot y^2 = 3x^7y^2$

3. $x^{(2n+1)-(n-1)} = x^{n+2}, \quad x \neq 0$

4. $x^{8+4} = x^{12}$

5. $\frac{4x^3y(xy)}{xy} = 4x^3y, x \neq 0, \quad y \neq 0$

B. **6.** $(x^2 + 1)(x^2 - 1) = (x^2 + 1)(x + 1)(x - 1)$

7. $(2x + 1)(x - 5)$

8. $x^6(16x^2 + 6x - 7) = x^6(2x - 1)(8x + 7)$

9. $3(x^3 + 1) = 3(x + 1)(x^2 - x + 1)$

10. $x^2(3x + 2) - 4(3x + 2) = (3x + 2)(x^2 - 4)$
$= (3x + 2)(x + 2)(x - 2)$

11. $9(x^2 - 4) = 9(x + 2)(x - 2)$

12. $x(y + z)$

13. $[2y^2 + (x - y)][2y^2 - (x - y)] = (2y^2 + x - y)(2y^2 - x + y)$

14. $(3x^2 - 2)(9x^4 + 6x^2 + 4)$

(40)

SOLUTIONS TO STUDY EXERCISES

Study Exercise One (Frame 4)

1. $-10a^3$

2. $-2x^2y^4$

3. $a^4b - a^2b^2 + ab^2$

4. $x^{n+n} = x^{2n}$

5. $x^{2n+2n} - 4 = x^{4n} - 4$

(4A)

Study Exercise Two (Frame 9)

1. $x^{8-2} = x^6, \quad x \neq 0$

2. $x^{(3n+1)-2n} = x^{n+1}, \quad x \neq 0$

3. $12b, a \neq 0, \quad b \neq 0$

4. $-a^2, a \neq 0, \quad b \neq 0$

5. $y^{2n-(n+1)} = y^{n-1}, \quad y \neq 0$

6. $x^{10-2} \cdot x^{5-3} = x^8 \cdot x^2$
$= x^{8+2}$
$= x^{10}, \quad x \neq 0$

7. $5x, x \neq 0$

(9A)

SOLUTIONS TO STUDY EXERCISES, CONTD.

Study Exercise Three (Frame 12)

1. $3x - 3y + 1$ **2.** $1 + 2x$ **3.** $x^a - 2$
4. $2x + 3$ **5.** $-x^{2n}$ **6.** $x - 1$

12A

Study Exercise Four (Frame 17)

1. $x^2 + 2x - 15$ **2.** $3x^2 + 14x + 8$ **3.** $2x^2 + 2xy - 4y^2$
4. $x^2 - 16$ **5.** $x^2 + 6xy + 9y^2$

17A

Study Exercise Five (Frame 27)

1. $5a(x + 3y)$ **2.** $xy(x + y)(x - y)$
3. $(3a + 2b)(3a - 2b)$ **4.** $9(x + 2)(x - 2)$
5. $(a + bc)(a^2 - abc + b^2c^2)$ **6.** $(2x - 3y)(4x^2 + 6xy + 9y^2)$
7. $(a^2b + 2)(a^2b - 2)$ **8.** $2(x^4 - 9) = 2(x^2 + 3)(x^2 - 3)$

27A

Study Exercise Six (Frame 32)

1. $(x + 2)(x - 1)$ **2.** $(3x + 1)(2x - 1)$ **3.** $(2x - 1)(x - 5)$
4. $(y - 4)^2$ **5.** $(6x - 1)^2$ **6.** $(5x + 6)(2x - 3)$
7. $(x^2 - 9)(x^2 + 3) = (x + 3)(x - 3)(x^2 + 3)$
8. $(x^2 + 4)(x^2 - 1) = (x^2 + 4)(x + 1)(x - 1)$

32A

Study Exercise Seven (Frame 35)

1. $(x + 1 + y)(x + 1 - y)$
2. $[1 - (x - y)][1 + (x - y) + (x - y)^2] =$
$(1 - x + y)(1 + x - y + x^2 - 2xy + y^2)$
3. $[(x - y - 1) + (2x + y + 1)][(x - y - 1) - (2x + y + 1)] =$
$(3x)(-x - 2y - 2)$ or $-(3x)(x + 2y + 2)$
4. $(x + 2)(x^2 - 4) = (x + 2)(x + 2)(x - 2)$

35A

Study Exercise Eight (Frame 38)

1. $y^2(y - 2) + 5(y - 2) = (y - 2)(y^2 + 5)$
2. $x(x^2 + 3) - 7(x^2 + 3) = (x^2 + 3)(x - 7)$
3. $2(x - 3) - y^2(x - 3) = (x - 3)(2 - y^2)$
4. $4x^2(x - 3) - (x - 3) = (x - 3)(4x^2 - 1)$
$= (x - 3)(2x + 1)(2x - 1)$

38A

UNIT 5—SUPPLEMENTARY PROBLEMS

A. Simplify:

1. $\dfrac{x^{14}}{x^2}$

2. $\dfrac{x^{3n+1}}{x^2}, n \in N$

3. $\dfrac{x^{2n+2}}{x^n}, n \in N$

4. $2x^4y(-3x^3y^4)$

5. $-2x^2y^3(2x^4y^6)$

6. $x^4(x^6)(x^2)$

B. Factor completely over the integers:

7. $36x^6 - 13x^4 + x^2$

8. $y^6 + 1$

9. $4 - 12x + 9x^2$

10. $4a(x + y) - (x + y)$

11. $x + y + 7ax + 7ay$

12. $ax^2 + bx - ax - b$

13. $2x^2 + 11x - 21$

14. $6x^3 - x^2 - 6x + 1$

15. $3x^2 - 12y^2$

16. $(s + t)^2 - 9a^2$

17. $9x^2 - y^2$

18. $2xy + 2y^2 + y + x$

19. $2x^3 - 4x^2 + 6x$

20. $(x + 1)^2 - 36y^2$

21. $(5x + 2y)^2 - (3x - 7y)^2$

22. $(x - 2)^3 + 8y^3$

23. $x^3y^3 - y^3 + 8x^3 - 8$

24. $3x^2 + 7ax - 6a^2$

25. $y^2 + 2yz + z^2 - 4x^2$

26. $8a^2c - 18\ c^3$

27. $8x^3 - 125y^3$

unit

Fractions

6

Objectives

1. Know the Fundamental Principle of Fractions.
2. Be able to add, subtract, multiply, and divide algebraic fractions and express the results in lowest terms.
3. Be able to use the division algorithm on polynomials.

Terms complex fraction

(1)

A *fraction* is an expression which appears in the form shown below.

A fraction denotes a quotient.

$\frac{a}{b}$ — a ⟵ numerator, b ⟵ denominator, where $b \neq 0$

$\frac{a}{b}$ — a ⟵ dividend, b ⟵ divisor, where $b \neq 0$

(2)

Theorem. If $a, b, c, d \in R$; $b, d \neq 0$, and $a/b = c/d$, then $ad = bc$

Proof:

1. $a, c, b, d \in R$; $b, d \neq 0$, $a/b = c/d$	1. **given**
2. $(a/b)(bd) = (c/d)(bd)$	2. **multiplication law of equality**
3. $(a \cdot 1/b)(bd) = (c \cdot 1/d)(bd)$	3. **definition of quotient**
4. $(ad)(b \cdot 1/b) = (bc)(d \cdot 1/d)$	4. **repeated use of cpm and apm**
5. $(ad) \cdot 1 = (bc) \cdot 1$	5. **mult. inv.**
6. $ad = bc$	6. **mult. ident.**

(3)

Theorem. If $a, b, c, d \in R$; $b, d \neq 0$ and $ad = bc$, then $a/b = c/d$

Proof:

1. $a, b, c, d \in R$; $b, d \neq 0$, $ad = bc$	1. **given**
2. $1/b$, $1/d$ exists	2. **mult. inv.**
3. $(ad)(1/b)(1/d) = (bc)(1/b)(1/d)$	3. **multiplication law for equality**
4. $a(1/b)(d \cdot 1/d) = c(1/d)(b \cdot 1/b)$	4. **repeated use of cpm and apm**
5. $a(1/b)(1) = c(1/d)(1)$	5. **mult inv.**
6. $a(1/b) = c(1/d)$	6. **mult ident.**
7. $a/b = c/d$	7. **definition of quotient**

(4)

$a, b, c, d \in R, b, d \neq 0, a/b = c/d$ if and only if $ad = bc$ means if $a/b = c/d$, then $ad = bc$ *and* if $ad = bc$, then $a/b = c/d$.

(5)

Fundamental Principle of Fractions

Let $a, b, c \in R, b \neq 0, c \neq 0$, then;

Line (1)	$abc = abc$
Line (2)	$abc = bac$
Line (3)	$a(bc) = b(ac)$
Line (4)	$a/b = ac/bc$

Fundamental principle of fractions

(6)

Fundamental Principle of Fractions

If $a, b, c \in R$ and $b, c \neq 0$, then $a/b = ac/bc$ or $ac/bc = a/b$

The fundamental principle states,

1. You may multiply numerator and denominator of a fraction by the same non zero number and the resulting fraction is equivalent to the original fraction.
2. If a fraction contains a common factor in its numerator and denominator, the fraction without the common factor is equivalent to the original fraction.

(7)

Example 1: $\dfrac{3x}{2} = \dfrac{?}{6}$

Solution: $\dfrac{3x}{2} = \dfrac{(3x)\cdot 3}{(2)\cdot 3}$

$= \dfrac{9x}{6}$

Example 2: $\dfrac{ay}{b} = \dfrac{?}{bx}$

Solution: $\dfrac{(ay)}{b} = \dfrac{(ay)\cdot x}{b \cdot x}$

$= \dfrac{axy}{bx}, b \neq 0, \quad x \neq 0$

Example 3: $\dfrac{5}{x+2} = \dfrac{?}{x^2 - 4}$

Solution: $\dfrac{5}{x+2} = \dfrac{(5)(x-2)}{(x+2)(x-2)}$

$= \dfrac{5x - 10}{x^2 - 4}, \quad x \neq 2, \neq -2$

(8)

Further examples:

Example 1: Write a fraction equivalent to $\dfrac{x^2 - 1}{x + 1}$

Solution: $\dfrac{x^2 - 1}{x + 1} = \dfrac{(x+1)(x-1)}{(x+1)\cdot 1} = \dfrac{x-1}{1} = x - 1, \quad x \neq -1$

Example 2: Write a fraction equivalent to $\dfrac{(a-b)^2}{a^2 - b^2}$

Solution: $\dfrac{(a-b)^2}{a^2 - b^2} = \dfrac{(a-b)(a-b)}{(a+b)(a-b)}$

$= \dfrac{a-b}{a+b}, a \neq -b, \quad a \neq b$

(9)

Lowest Terms

A fraction is reduced or is in lowest terms if its numerator and denominator do not contain a common factor.

In order to reduce a fraction:
1. factor the numerator
2. factor the denominator
3. apply the fundamental principle of fractions.

(10)

Example 1: Reduce $\dfrac{2x + 2y}{x + y}$

Solution: $$\frac{2x + 2y}{x + y} = \frac{2(x + y)}{(x + y) \cdot 1} = 2/1 = 2, \quad x \neq -y$$

Example 2: Reduce $\dfrac{x^2 - 1}{x + 1}$

Solution: $$\frac{x^2 - 1}{x + 1} = \frac{(x + 1)(x - 1)}{(x + 1) \cdot 1} = \frac{x - 1}{1} = x - 1, \quad x \neq -1$$

(11)

Study Exercise One

A. Which of the following fractions are given in lowest terms?

1. $\dfrac{2x + 3}{2}$ **2.** $\dfrac{(a - 2)^2}{2}$ **3.** $\dfrac{x^2 + xy}{y}, y \neq 0$ **4.** $\dfrac{1 + x}{x}, x \neq 0$

B. Reduce to lowest terms:

5. $\dfrac{x^2 + x - 6}{x - 2}$ **6.** $\dfrac{2x^2 - 2x}{2x}$ **7.** $\dfrac{y^2 - y - 2}{y + 1}$ **8.** $\dfrac{ax - a}{a}$

(12)

The Signs of a Fraction

There are three signs associated with a fraction:

(1) a sign for the numerator
(2) a sign for the denominator
(3) a sign for the fraction itself

$$-\frac{-2}{+3}$$

(13)

The fraction $-a/b$ may also be written:

(1) $a/-b$ by multiplying numerator and denominator by -1.

(2) $-(a/b)$, since $-a/b = (-a) \cdot 1/b = -(a)(1/b) = -a/b$

(3) $-(-a/-b)$, since $-a/b = a/-b = -a/b = - -a/-b$

(14)

Furthermore, the fraction a/b which is equivalent to $+ \ +a/+b$ may also be written $- \ -a/+b$ since

$$+\frac{+a}{+b} = +\frac{-a}{-b}$$
$$= -\frac{+a}{-b}$$
$$= -\frac{-a}{+b}$$

(15)

Any two of the three signs of a fraction may be changed without changing the fraction.

Example 1: $+\frac{-2}{-3} = -\frac{-(-2)}{-3} = -\frac{2}{-3}$

Example 2: $-\frac{+7}{3} = +\frac{-7}{3} = +\frac{+7}{-3}$

(16)

To change the sign if the numerator or denominator consists of more than one term, group all the terms in a grouping symbol and change the sign in front.

Example: $+\frac{x-2}{3} = \frac{-(x-2)}{-3} = \frac{-x+2}{-3}$

(17)

Example: Reduce to lowest terms: $\frac{x^2-4}{2-x}$

Solution:
$$\frac{x^2-4}{2-x} = \frac{(x+2)(x-2)}{(2-x)}$$
$$= -\frac{(x+2)(x-2)}{-(2-x)}$$
$$= -\frac{(x+2)(x-2)}{(x-2)}$$
$$= -(x+2),\ x \neq 2$$

(18)

Example: Reduce to lowest terms: $\frac{x^2-xy}{3y-3x}$

Solution:
$$\frac{x^2-xy}{3y-3x} = \frac{x(x-y)}{3(y-x)}$$
$$= -\frac{x(x-y)}{-3(y-x)}$$
$$= -\frac{x(x-y)}{3(x-y)}$$
$$= -\frac{x}{3},\ x \neq y$$

(19)

Study Exercise Two

Reduce to lowest terms:

1. $\dfrac{(a-b)^2}{(b-a)}$ 2. $\dfrac{(a-b)}{(b-a)}$ 3. $\dfrac{x^3-1}{1-x^2}$ 4. $\dfrac{y^2+5y-14}{2-y}$

(20)

Multiplication of Fractions

Let the product of a/b and c/b be p.

Line (1) $\dfrac{a}{b}\cdot\dfrac{c}{d}=p$

Line (2) $a(1/b)\,(c\cdot 1/d)=p$

Line (3) $(a\cdot 1/b)\,(c\cdot 1/d)\,(bd)=p(bd)$

Line (4) $(b\cdot 1/b)\,(d\cdot 1/d)\,(ac)=p(bd)$

Line (5) $1\cdot 1\cdot ac=p(bd)$

Line (5) $ac=p(bd)$

Line (7) $ac/bd=p$

Thus, $a/b\cdot c/d=ac/bd,\quad b,\quad d\neq 0$

(21)

In order to multiply fractions;

(1) factor each numerator
(2) factor each denominator
(3) apply the fundamental principle of fractions
(4) multiply according to $a/b\cdot c/d=ac/bd,\ b,\ d\neq 0$

(22)

Example: Multiply and express in simplest form: $\dfrac{(x-2)^2}{3}\cdot\dfrac{9}{x^2-4}$

Solution: $\dfrac{(x-2)^2}{3}\cdot\dfrac{9}{x^2-4}=\dfrac{(x-2)\cancel{(x-2)}}{\cancel{3}}\cdot\dfrac{3\cdot\cancel{3}}{(x+2)\cancel{(x-2)}}$

$=\dfrac{3(x-2)}{x+2},\quad x\neq 2,\quad x\neq -2$

(23)

Example: Multiply and express in simplest form: $\dfrac{x^2+2x}{x}\cdot\dfrac{x^2}{x^2+2x}$

Solution: $\dfrac{x^2+2x}{x}\cdot\dfrac{x^2}{x^2+2x}=\dfrac{x\cancel{(x+2)}}{\cancel{x}}\cdot\dfrac{x\cdot\cancel{x}}{x\cancel{(x+2)}}$

$=\dfrac{x}{1}$

$=x,\quad x\neq 0,\quad x\neq -2$

(24)

Study Exercise Three

Express each product as a single fraction in lowest terms:

1. $\dfrac{x^3y-y^3x}{x^2y-xy^2}\cdot\dfrac{1}{x+y}$ 2. $\dfrac{8x-8y}{16x-16y}\cdot\dfrac{(x-y)^2}{2}$

3. $\dfrac{x^2-4}{xy^2}\cdot\dfrac{2xy}{x^2-4x+4}$ 4. $\dfrac{x^2-x-6}{x^2-2x+1}\cdot\dfrac{x^2+3x-4}{9x-x^3}$

(25)

Division of Fractions

Let $\frac{a}{b} \div \frac{c}{d} = q$

Line (1) $\quad \dfrac{a/b}{c/d} = q$

Line (2) $\quad (c/d) \cdot q = a/b$

Line (3) $\quad (d/c)\,(c/d)\,q = (d/c)\,(a/b)$

Line (4) $\quad q = ad/bc$

(26)

Theorem.

If $a, b, c, d \in R$, $b, c, d \neq 0$, then $\frac{a}{b} \div \frac{c}{d} = \frac{ad}{bc}$.

Since $\frac{a}{b} \div \frac{c}{d} = \frac{ad}{bc}$ and since $\frac{a}{b} \cdot \frac{d}{c} = \frac{ad}{bc}$, we get the familiar rule invert and multiply.

$$\frac{a}{b} \div \frac{c}{d} = \frac{a}{b} \cdot \frac{d}{c} = \frac{ad}{bc},\ b, c, d \neq 0$$

(27)

Example: Perform the indicated operation and simplify $\dfrac{x + 2xy}{3x^2} \div \dfrac{2y + 1}{6x}$

Solution:

$$\frac{x + 2xy}{3x^2} \div \frac{2y + 1}{6x} = \frac{x + 2xy}{3x^2} \cdot \frac{6x}{2y + 1}$$

$$= \frac{\cancel{x}\cancel{(1 + 2y)}}{\cancel{3} \cdot \cancel{x} \cdot \cancel{x}} \cdot \frac{2 \cdot \cancel{3x}}{\cancel{(2y + 1)}}$$

$$= \frac{2}{1}$$

$$= 2,\ x \neq 0,\ y \neq -\frac{1}{2}$$

(28)

Study Exercise Four

Perform the indicated operations and simplify:

1. $\dfrac{xy^3}{yz} \div x^2z$ **2.** $\dfrac{x^2 - y^2}{x - y} \div \dfrac{x + y}{x}$ **3.** $\dfrac{x^2}{x^2 - 1} \div \dfrac{x^3}{(1 - x)^2}$

(29)

Complex Fractions

A fraction that contains a fraction in either the numerator or denominator or both is called a *complex fraction.*

Some examples of complex fractions include:

$$\frac{\frac{3}{5}}{6},\quad \frac{x + \frac{1}{x}}{1 - \frac{x}{2}},\quad \text{and } \frac{\frac{5}{2}}{\frac{3}{4}}$$

(30)

One way to simplify complex fractions is to multiply numerator and denominator by the same nonzero number.

Example:
$$\frac{\frac{5}{2}}{\frac{3}{4}} = \frac{4\cdot(\frac{5}{2})}{4\cdot(\frac{3}{4})} = \frac{2(5)}{3} = \frac{10}{3}$$

(31)

Example: Simplify $\dfrac{1+\frac{1}{x}}{1+\frac{2}{x}}$

Solution:
$$\frac{1+\frac{1}{x}}{1+\frac{2}{x}} = \frac{x(1+\frac{1}{x})}{x(1+\frac{2}{x})} = \frac{x\cdot 1 + x(\frac{1}{x})}{x\cdot 1 + x(\frac{2}{x})} = \frac{x+1}{x+2},\ x \neq 0,\ x \neq -2$$

(32)

Example: Simplify $\dfrac{\frac{1}{x^2}}{1-\frac{1}{x}}$

Solution:
$$\frac{\frac{1}{x^2}}{1-\frac{1}{x}} = \frac{x^2\left(\frac{1}{x^2}\right)}{x^2\left(1-\frac{1}{x}\right)} = \frac{1}{x^2-x},\ x \neq 0,\ x \neq 1$$

(33)

Study Exercise Five

Write as a single fraction in lowest terms:

1. $\dfrac{\frac{3}{5}}{5-\frac{1}{5}}$

2. $\dfrac{x}{\frac{x}{x-1}}$

3. $\dfrac{x-1}{1-\dfrac{2}{x^2}}$

4. $\dfrac{1+\frac{1}{x}}{1-\frac{1}{x}}$

5. $\dfrac{\frac{2}{x}+\frac{3}{2x}}{5+\frac{1}{x}}$

6. $\dfrac{\frac{1}{2}+\frac{1}{3}}{\frac{1}{6}-\frac{3}{2}}$

(34)

Addition of Fractions

$$\frac{a}{b} + \frac{c}{b} = a\left(\frac{1}{b}\right) + c\left(\frac{1}{b}\right)$$

$$= (a + c)\cdot\frac{1}{b}$$

$$= \frac{a + c}{b}$$

Thus, $\frac{a}{b} + \frac{c}{b} = \frac{a + c}{b}, b \neq 0$

(35)

Subtraction of Fractions

$$\frac{a}{b} - \frac{c}{b} = \frac{a}{b} + \frac{-c}{b}$$

$$= \frac{a + (-c)}{b}$$

$$= \frac{a - c}{b}$$

Thus, $\frac{a}{b} - \frac{c}{b} = \frac{a - c}{b}, b \neq 0$

(36)

In order to add fractions

1. factor each numerator and each denominator.
2. make all the denominators the same by the fundamental principle of fractions.
3. add according to $\frac{a}{b} + \frac{c}{b} = \frac{a + c}{b}, b \neq 0$.
4. reduce result to lowest terms.

(37)

Example 1: Find the sum of $\frac{5}{18} + \frac{5}{6}$

Solution: $\frac{5}{18} + \frac{5}{6} = \frac{5}{18} + \frac{15}{18}$

$$= \frac{20}{18}$$

$$= \frac{10}{9}$$

Example 2: Find the sum of $\frac{3}{x - 2} + \frac{x}{x^2 - 4}$

Solution: $\frac{3}{x - 2} + \frac{x}{x^2 - 4} = \frac{3}{x - 2} + \frac{x}{(x - 2)(x + 2)}$

$$= \frac{3(x + 2)}{(x - 2)(x + 2)} + \frac{x}{(x - 2)(x + 2)}$$

$$= \frac{3x + 6 + x}{(x - 2)(x + 2)}$$

$$= \frac{4x + 6}{(x - 2)(x + 2)}$$

$$= \frac{2(2x + 3)}{(x - 2)(x + 2)}, x \neq 2, x \neq -2$$

(38)

Example: Perform the indicated operation and simplify: $\frac{3}{2x} - \frac{x-1}{x+1}$

Solution:
$$\frac{3}{2x} - \frac{x-1}{x+1} = \frac{3}{2x} + \frac{-(x-1)}{x+1}$$
$$= \frac{3(x+1)}{2x(x+1)} + \frac{-(x-1) \cdot 2x}{2x(x+1)}$$
$$= \frac{3x+3}{2x(x+1)} + \frac{-2x^2+2x}{2x(x+1)}$$
$$= \frac{-2x^2+5x+3}{2x(x+1)}, x \neq 0, x \neq -1$$

(39)

Study Exercise Six

Perform the indicated operations and write as a single fraction in lowest terms.

1. $\frac{x-2}{2} + \frac{4-x}{6}$

2. $\frac{5}{2x-6} - \frac{3}{x+3}$

3. $\frac{1}{x^2-1} - \frac{1}{x^2+2x+1}$

4. $x + \frac{1}{x-1} - \frac{2x+1}{x+1}$

5. $\frac{2}{3-x} - \frac{1}{x-3}$

(40)

Division

$$\frac{9}{2} = 4 + \frac{1}{2}$$

```
                 4  ←— Quotient
divisor ——→ 2 )9  ←— dividend
                 8
                 1  ←— remainder
```

$$\frac{\text{dividend}}{\text{divisor}} = \text{Quotient} + \frac{\text{remainder}}{\text{divisor}}$$

(41)

A similar procedure, known as the *division algorithm* (long division), can be employed with polynomials.

Example: $(x^2 + 3x + 5) \div (x + 1)$

Solution:

```
              x + 2
      x + 1 ) x² + 3x + 5
              x² + x
              -------
                   2x + 5
                   2x + 2
                   ------
                        3
```

Answer: $\frac{x^2+3x+5}{x+1} = x + 2 + \frac{3}{x+1}, x \neq -1$

(42)

Example: Use long division to simplify $\dfrac{2x^3 + 3x - 1}{x - 2}$

Solution:

$$
\begin{array}{r}
2x^2 + 4x + 11 \\
x - 2 \overline{)\, 2x^3 \qquad\quad + 3x - 1} \\
\underline{2x^3 - 4x^2 \qquad\qquad\quad} \\
4x^2 + 3x - 1 \\
\underline{4x^2 - 8x \quad\;\;} \\
11x - 1 \\
\underline{11x - 22} \\
21
\end{array}
$$

Answer: $\dfrac{2x^3 + 3x - 1}{x - 2} = 2x^2 + 4x + 11 + \dfrac{21}{x - 2}, \qquad x \neq 2$

(43)

Study Exercise Seven

Perform the following by long division and express the result in the form of quotient plus remainder over divisor.

1. $\dfrac{2x^2 + 13x - 7}{2x - 1}$

2. $(2x^4 - x + 6) \div (x - 5)$

3. $(4x^2 + 12x + 5) \div (2x + 1)$

4. $\dfrac{x^5 - 1}{x - 1}$

(44)

REVIEW EXERCISES

A. Reduce to lowest terms.

1. $\dfrac{2x + 4}{4}$

2. $\dfrac{x^2 - 1}{x + 1}$

3. $\dfrac{a^2 - 9}{3 - a}$

4. $\dfrac{x^2 - 3x - 4}{x^2 - x - 12}$

B. Perform the indicated operations and express the results in lowest terms or simplest form.

5. $\dfrac{3x - 1}{5} + \dfrac{4 - 5x}{6}$

6. $\dfrac{5a}{a - b} + \dfrac{ab}{a^2 - b^2}$

7. $\dfrac{7a}{12b^2} \cdot \dfrac{20b^3}{35a^3}$

8. $\dfrac{(x - y)^2}{x + y} \cdot \dfrac{3x + 3y}{x^2 - y^2}$

9. $\dfrac{2a}{5a - 7} \div \dfrac{a + 4}{7 - 5a}$

10. $\dfrac{3 + \frac{2}{x}}{5 - \frac{1}{x}}$

11. $\dfrac{\frac{b}{a} - \frac{a}{b}}{\frac{1}{b} + \frac{1}{a}}$

12. $(4x^3 - 2x + 5) \div (x + 3)$ (Use long division)

(45)

SOLUTIONS TO REVIEW EXERCISES

A. **1.** $\dfrac{\cancel{2}(x + 2)}{\cancel{4}_2} = \dfrac{x + 2}{2}$

2. $\dfrac{(x + 1)(x - 1)}{(x + 1)} = x - 1, x \neq -1$

3. $\dfrac{(a + 3)\overset{-1}{\cancel{(a - 3)}}}{\cancel{(3 - a)}} = -(a + 3)$ or $-a - 3, \quad a \neq 3$

4. $\dfrac{\cancel{(x - 4)}(x + 1)}{\cancel{(x - 4)}(x + 3)} = \dfrac{x + 1}{x + 3}, x \neq 4, x \neq -3$

B. **5.** $\dfrac{(3x - 1)\cdot 6}{5 \cdot 6} + \dfrac{(4 - 5x)\cdot 5}{6 \cdot 5} = \dfrac{18x - 6 + 20 - 25x}{30}$

$= \dfrac{14 - 7x}{30}$

6. $\dfrac{5a(a + b)}{(a - b)(a + b)} + \dfrac{(ab)}{(a + b)(a - b)} = \dfrac{5a^2 + 5ab + ab}{(a + b)(a - b)}$

$= \dfrac{5a^2 + 6ab}{(a + b)(a - b)}, a \neq b, \quad a \neq -b$

7. $\dfrac{\cancel{7a}}{\cancel{4}\cdot 3 \cdot \cancel{b^2}} \cdot \dfrac{\cancel{4}\cdot\cancel{5}\cdot\cancel{b^2}\cdot b}{\cancel{7}\cdot\cancel{5}\cdot\cancel{a}\cdot a^2} = \dfrac{b}{3a^2}, a \neq 0, b \neq 0$

8. $\dfrac{(x - y)\cancel{(x - y)}}{(x + y)} \cdot \dfrac{3\cancel{(x + y)}}{\cancel{(x + y)}\cancel{(x - y)}} = \dfrac{3(x - y)}{x + y}, x \neq y, x \neq -y$

9. $\dfrac{2a}{\cancel{5a - 7}} \cdot \dfrac{\overset{-1}{\cancel{7 - 5a}}}{a + 4} = -\dfrac{2a}{a + 4}, \quad a \neq -4, a \neq \dfrac{7}{5}$

10. $\dfrac{x(3 + \frac{2}{x})}{x(5 - \frac{1}{x})} = \dfrac{3x + 2}{5x - 1}, x \neq 0, x \neq \dfrac{1}{5}$

SOLUTIONS TO REVIEW EXERCISES, CONTD.

(Frame 46, contd.)

11. $$\frac{ab(\frac{b}{a} - \frac{a}{b})}{ab(\frac{1}{b} + \frac{1}{a})} = \frac{b^2 - a^2}{a + b}$$

$$= \frac{(b + a)(b - a)}{(a + b)} = b - a, a \neq 0, b \neq 0, a \neq -b$$

12.
$$\begin{array}{r} 4x^2 - 12x + 34 \\ x + 3 \overline{\big)\, 4x^3 \qquad\qquad - 2x + 5} \\ 4x^3 + 12x^2 \qquad\qquad\quad \\ \hline -12x^2 - 2x + 5 \\ -12x^2 - 36x \qquad \\ \hline 34x + 5 \\ 34x + 102 \\ \hline -97 \end{array}$$

$$4x^2 - 12x + 34 + \frac{-97}{x + 3}, \quad x \neq -3$$

(46)

SOLUTIONS TO STUDY EXERCISES

Study Exercise One (Frame 12)

A. **1—4.** Each of the fractions in exercise 1 through 4 is given in lowest terms since there are no common factors in numerator and denominator.

B. **5.** $\frac{(x + 3)\cancel{(x - 2)}}{\cancel{(x - 2)} \cdot 1} = x + 3, x \neq 2$

6. $\frac{\cancel{2x}(x - 1)}{\cancel{2x}(1)} = x - 1, x \neq 0$

7. $\frac{(y - 2)\cancel{(y + 1)}}{\cancel{(y + 1)}(1)} = y - 2, y \neq -1$

8. $\frac{\cancel{a}(x - 1)}{\cancel{a}(1)} = x - 1, a \neq 0$

(12A)

Study Exercise Two (Frame 20)

1. $$\frac{(a - b)(a - b)}{(b - a)} = \frac{-(a - b)(a - b)}{-(b - a)}$$

$$= \frac{-(a - b)\cancel{(a - b)}}{\cancel{(a - b)}}$$

$$= -(a - b), a \neq b$$

2. $$-\frac{(a - b)}{-(b - a)} = -\frac{\cancel{(a - b)}}{\cancel{(a - b)}}$$

$$= -1, a \neq b$$

3. $$\frac{(x - 1)(x^2 + x + 1)}{(1 + x)(1 - x)} = -\frac{-(x - 1)(x^2 + x + 1)}{(1 + x)(1 - x)}$$

$$= -\frac{\cancel{(1 - x)}(x^2 + x + 1)}{(1 + x)\cancel{(1 - x)}}$$

$$= -\frac{x^2 + x + 1}{1 + x}, x \neq 1, x \neq -1$$

4. $$\frac{(y + 7)(y - 2)}{(2 - y)} = -\frac{(y + 7)(y - 2)}{-(2 - y)}$$

$$= -\frac{(y + 7)\cancel{(y - 2)}}{\cancel{(y - 2)} \cdot 1}$$

$$= -(y + 7), y \neq 2$$

(20A)

SOLUTIONS TO STUDY EXERCISES, CONTD.

Study Exercise Three (Frame 25)

1. $\dfrac{\cancel{xy}\cancel{(x+y)}\cancel{(x-y)}}{\cancel{xy}\cancel{(x-y)}}\cdot\dfrac{1}{\cancel{(x+y)}} = 1, x \neq 0, y \neq 0, x \neq y, x \neq -y$

2. $\dfrac{8\cancel{(x-y)}}{\cancel{8}\cdot 2\cancel{(x-y)}}\cdot\dfrac{(x-y)(x-y)}{2} = \dfrac{(x-y)(x-y)}{4}, x \neq y$

3. $\dfrac{(x+2)\cancel{(x-2)}}{\cancel{x}\cdot\cancel{y}\cdot y}\cdot\dfrac{2\cancel{xy}}{(x-2)\cancel{(x-2)}} = \dfrac{2(x+2)}{y(x-2)}, x \neq 0, y \neq 0, x \neq 2$

4. $\dfrac{(x-3)(x+2)}{(x-1)\cancel{(x-1)}}\cdot\dfrac{(x+4)\cancel{(x-1)}}{x(3-x)(3+x)} = -\dfrac{-(x-3)(x+2)(x+4)}{x(x-1)(3-x)(3+x)}$

$$= -\dfrac{\cancel{(3-x)}(x+2)(x+4)}{x(x-1)\cancel{(3-x)}(3+x)}$$

$$= -\dfrac{(x+2)(x+4)}{x(x-1)(3+x)},$$

$x \neq 1, x \neq 0, x \neq 3, x \neq -3$

25A

Study Exercise Four (Frame 29)

1. $\dfrac{\cancel{x}\cancel{y}y^2}{\cancel{y}z}\cdot\dfrac{1}{\cancel{x}xz} = \dfrac{y^2}{xz^2}, x \neq 0, y \neq 0, z \neq 0$

2. $\dfrac{\cancel{(x+y)}\cancel{(x-y)}}{\cancel{(x-y)}}\cdot\dfrac{x}{\cancel{(x+y)}} = x, x \neq y, x \neq -y, x \neq 0$

3. $\dfrac{x^2}{(x+1)(x-1)}\cdot\dfrac{(1-x)(1-x)}{xx^2} = -\dfrac{1}{(x+1)(x-1)}\cdot\dfrac{-(1-x)(1-x)}{x}$

$$= -\dfrac{1}{(x+1)\cancel{(x-1)}}\cdot\dfrac{\cancel{(x-1)}(1-x)}{x}$$

$$= -\dfrac{1-x}{x(x+1)} \text{ or } \dfrac{x-1}{x(x+1)}, x \neq 0, x \neq 1, x \neq -1$$

29A

Study Exercise Five (Frame 34)

1. $\dfrac{5(3/5)}{5\left(5-\dfrac{1}{5}\right)} = \dfrac{3}{25-1}$

$$= 3/24$$

$$= 1/8$$

2. $\dfrac{(x-1)(x)}{(x-1)\left(\dfrac{x}{x-1}\right)} = \dfrac{x^2-x}{x}$

$$= \dfrac{\cancel{x}(x-1)}{\cancel{x}}$$

$$= x-1, x \neq 0, x \neq 1$$

3. $\dfrac{x^2(x-1)}{x^2\left(1-\dfrac{2}{x^2}\right)} = \dfrac{x^3-x^2}{x^2-2}, x \neq 0, x^2 \neq 2$

Study Exercise Five (Frame 34, contd.)

4. $\dfrac{x(1 + \frac{1}{x})}{x(1 - \frac{1}{x})} = \dfrac{x + 1}{x - 1}, x \neq 0, x \neq 1$

5. $\dfrac{2x(\frac{2}{x} + \frac{3}{2x})}{2x(5 + \frac{1}{x})} = \dfrac{4 + 3}{10x + 2}$

$$= \frac{7}{10x + 2}, x \neq 0, x \neq -\frac{1}{5}$$

6. $\dfrac{6(\frac{1}{2} + \frac{1}{3})}{6(\frac{1}{6} - \frac{3}{2})} = \dfrac{3 + 2}{1 - 9}$

$$= -\frac{5}{8}$$

34A

Study Exercise Six (Frame 40)

1. $\dfrac{3(x - 2)}{3 \cdot 2} + \dfrac{(4 - x)}{6} = \dfrac{3x - 6 + 4 - x}{6}$

$$= \frac{2x - 2}{6}$$

$$= \frac{\cancel{2}(x - 1)}{\cancel{2} \cdot 3}$$

$$= \frac{x - 1}{3}$$

2. $\dfrac{5(x + 3)}{2(x - 3)(x + 3)} - \dfrac{2 \cdot 3(x - 3)}{2(x + 3)(x - 3)} = \dfrac{5x + 15 - (6x - 18)}{2(x - 3)(x + 3)}$

$$= \frac{-x + 33}{2(x - 3)(x + 3)}, x \neq 3, x \neq -3$$

3. $\dfrac{1(x + 1)}{(x + 1)(x - 1)(x + 1)} + \dfrac{-1(x - 1)}{(x + 1)(x + 1)(x - 1)} = \dfrac{x + 1 - x + 1}{(x + 1)(x + 1)(x - 1)}$

$$= \frac{2}{(x + 1)(x + 1)(x - 1)},$$

$$x \neq 1, x \neq -1$$

4. $\dfrac{x(x + 1)(x - 1)}{1(x + 1)(x - 1)} + \dfrac{1(x + 1)}{(x - 1)(x + 1)} - \dfrac{(2x + 1)(x - 1)}{(x + 1)(x - 1)}$

$$= \frac{x^3 - x + x + 1 - [2x^2 - x - 1]}{(x + 1)(x - 1)}$$

$$= \frac{x^3 - 2x^2 + x + 2}{(x + 1)(x - 1)}, x \neq 1, x \neq -1$$

5. $\dfrac{-2}{x - 3} + \dfrac{-1}{x - 3} = \dfrac{-3}{x - 3}, x \neq 3$

40A

Study Exercise Seven (Frame 44)

1.
$$\begin{array}{r|l} & x + 7 \\ \hline 2x - 1 & 2x^2 + 13x - 7 \\ & 2x^2 - x \\ \hline & 14x - 7 \\ & 14x - 7 \\ \hline & 0 \end{array}$$

Answer: $x + 7, x \neq 1/2$

2.
$$\begin{array}{r|l} & 2x^3 + 10x^2 + 50x + 249 \\ \hline x - 5 & 2x^4 \qquad - x + 6 \\ & 2x^4 - 10x^3 \\ \hline & 10x^3 \qquad - x + 6 \\ & 10x^3 - 50x^2 \\ \hline & 50x^2 - x + 6 \\ & 50x^2 - 250x \\ \hline & 249x + 6 \\ & 249x - 1245 \\ \hline & 1251 \end{array}$$

Answer: $2x^3 + 10x^2 + 50x + 249 + \dfrac{1251}{x - 5}, x \neq 5$

3.
$$\begin{array}{r|l} & 2x + 5 \\ \hline 2x + 1 & 4x^2 + 12x + 5 \\ & 4x^2 + 2x \\ \hline & 10x + 5 \\ & 10x + 5 \\ \hline & 0 \end{array}$$

Answer: $2x + 5, x \neq -\dfrac{1}{2}$

4.
$$\begin{array}{r|l} & x^4 + x^3 + x^2 + x + 1 \\ \hline x - 1 & x^5 \qquad - 1 \\ & x^5 - x^4 \\ \hline & x^4 \qquad - 1 \\ & x^4 - x^3 \\ \hline & x^3 \qquad - 1 \\ & x^3 - x^2 \\ \hline & x^2 \qquad - 1 \\ & x^2 - x \\ \hline & x - 1 \\ & x - 1 \\ \hline & 0 \end{array}$$

Answer: $x^4 + x^3 + x^2 + x + 1, x \neq 1$

44A

UNIT 6—SUPPLEMENTARY PROBLEMS

A. Write a fraction equivalent to the given fraction with no minus sign in the numerator or denominator.

1. $-7/-5$ **2.** $3/-5$ **3.** $\dfrac{-(2x + 1)}{3}$ **4.** $\dfrac{-(4x + y)}{x + y}$

B. Reduce to lowest terms. List any restrictions on the variables.

5. $\dfrac{1 + x}{x^2 - 1}$ **6.** $\dfrac{8x^2 - 4x^3}{x(2 - x)}$ **7.** x/x

8. $\dfrac{x^2 - 4xy + 3y^2}{y^2 - x^2}$ **9.** $\dfrac{4x^2 - 20x}{x^2 - 4x - 5}$ **10.** $\dfrac{6x^2 - 3xy}{-4x^2y + 2xy^2}$

11. $\dfrac{y - x}{x^2 - y^2}$ **12.** $\dfrac{x^3 + y^3}{x^2 - y^2}$ **13.** $\dfrac{(x - 1)(x - 3)}{(1 - x)(3 + x)}$

C. Perform the indicated operations and write each as a single fraction in reduced form. List any restrictions on the variable.

14. $\dfrac{x^2 - y^2}{y} \div \dfrac{x + y}{y^2}$ **15.** $\dfrac{x - y}{x^2 + xy} \cdot \dfrac{x^2}{y^2 - xy}$

16. $\dfrac{2y-6}{y+2x} \div \dfrac{4y-12}{2y+4x}$

17. $\dfrac{1}{1-\frac{4}{x}}$

18. $\dfrac{x^2-\frac{1}{x}}{x^2-\dfrac{1}{x^2}}$

19. $\dfrac{3}{a-1}+\dfrac{2}{a+1}$

20. $\dfrac{1}{t}+\dfrac{5}{t^2}-\dfrac{t+3}{t^2-t}$

21. $8x+\dfrac{x-9}{x+7}$

22. $1+\dfrac{1}{x-3}+\dfrac{3}{x^2+3x}$

23. $\dfrac{x}{x+y}-\dfrac{y}{x-y}$

24. $\dfrac{a+b}{b^2}+\dfrac{1-a^2}{a-b}$

25. $\dfrac{xy^2}{2x-2y}\cdot\dfrac{x^2-y^2}{x^3y^2}$

26. $\dfrac{6x-12}{4xy+4x}\cdot\dfrac{y^2-1}{2-3x+x^2}$

27. $\dfrac{x+2xy}{3x^2}\div\dfrac{2y+1}{6x}$

28. $\dfrac{3x-1}{10}+\dfrac{5-2x}{15}$

29. $\dfrac{3}{x-2}-\dfrac{2}{x-2}-\dfrac{x}{x^2-4}$

30. $\dfrac{(x+y)/3x^2}{(x-y)/x}$

31. $\dfrac{x+y}{\frac{1}{x}+\frac{1}{y}}$

32. $\dfrac{3}{2(a-b)}-\dfrac{2}{5(a-b)}$

33. $\dfrac{1}{3x-3}-\dfrac{x+6}{x^2+3x-4}$

D. Perform the following divisions by the long division method. Give answer in the form quotient plus remainder over divisor.

34. $(x^4+x^2+2x-1)\div(x+3)$

35. $(3x^3-x+1)\div(x-1)$

36. $(x^4+1)\div(x+1)$

E. Find the fallacy in the following.

37. Given that $a = 1$

$$\begin{aligned} a\cdot a &= 1\cdot a \\ a^2-1 &= 1\cdot a-1 \\ (a+1)(a-1) &= (a-1)\cdot 1 \\ (a+1) &= 1 \\ \text{since } a &= 1, \\ (1+1) &= 1 \\ 2 &= 1 \end{aligned}$$

unit

7

Exponents

Objectives

1. Know the five laws of exponents.
2. Be able to simplify expressions with rational exponents.
3. Be able to write numbers in scientific notation form.

Terms root

(1)

Review

1. $(x^2)^3 = (x^2)(x^2)(x^2)$
   $= (x \cdot x)(x \cdot x)(x \cdot x)$
   $= x^6$

2. $(3x)^4 = (3x)(3x)(3x)(3x)$
   $= 3^4 \cdot x^4$ or $81x^4$

3. $(x/y)^3 = (x/y)(x/y)(x/y)$
   $= \dfrac{x^3}{y^3}, y \neq 0$

(2)

The three previous examples illustrate these three properties of exponents:

1. $(x^m)^n = x^{mn}, m, n \in N$ **2.** $(xy)^m = x^m \cdot y^m$ **3.** $(x/y)^m = \dfrac{x^m}{y^m}, y \neq 0$

(3)

Laws of Natural Number Exponents

$m, n \in N$

I. $x^m \cdot x^n = x^{m+n}$

II. **(a)** $\dfrac{x^m}{x^n} = x^{m-n}$ if $m > n, x \neq 0$ **(b)** $\dfrac{x^m}{x^n} = 1$ if $m = n, x \neq 0$

(c) $\dfrac{x^m}{x^n} = \dfrac{1}{x^{n-m}}$ if $n > m, x \neq 0$

III. $(x^m)^n = x^{mn}$

IV. $(x \cdot y)^m = x^m \cdot y^m$

V. $\left(\dfrac{x}{y}\right)^m = \dfrac{x^m}{y^m}, y \neq 0$

(4)

Study Exercise One

State which of the laws of exponents I through V justifies the given statement.

1. $\left(\frac{2x^2}{y}\right)^3 = \frac{(2x^2)^3}{(y)^3}, y \neq 0$
2. $(x^2y)^4 = (x^2)^4 \cdot (y)^4$
3. $(x^4)^2 = x^8$
4. $(x^2)^3 (y^3)^3 = (x^2 \cdot y^3)^3$
5. $x^8 \cdot x^2 = x^{10}$

(5)

Example: Simplify $(2x^2y)^2$

Solution:

Line (1)	$(2x^2y)^2 = (2x^2)^2 \cdot y^2;$	Law IV
Line (2)	$= 2^2(x^2)^2 \cdot y^2;$	Law IV
Line (3)	$= 2^2x^4y^2;$	Law III
	or $4x^4y^2$	

(6)

Example: Simplify $\left(\frac{x^2}{y^3}\right)^3$

Solution:

Line (1)	$\left(\frac{x^2}{y^3}\right)^3 = \frac{(x^2)^3}{(y^3)^3},$	Law V
Line (2)	$= \frac{x^6}{y^9}, y \neq 0;$	Law III

(7)

Example: Simplify $\frac{(x^{n+1} \cdot x^{2n})^2}{x^{3n}}, n \in N$

Solution:

Line (1)	$\frac{(x^{n+1} \cdot x^{2n})^2}{x^{3n}} = \frac{(x^{3n+1})^2}{x^{3n}};$	Law I
Line (2)	$= \frac{x^{6n+2}}{x^{3n}}$	Law III
Line (3)	$= x^{(6n+2)-(3n)};$	Law II
Line (4)	$= x^{3n+2}, x \neq 0$	

(8)

Study Exercise Two

Simplify using the laws of exponents:

1. $(-3xy^2)^3$
2. $(3y)^2 (2y^4)$
3. $\frac{2^6}{2^4}$
4. $\frac{2x^4 \cdot x^3}{3y^4 \cdot y^3}$
5. $\left(\frac{3y^2}{4}\right)^3$
6. $\frac{1}{[(x)^2]^3}$
7. $(3 \cdot 10^2)^3$
8. $3^4 \cdot 3^6$
9. $a^{n+1} \cdot a^{n-2}, n \in N, n > 2$
10. $-(x^2)^2$

(9)

Extension of the Laws of Exponents

It is desirable to maintain the five laws of exponents as previously stated.

Under these conditions, what sort of meaning could then be attached to expressions such as

$$2^{1/2}, 3^0, (-1)^{-1}, 5^{-3/4}\ ?$$

(10)

Zero Exponents

If law I, $x^m \cdot x^n = x^{m+n}$ is to hold, then $x^n \cdot x^0 = x^{n+0}$.

Thus, $x^n \cdot x^0 = x^n$ and x^0 must equal 1 by the uniqueness of the identity element.

Further, $\frac{x^n}{x^n} = x^n \cdot \frac{1}{x^n} = 1, x \neq 0$ and if we want law II to hold, $\frac{x^n}{x^n} = x^{n-n}$

$$= x^0$$
$$= 1, x \neq 0$$

Thus we define $x^0 = 1, x \neq 0$.

(11)

Zero Exponent Property $x^0 = 1, x \neq 0$

Examples:

1. $3^0 = 1$ **2.** $(125 \cdot 23)^0 = 1$ **3.** $(a + b)^0 = 1, a \neq -b$

(12)

Integral Exponents

Let $n \in I$ and $n > 0$,

$$\begin{aligned} x^n \cdot x^{-n} &= x^{n+(-n)} \\ &= x^0 \\ &= 1, x \neq 0 \end{aligned}$$

If $x^n \cdot x^{-n} = 1$ and $x \neq 0$

Then x^{-n} must be the reciprocal of x^n.

$$x^{-n} = \frac{1}{x^n}, x \neq 0$$

(13)

Negative Exponent Property $x^{-n} = \frac{1}{x^n}, x \neq 0$

Examples:

1. $2^{-2} = \frac{1}{2^2}$

2. $x^{-1} = \frac{1}{x}, x \neq 0$

3. $(a + b)^{-2} = \frac{1}{(a + b)^2}, a \neq -b$

(14)

If $m, n \in I$,

$$\begin{aligned} x^{m-n} &= \frac{1}{x^{-(m-n)}} \\ &= \frac{1}{x^{n-m}}, x \neq 0 \end{aligned}$$

We can now do away with laws II(b), and II(c) of the laws of exponents. All we need is

II(a) $\frac{x^m}{x^n} = x^{m-n}, x \neq 0$

(Frame 15, contd.)

If the exponents come out negative, we now know how to change them to positive exponents.

Example:

(1) $\dfrac{x^6}{x^{10}} = x^{6-10}$
$= x^{-4}$
$= \dfrac{1}{x^4}, x \neq 0$

(2) $\dfrac{2^4}{2^8} = 2^{4-8}$
$= 2^{-4}$
$= \dfrac{1}{2^4}$ or $\dfrac{1}{16}$

(15)

Inverses and Sums

Line (1) $x^{-1} = \dfrac{1}{x}, x \neq 0$ and $y^{-1} = \dfrac{1}{y}, y \neq 0$

Line (2) $x^{-1} + y^{-1} = \dfrac{1}{x} + \dfrac{1}{y}, x \neq 0, y \neq 0$

Line (3) $(x + y)^{-1} = \dfrac{1}{x + y}, x \neq -y$

Line (4) $\dfrac{1}{x + y} \neq \dfrac{1}{x} + \dfrac{1}{y}$

Line (5) Thus, $(x + y)^{-1} \neq x^{-1} + y^{-1}$

(16)

Examples

Simplify. Leave no answer with a zero or a negative exponent.

1. $2^{-3} = \dfrac{1}{2^3}$
$= \dfrac{1}{8}$

2. $5^{-1} + 3^0 = \dfrac{1}{5} + 1$
$= \dfrac{6}{5}$

3. $\left(\dfrac{x+1}{x}\right)^{-1} = \dfrac{1}{\frac{x+1}{x}}$
$= \dfrac{x \cdot 1}{x(\frac{x+1}{x})}$
$= \dfrac{x}{x + 1}, x \neq 0, x \neq -1$

4. $(x^2y^{-4})^{-1} = (x^2)^{-1}(y^{-4})^{-1}$, Law IV
$= x^{-2} \cdot y^4$, Law III
$= \dfrac{1}{x^2} \cdot y^4$
$= \dfrac{y^4}{x^2}, x \neq 0, y \neq 0$

5. $\dfrac{x^{-1} + y^{-1}}{x^{-1}} = \dfrac{\frac{1}{x} + \frac{1}{y}}{\frac{1}{x}}$
$= \dfrac{xy(\frac{1}{x} + \frac{1}{y})}{xy(\frac{1}{x})}$
$= \dfrac{y + x}{y}, x \neq 0, y \neq 0$

(17)

Study Exercise Three

Simplify using the laws of exponents. Leave no answer with a zero or a negative exponent.

1. $-5(5)^{-2}$ **2.** $-2x^{-2}$ **3.** $(-2x)^{-2}$ **4.** $\dfrac{ax^{-3}}{a^{-2}x}$ **5.** $3x^0$ **6.** $(3x)^0$

7. $\dfrac{x^{m+3}}{x^{m-1}}, m \in N$ **8.** $x^{-2} + \dfrac{1}{x^{-2}}$ **9.** $x^{-1} + y^{-1}$ **10.** $\left(\dfrac{x^{-1}y^3}{2x^0y^{-5}}\right)^{-2}$

(18)

Roots

One of two equal factors of a number is called a *square root* of the number.

One of three equal factors of a number is called a *cube root* of the number.

One of four equal factors of a number is called a *fourth root* of the number.

One of n equal factors of a number is called an *nth root* of the number, $n \in N, n > 2$.

(19)

Since $4 = 2 \cdot 2$, 2 is a square root of 4. Further, since $4 = (-2)(-2)$, -2 is a square root of 4.

Since $25 = (5)(5)$ and $25 = (-5)(-5)$, 5 and -5 are square roots of 25.

(20)

---

0 is the only square root of 0 because

$$0 = 0 \cdot 0$$

These are the only two equal factors whose product is zero.

(21)

-36 has no square roots in the real number system since no two equal factors will give a product which is negative.

Square roots of negative numbers do not exist.

(22)

Summary of Square Roots

Each positive number has two square roots.

Zero has only one square root, namely zero.

There are no square roots for negative numbers.

(23)

Principal Square Root

Since each positive number has two square roots, we will define the positive square root as the principal square root.

There are no principal square roots of negative numbers since square roots of negative numbers do not exist.

We also define the principal square root of zero to be zero.

(24)

Since the principal square root of a positive number is positive, the principal square root of zero is zero, and principal square roots of negative numbers are not defined, it follows that the principal square root of a number is never negative.

(25)

Examples:

1. principal square root of 64 is 8
2. principal square root of 100 is 10
3. principal square root of 0 is 0
4. principal square root of $(-2)^2$ is 2
5. principal square root of $(-4)^2$ is 4
6. principal square root of $(x)^2$ is ______?

What do you think the answer to number 6 is? Remember, the answer must not be negative.

(26)

The principal square root of $(x)^2$ is x when $x > 0$ and, $-x$ when $x < 0$

The principal square root of $(x)^2$ is $|x|$ if x is any real number.

(27)

Cube Roots

Since $8 = (2)(2)(2)$, 2 is a cube root of 8
Since $-8 = (-2)(-2)(-2)$, -2 is a cube root of -8
Since $0 = (0)(0)(0)$, 0 is a cube root of 0

Every real number has exactly one real cube root. A cube root may be positive, negative, or zero. There is no need for a principal cube root.

(28)

The cube root of a positive number is positive.

The cube root of zero is zero.

The cube root of a negative number is negative.

(29)

Fourth Roots

Since $16 = (2)(2)(2)(2)$ and $16 = (-2)(-2)(-2)(-2)$, 2 and -2 are fourth roots of 16. The number 2 is taken as the principal fourth root of 16.

The only fourth root of zero is zero.

There are no fourth roots of -16 since four equal factors will not give a product which is negative.

Fourth roots behave like square roots.

(30)

Fifth Roots

Since $32 = (2)(2)(2)(2)(2)$, 2 is a fifth root of 32.

Since $-32 = (-2)(-2)(-2)(-2)(-2)$, -2 is a fifth root of -32

Fifth roots behave like cube roots.
This process could be kept up and we would find sixth, eighth, tenth, etc., roots behave like square roots; and seventh, ninth, eleventh, etc., roots behave like cube roots.

(31)

Summary of Roots

Even Roots (2, 4, 6, ...)

Even roots of non negative numbers exist.
Even roots of negative numbers do not exist.
The principal even root of a positive number is the positive even root.

The principal even root of zero is zero.

Odd Roots (3, 5, 7, ...)

Odd roots of positive, negative, and zero numbers exist.
Principal odd roots are not needed since each odd root is unique.

(32)

Study Exercise Four

A. Find the square roots of the following numbers:

1. 36 **2.** 49 **3.** 0 **4.** -9

B. Find the principal square root of the following:

5. 64 **6.** 121 **7.** -49 **8.** $100x^2$

(Frame 33, contd.)

C. Find the cube root of each of the following:

9. -27 **10.** $64x^3$ **11.** 125 **12.** -8

D. Find the principal fourth root:

13. 16 **14.** -1 **15.** 81 **16.** 1

(33)

Rational Exponents

Meanings consistent with the five laws of exponents will have to be found for rational exponents.

What will $16^{1/2}$ have to mean?

$$\text{Since } (16^{1/2})(16^{1/2}) = 16^{1/2+1/2} = 16$$

$16^{1/2}$ is a square root of 16. But which one?

It is agreed that $16^{1/2}$ will be taken as the principal square root of 16. Thus, $x^{1/2}$ means the principal square root of x where $x \geqslant 0$.

(34)

Remember that when written $-x^{1/2}$, the expression is taken to mean $-(x^{1/2})$.

Thus, $-4^{1/2}$ means the opposite of the principal square root of 4 or -2, while $(-4)^{1/2}$ does not exist since there are no square roots of negative numbers.

(35)

Since $x^{1/4} \cdot x^{1/4} \cdot x^{1/4} \cdot x^{1/4} = x^{1/4+1/4+1/4+1/4}$ or x, $x^{1/4}$ is taken as the principal fourth root of x.

$x^{1/4}$ means the principal fourth root of x, $x \geqslant 0$.

$16^{1/4}$ represents the principal fourth root of 16 and is equal to 2.

(36)

$x^{1/n}$ represents the principal nth root of x where $x \geqslant 0$, $n \in N$, $n \geqslant 2$ and n is even.

$-x^{1/n}$ represents the opposite of the principal nth root of x where $x \geqslant 0$, $n \in N$, $n \geqslant 2$ and n is even.

$x^{-1/n}$ will be taken to mean $\dfrac{1}{x^{1/n}}$ where $x > 0$, $n \in N$, $n \geq 2$ and n is even.

(37)

Examples:

1. $4^{1/2} = 2$ **2.** $16^{1/4} = 2$ **3.** $81^{1/4} = 3$

4. $0^{1/2} = 0$ **5.** $1^{1/6} = 1$ **6.** $(x^8)^{1/4} = x^2$

7. $4^{-1/2} = \dfrac{1}{4^{1/2}} = \dfrac{1}{2}$ **8.** $0^{1/4} = 0$ **9.** $-25^{1/2} = -(25^{1/2}) = -5$

10. $(-9)^{1/2} =$ does not exist **11.** $-9^{1/2} = -(9^{1/2}) = -3$

12. $(x^4)^{1/2} = x^2$ **13.** $(x^2)^{1/2} = |x|$; the result must be non negative.

(38)

Study Exercise Five

Write as a product or quotient of powers in which each variable occurs but once and all exponents are positive.

1. $(16)^{1/4}$ **2.** $(16)^{-1/4}$ **3.** $(x^2y^{-1})^{-1/2}$

4. $(x^{1/2} \cdot x^{1/4})^4$ **5.** $(-1)^{1/4}$ **6.** $(4x^2)^{1/2}$

7. $-4^{1/2}$ **8.** $-(25x^4y^4)^{1/2}$

(39)

Since $x^{1/3} \cdot x^{1/3} \cdot x^{1/3} = x^{1/3+1/3+1/3}$ or x, $x^{1/3}$ will be taken as the cube root of x.

Since $x^{1/5} \cdot x^{1/5} \cdot x^{1/5} \cdot x^{1/5} \cdot x^{1/5} = x$, $x^{1/5}$ will be taken as the fifth root of x.

$8^{1/3}$ represents the cube root of 8 or 2

$(-32)^{1/5}$ represents the fifth root of -32 or -2

(40)

$x^{1/n}$ represents the nth root of x where $x \in R$, $n \in N$, $n \geqslant 3$ and n is odd.

$-x^{1/n}$ represents the opposite of the nth root of x where $x \in R$, $n \in N$, $n \geqslant 3$ and n is odd.

$x^{-1/n}$ will be taken to mean $\dfrac{1}{x^{1/n}}$ where $x \in R$, $x \neq 0$, $n \in N$, $n \geqslant 3$ and n is odd.

(41)

Study Exercise Six

Find the numerical value:

1. $27^{1/3}$ **2.** $-27^{1/3}$ **3.** $(-27)^{1/3}$

4. $0^{1/5}$ **5.** $8^{-1/3}$ **6.** $64^{1/3}$

7. $0^{1/3}$ **8.** $1^{1/7}$

(42)

If $a \geqslant 0$, $m, n \in N$, we define the general rational exponent to obey the third law of exponents

$$(a^{1/n})^m = a^{m/n} \text{ and } (a^m)^{1/n} = a^{m/n}$$

Thus, $a^{m/n}$ may be viewed as (1) the mth power of the nth root of a.
(2) the nth root of the mth power of a.

(43)

$$a^{m/n} = \begin{cases} (a^{1/n})^m \\ \text{or} \\ (a^m)^{1/n} \end{cases} \quad \text{provided } a^{1/n} \text{ exists and } m, n \in N$$

Example: $8^{2/3} = (8^{1/3})^2 = 2^2$ or 4

or

$(8^2)^{1/3} = 64^{1/3}$ or 4

(44)

A special rule must be given if both m and n are even and we wish to evaluate $(a^m)^{1/n}$.

$$(a^m)^{1/n} = |a|^{m/n}, \text{ both } m \text{ and } n \text{ are even.}$$

Example 1: $[(-2)^2]^{1/2} = |-2|^{2(1/2)}$
$= |-2|$ or 2

Note: $[(-2)^2]^{1/2} \neq (-2)^1$

Example 2: $(x^2)^{1/2} = |x|^{2/2}$
$= |x|$

(45)

$$a^{-m/n} = \frac{1}{a^{m/n}}, a \in R, a \neq 0, m, n \in N$$

Example: $8^{-2/3} = \frac{1}{8^{2/3}}$

$= \frac{1}{(8^{1/3})^2}$

$= \frac{1}{2^2}$

$= \frac{1}{4}$

(46)

Study Exercise Seven

Simplify:

1. $27^{2/3}$ **2.** $64^{-2/3}$ **3.** $16^{5/4}$

4. $\left(\frac{1}{4}\right)^{3/2}$ **5.** $\left(\frac{1}{4}\right)^{-1/2}$ **6.** $\left(\frac{4}{9}\right)^{1/2}$

7. $\left(\frac{1}{16}\right)^{-3/4}$ **8.** $[(-4)^4]^{1/4}$

(47)

Scientific Notation

Consider products of powers of ten:

$10^0(2.314) = 2.314$ $10^{-1}(2.31) = .231$
$10^1(1.234) = 12.34$ $10^{-2}(1.2) = .012$
$10^2(1.45) = 145$ $10^{-3}(1.2) = .0012$
$10^3(.2834) = 283.4$ $10^{-4}(31.1) = .00311$

Notice that the power on ten tells how many places the decimal point is moved. A positive exponent indicates that the decimal point is moved to the right. A negative exponent indicates that the decimal point is moved to the left. A zero exponent on ten leaves the number unchanged.

(48)

Any positive number N can be written as $N = p \cdot 10^k$ where p is the number made by placing the decimal point to form a number between 1 and 10 ($1 \leqslant p < 10$).

k is the number of places the decimal point has to be moved from its position in p so that p becomes N. k is positive if the movement is to the right, negative if the movement is to the left, and zero if there is no movement.

(49)

Scientific notation is a process of writing a number as an integral power of ten times a number between 1 and 10.

$$N = p \cdot 10^k, 1 \leqslant p < 10, k \in I$$

Example: $12.3 = (1.23)(10^1)$

(50)

Examples: Write in scientific notation form:

1. $.00134 = 1.34 \cdot 10^{-3}$
2. $3142 = 3.142 \cdot 10^{3}$
3. $2.134 = 2.134 \cdot 10^{0}$
4. $2{,}123{,}000 = 2.123 \cdot 10^{6}$
5. $.890 = 8.90 \cdot 10^{-1}$

(51)

Study Exercise Eight

Write in scientific notation:

1. 3,423
2. .00789
3. 23.58
4. .896
5. 7,899,100
6. .008345

(52)

Computation Using Scientific Notation

Example 1:
$$\frac{248{,}000}{.0124} = \frac{2.48 \cdot 10^{5}}{1.24 \cdot 10^{-2}} = 2 \cdot 10^{5-(-2)} = 2 \cdot 10^{7}$$

Example 2:
$$\frac{(4 \cdot 10^{3})(10^{5})}{2(10^{2})} = \frac{4 \cdot 10^{3+5}}{2 \cdot 10^{2}} = 2 \cdot 10^{8-2} = 2 \cdot 10^{6}$$

(53)

Study Exercise Nine

Compute using scientific notation:

1. $\dfrac{(4 \cdot 10^{3})(6 \cdot 10^{-3})}{3 \cdot 10^{-7}}$
2. $\dfrac{(.065)(2.2)(50)}{(1.30)(.011)(.05)}$

(54)

REVIEW EXERCISES

A. Evaluate and/or simplify:

1. -2^3 **2.** $(2x)^5$ **3.** $\dfrac{10^{m+6}}{10^{m+4}}$

4. $\left(\dfrac{x^3}{x^2}\right)^2$ **5.** $\left(\dfrac{3}{7}\right)^{-1}$ **6.** $(10x^{-1}y^2)^3$

7. $(3^{-1} + 6^{-1})^{-2}$ **8.** $-(-1)^{-3/5}$ **9.** $\dfrac{7x}{(5)^0}$

10. $\dfrac{x - x^{-1}}{x^{-1}}$ **11.** $(x^2)^{1/2}$ **12.** $64^{2/3}$

13. $3^{-2} + 5(2)^0$ **14.** $\dfrac{(2^{-8} \cdot 3^4)^{1/4}}{5^{-4}}$ **15.** $-16^{1/2}$

B. Write in scientific notation:

16. .078 **17.** 36.120

C. Compute using scientific notation:

18. $\dfrac{.078}{.00012}$

(55)

SOLUTIONS TO REVIEW EXERCISES

A. **1.** -8

2. $32x^5$

3. $10^{(m+6)-(m+4)} = 10^2$

4. $(x^{3-2})^2 = x^2, x \neq 0$

5. $7/3$

6. $\dfrac{1000y^6}{x^3}, x \neq 0$

7. $(1/3 + 1/6)^{-2} = (3/6)^{-2}$

$$= \frac{1}{(1/2)^2}$$

$$= 4$$

8. $-\left(\dfrac{1}{(-1)^{3/5}}\right) = -\left(\dfrac{1}{(-1)^3}\right)$

$$= -\left(\frac{1}{-1}\right)$$

$$= 1$$

9. $7x$

10. $\dfrac{x - \frac{1}{x}}{1/x} = x^2 - 1, x \neq 0$

11. $|x|$

12. 16

13. $\frac{1}{9} + 5(1) = 46/9$

14. $\dfrac{2^{-2} \cdot 3^1}{\frac{1}{5^4}} = \dfrac{3 \cdot 5^4}{4}$

$$= \frac{1875}{4}$$

15. -4

B. **16.** $7.8 \cdot 10^{-2}$

17. $3.612 \cdot 10^1$

C. **18.** $\dfrac{7.8 \cdot 10^{-2}}{1.2 \cdot 10^{-4}} = 6.5 \cdot 10^2$

$$= 650$$

(56)

SOLUTIONS TO STUDY EXERCISES

Study Exercise One (Frame 5)

1. Law V, $\left(\frac{x}{y}\right)^m = \frac{x^m}{y^m}, y \neq 0$

2. Law IV, $(x \cdot y)^m = x^m \cdot y^m$

3. Law III, $(x^m)^n = x^{mn}$

4. Law IV, $(x \cdot y)^m = x^m \cdot y^m$

5. Law I, $x^m \cdot x^n = x^{m+n}$

5A

Study Exercise Two (Frame 9)

1. $-27x^3y^6$

2. $18y^6$

3. 2^2 or 4

4. $\frac{2x^7}{3y^7}, y \neq 0$

5. $\frac{27y^6}{64}$

6. $\frac{1}{x^6}, x \neq 0$

7. $27 \cdot 10^6$

8. 3^{10}

9. a^{2n-1}

10. $-x^4$

9A

Study Exercise Three (Frame 18)

1. $-5\left(\frac{1}{5^2}\right) = \frac{-5}{25}$
$$= -\frac{1}{5}$$

2. $-2\left(\frac{1}{x^2}\right) = \frac{-2}{x^2}, x \neq 0$

3. $\frac{1}{(-2x)^2} = \frac{1}{4x^2}, x \neq 0$

4. $\frac{a \cdot \frac{1}{x^3}}{\frac{1}{a^2} \cdot x} = \frac{a^2x^3\left(a \cdot \frac{1}{x^3}\right)}{a^2x^3\left(\frac{1}{a^2} \cdot x\right)}$
$$= \frac{a^3}{x^4}, a \neq 0, x \neq 0$$

5. $3, x \neq 0$

6. $1, x \neq 0$

7. $x^{(m+3)-(m-1)} = x^4, x \neq 0$

8. $\frac{1}{x^2} + \frac{1}{\frac{1}{x^2}} = \frac{1}{x^2} + \frac{x^2(1)}{x^2\left(\frac{1}{x^2}\right)}$
$$= \frac{1}{x^2} + x^2$$
$$= \frac{1}{x^2} + \frac{x^2(x^2)}{x^2}$$
$$= \frac{1 + x^4}{x^2}, x \neq 0$$

9. $x^{-1} + y^{-1} = \frac{1}{x} + \frac{1}{y}$
$$= \frac{y + x}{xy}, x \neq 0, y \neq 0$$

10. $\frac{x^2y^{-6}}{2^{-2} \cdot x^0 \cdot y^{10}} = \frac{x^2}{\frac{1}{2^2} \cdot 1 \cdot y^{10}} \cdot \frac{1}{y^6}$
$$= \frac{4x^2}{y^{16}}, x \neq 0, y \neq 0$$

18A

SOLUTIONS TO STUDY EXERCISES, CONTD.

Study Exercise Four (Frame 33)

A. **1.** 6 and -6 **2.** 7 and -7 **3.** 0 **4.** does not exist

B. **5.** 8 **6.** 11 **7.** does not exist **8.** $10|x|$

C. **9.** -3 **10.** $4x$ **11.** 5 **12.** -2

D. **13.** 2 **14.** does not exist **15.** 3 **16.** 1

33A

Study Exercise Five (Frame 39)

1. 2

2. $\frac{1}{16^{1/4}} = \frac{1}{2}$

3. $$\frac{1}{(x^2y^{-1})^{1/2}} = \frac{1}{\left(\frac{x^2}{y}\right)^{1/2}} = \frac{y^{1/2}(1)}{y^{1/2}\left(\frac{|x|}{y^{1/2}}\right)} = \frac{y^{1/2}}{|x|},\ x \neq 0,\ y > 0$$

4. $x^2 \cdot x = x^3,\ x \geqslant 0$

5. does not exist

6. $2|x|$

7. -2

8. $-5x^2y^2$

39A

Study Exercise Six (Frame 42)

1. 3 **2.** -3 **3.** -3 **4.** 0

5. $\frac{1}{2}$ **6.** 4 **7.** 0 **8.** 1

42A

Study Exercise Seven (Frame 47)

1. $$(27^{1/3})^2 = 3^2 = 9$$

2. $$\frac{1}{64^{2/3}} = \frac{1}{(64^{1/3})^2} = \frac{1}{4^2} = \frac{1}{16}$$

3. $$(16^{1/4})^5 = 2^5 = 32$$

4. $$\left[\left(\frac{1}{4}\right)^{1/2}\right]^3 = \left(\frac{1}{2}\right)^3 = \frac{1}{8}$$

5. $$\frac{1}{\left(\frac{1}{4}\right)^{1/2}} = \frac{1}{\frac{1}{2}} = 2$$

6. $\frac{2}{3}$

7. $$\frac{1}{\left(\frac{1}{16}\right)^{3/4}} = \frac{1}{\left(\frac{1}{2}\right)^3} = \frac{1}{\frac{1}{8}} = 8$$

8. $|-4| = 4$

47A

Study Exercise Eight (Frame 52)

1. $3.423 \cdot 10^3$ **2.** $7.89 \cdot 10^{-3}$ **3.** $2.358 \cdot 10^1$

4. $8.96 \cdot 10^{-1}$ **5.** $7.8991 \cdot 10^6$ **6.** $8.345 \cdot 10^{-3}$

52A

Study Exercise Nine (Frame 54)

1. $\frac{24 \cdot 10^0}{3 \cdot 10^{-7}} = \frac{8 \cdot 1}{10^{-7}}$
$= 8 \cdot 10^7$

2. $\frac{6.5 \cdot 10^{-2} \cdot 2.2 \cdot 5 \cdot 10^1}{(1.3)(1.1) \cdot 10^{-2} \cdot 5 \cdot 10^{-2}} = \frac{(5)(2)(10)}{10^{-2}}$
$= \frac{10^2}{10^{-2}}$
$= 10^4$

54A

UNIT 7—SUPPLEMENTARY PROBLEMS

A. Find the numerical value:

1. $27^{1/3}$ **2.** 1^{-17} **3.** $\left(\frac{1}{27}\right)^{-1/3}$

4. $-8^{2/3}$ **5.** $16^{3/4}$ **6.** $(9 + 16)^{1/2}$

7. $-4^{1/2}$ **8.** $[8^{1/3} + (-8)^{1/3}]^2$ **9.** $4 - (5^2 - 3^2)^{1/2} + (17)^0$

10. $16^{-1/2}$

B. Simplify or find the numerical value. Leave no answer with a negative or zero exponent. List restrictions on the variables.

11. $4^{3/2} + (-2)^{-3}$ **12.** $16^{-1/2} - 16^{1/4}$ **13.** $(x^2y^{-3})^{-1}$

14. $(x^{-2}y^3)^0$ **15.** $\frac{4^{2/3}}{4^{1/6}}$ **16.** $\left(\frac{x^{-1}y^{-2}}{x^0xy}\right)^{-1}$

17. $x^{-1} + \frac{1}{x^{-1}}$ **18.** $\left(\frac{x^{1-n}}{x^{2-n}}\right)^2$ **19.** $x^6 \cdot x^2$

20. $x^{-3} \cdot x^{-3}$ **21.** $\frac{x^{1/2}}{x^{1/2}}$ **22.** $\left(\frac{2}{5}\right)^{-2} - \left(\frac{8}{27}\right)^{-2/3}$

23. $(-27)^{2/3} + 9^{3/2}$ **24.** $\frac{2^{-2} + 3^{-1}}{2}$ **25.** $\frac{2^{-2} + 3^{-1}}{2^{-2}}$

26. $\frac{x^{-1}}{y^{-1}}$ **27.** $\frac{x^{-1}}{y^{-1}} + \frac{y^{-1}}{x}$ **28.** $\frac{x^{-1} - y^{-1}}{x - y}$

29. $(x^3)^{-1/3}$ **30.** $[(-3)^2]^{1/2}$ **31.** $(x^2)^{1/2}$ **32.** $(x^{1/2})^2$

C. Miscellaneous:

33. Write in scientific notation:
(a) 233 **(b)** .0478 **(c)** 123.4 **(d)** .18

34. Simplify: $\frac{(4 \cdot 10^3)(6 \cdot 10^{-6})}{8 \cdot 10^{-2}}$ **35.** Simplify: $\frac{(.003)(4{,}000)}{.0000012}$

36. Select the number(s) that is (are) negative from:
$16^{-1/4}, 3^{1/2}, 8^{-2/3}, (-1)^{-3}$.

37. $y^{1/2} + y = y(\_\_\_\_\_\_)$ **38.** $x^2 - x^{1/2} = x^{1/2}(\_\_\_\_\_\_)$

39. $x^{3/2} + x = x(\_\_\_\_\_\_)$

Radicals

unit **8**

Objectives

1. Know what a radical is and the properties of radicals.
2. Be able to simplify expressions with radicals.

Terms

radical
index
radicand
rationalizing the denominator

(1)

An alternate notation for roots will be given in this unit.
Recall that $x^{1/2}$ designates the principal square root of x.

$$x^{1/2} \text{ exists if } x \geqslant 0$$
$$x^{1/2} \text{ does not exist if } x < 0$$

(2)

$x^{1/2}$ will also be written $\sqrt[2]{x}$. $\sqrt[2]{x}$ is known as the *radical* form.

index → $\sqrt[2]{x}$ ← *radical*, *radicand*

$$x^{1/2} = \sqrt[2]{x}$$

Most of the time $\sqrt[2]{x}$ is written $\sqrt{x}$.

(3)

$$\sqrt[2]{x} = \sqrt{x}$$

$\sqrt{x}$ designates the principal square root of x. Hence $\sqrt{x}$ exists if $x \geqslant 0$.
$-\sqrt{x}$ designates the opposite of the principal square root of x.

(4)

Examples:

1. $\sqrt{9} = 3$ (not $+3$ and -3)

2. $\sqrt{25} = 5$

3. $-\sqrt{36} = -6$

4. $-\sqrt{100} = -10$

5. $\sqrt{-4} =$ does not exist

(5)

Since $(x + y)^{1/2} \neq x^{1/2} + y^{1/2}$, it follows that $\sqrt{x + y} \neq \sqrt{x} + \sqrt{y}$.

Thus, $\sqrt{4 + 9} = \sqrt{13}$ and not $2 + 3$.

(6)

Study Exercise One

A. Write the following in radical form.

1. $5^{1/2}$ **2.** $-46^{1/2}$ **3.** $(x^2 + y^2)^{1/2}$ **4.** $(1 + 4x)^{1/2}$

B. Find the value.

5. $\sqrt{64}$ **6.** $-\sqrt{25}$ **7.** $\sqrt{-81}$

(7)

$x^{1/3}$ will be written $\sqrt[3]{x}$. $x^{1/4}$ will be written $\sqrt[4]{x}$.

In general, $\sqrt[n]{x}$ designates $x^{1/n}$, $n \in N, n \geqslant 2$.

(8)

$\sqrt[n]{x}$ exists

1. when n is even ($n \in N, n \geqslant 2$) and $x \geqslant 0$ (for even roots of non negative numbers).
2. when n is odd ($n \in N, n \geqslant 3$) for all x (for odd roots of all numbers).

(9)

Since $(x^{1/n})^n = x$ if the nth root of x exists, then $(\sqrt[n]{x})^n = x$ if the nth root of x exists.

Example: **1.** $(\sqrt{5})^2 = 5$

2. $(\sqrt[3]{7})^3 = 7$

(10)

Recall that, $(x^n)^{1/n} = \begin{cases} |x| \text{ if } n \text{ is even} \\ x \text{ if } n \text{ is odd.} \end{cases}$

Thus it follows that, $\sqrt[n]{x^n} = \begin{cases} |x| \text{ if } n \text{ is even} \\ x \text{ if } n \text{ is odd} \end{cases}$

(11)

Example 1: $\sqrt{(2)^2} = |2|$ or 2

Example 2: $\sqrt{(-2)^2} = |-2|$ or 2

Example 3: $\sqrt{x^2} = |x|$

Example 4: $\sqrt[3]{x^3} = x$

(12)

$$\text{Since } x^{m/n} = (x^m)^{1/n} \quad \text{if } x^{1/n} \text{ exists}$$
$$= (x^{1/n})^m$$

$$\text{then } x^{m/n} = \sqrt[n]{x^m} \quad \text{if } \sqrt[n]{x} \text{ exists}$$
$$= (\sqrt[n]{x})^m$$

(13)

Example 1: $x^{2/3} = \sqrt[3]{x^2}$

$= \left(\sqrt[3]{x}\right)^2$

Example 2: $8^{2/3} = \sqrt[3]{8^2}$ or $8^{2/3} = \left(\sqrt[3]{8}\right)^2$

$= \sqrt[3]{64}$ $= 2^2$

$= 4$ $= 4$

(14)

Study Exercise Two

Simplify:

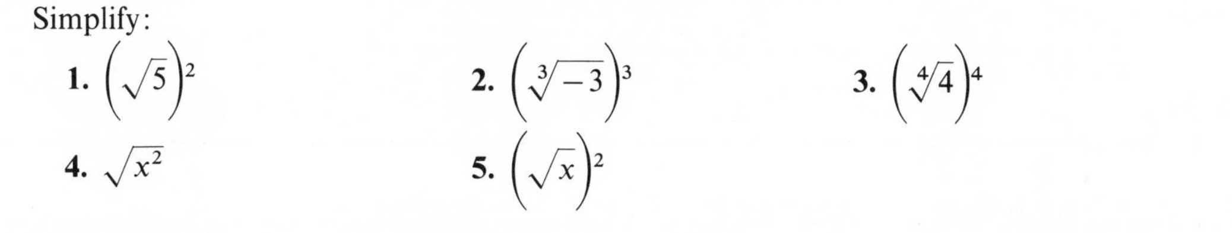

(15)

Properties of the Radical

Property I. $\sqrt[n]{a} \cdot \sqrt[n]{b} = \sqrt[n]{ab}$ provided that both $\sqrt[n]{a}$ and $\sqrt[n]{b}$ exist.

$$\begin{aligned} \textit{Proof: } \sqrt[n]{a} \cdot \sqrt[n]{b} &= a^{1/n} \cdot b^{1/n} \\ &= (ab)^{1/n} \\ &= \sqrt[n]{ab} \end{aligned}$$

(16)

Property II. $\dfrac{\sqrt[n]{a}}{\sqrt[n]{b}} = \sqrt[n]{\dfrac{a}{b}}$ provided that both $\sqrt[n]{a}$ and $\sqrt[n]{b}$ exist and $b \neq 0$.

$$\begin{aligned} \textit{Proof: } \frac{\sqrt[n]{a}}{\sqrt[n]{b}} &= \frac{a^{1/n}}{b^{1/n}} \\ &= \left(\frac{a}{b}\right)^{1/n} \\ &= \sqrt[n]{\frac{a}{b}}, \ b \neq 0 \end{aligned}$$

(17)

Examples: Apply either property I or II.

1. $\sqrt{2} \cdot \sqrt{3} = \sqrt{6}$ Property I
2. $\sqrt[3]{3} \cdot \sqrt[3]{3} = \sqrt[3]{9}$ Property I
3. $\sqrt[5]{2x} \cdot \sqrt[5]{5} = \sqrt[5]{10x}$ Property I
4. $\sqrt{\dfrac{5x}{2}} = \dfrac{\sqrt{5x}}{\sqrt{2}}, x \geqslant 0$ Property II
5. $\dfrac{\sqrt{15}}{\sqrt{3}} = \sqrt{\dfrac{15}{3}}$

   $= \sqrt{5}$ Property II

(18)

A radical is said to be in simplest form provided;

1. there are no factors which are perfect nth powers under a radical whose index is n.
2. there are no fractions under the radical sign.
3. there are no radicals in the denominator of a fraction.
4. the smallest possible index is used.

(19)

The following are examples of radical expressions that are *not* in simplest form.

$$\sqrt{\frac{1}{2}}, \quad \frac{1}{1+\sqrt{2}}, \quad \sqrt[3]{(x+1)^3}, \quad \sqrt{8}$$

(20)

Example 1: Write $\sqrt{50}$ in simplest form:

Solution:

Line (1) $\sqrt{50} = \sqrt{25 \cdot 2}$

Line (2) $= \sqrt{25} \cdot \sqrt{2}$

Line (3) $= 5\sqrt{2}$

Example 2: Write $\sqrt[3]{40}$ in simplest form:

Solution:

Line (1) $\sqrt[3]{40} = \sqrt[3]{8 \cdot 5}$

Line (2) $= \sqrt[3]{8} \cdot \sqrt[3]{5}$

Line (3) $= 2\sqrt[3]{5}$

(21)

To simplify a radical with a fraction under the radical multiply numerator and denominator by the same non-zero quantity to eliminate the radical in the denominator.

Example: Simplify $\sqrt{1/2}$

Solution:

Line (1) $\sqrt{1/2} = \dfrac{\sqrt{1}}{\sqrt{2}}$

Line (2) $= \dfrac{\sqrt{1}\sqrt{2}}{\sqrt{2}\sqrt{2}}$

Line (3) $= \dfrac{\sqrt{2}}{\sqrt{4}}$

Line (4) $= \dfrac{\sqrt{2}}{2}$

(22)

Example 1: Simplify $\sqrt{12x^5y^4}$

Solution:

Line (1) $\sqrt{12x^5y^4} = \sqrt{4 \cdot x^4 \cdot y^4 \cdot 3 \cdot x}$

Line (2) $= \sqrt{4x^4y^4}\sqrt{3x}$

Line (3) $= 2x^2y^2\sqrt{3x},\ x \geqslant 0$

Example 2: Simplify $\sqrt{1/9 + 1/4}$

Solution:

Line (1) $\sqrt{\dfrac{1}{9} + \dfrac{1}{4}} = \sqrt{\dfrac{4}{36} + \dfrac{9}{36}}$

Line (2) $= \sqrt{\dfrac{13}{36}}$

Line (3) $= \dfrac{\sqrt{13}}{6}$

(23)

Observe from the previous example that, $\sqrt{x^2 + y^2} \neq x + y$.

Also note that $\sqrt{x^2} = |x|$ if x is real.

$$= \begin{cases} x \text{ if } x \geqslant 0 \\ -x \text{ if } x < 0 \end{cases}$$

(24)

Example: Simplify $\sqrt{48x^2}$

Solution:

Line (1) $\sqrt{48x^2} = \sqrt{16 \cdot x^2 \cdot 3}$

Line (2) $= \sqrt{16x^2}\sqrt{3}$

Line (3) $= 4|x|\sqrt{3}$

(25)

Study Exercise Three

Simplify:

1. $\sqrt{27x^4}$

2. $\sqrt{98}$

3. $\sqrt[3]{-125}$

4. $\sqrt[4]{32}$

5. $\sqrt{\frac{3}{7}}$

6. $\sqrt[3]{\frac{3}{8}}$

7. $\sqrt{\frac{r}{4\pi}}$

8. $5\sqrt{\frac{9}{20}}$

9. $\sqrt{\frac{x}{y^2}}$

10. $\sqrt{20x^2}$

11. $\frac{\sqrt[3]{27}}{\sqrt[3]{4}}$

12. $\frac{2}{\sqrt{3}}$

(26)

Multiplication of Radicals

To multiply $(\sqrt{3} + \sqrt{5})$ by $(\sqrt{3} - \sqrt{5})$, we apply the distributive property.

$$\begin{aligned}(\sqrt{3} + \sqrt{5})(\sqrt{3} - \sqrt{5}) &= \sqrt{9} + \sqrt{15} - \sqrt{15} - \sqrt{25} \\ &= 3 + 0 - 5 \\ &= -2\end{aligned}$$

(27)

Example 1: Simplify $\dfrac{2\sqrt{3}}{4\sqrt{5}}$

Solution:

Line (1) $\dfrac{2\sqrt{3}}{4\sqrt{5}} = \dfrac{\sqrt{3}}{2\sqrt{5}}$

Line (2) $= \dfrac{\sqrt{3}\cdot\sqrt{5}}{2\sqrt{5}\cdot\sqrt{5}}$

Line (3) $= \dfrac{\sqrt{15}}{2\sqrt{25}}$

Line (4) $= \dfrac{\sqrt{15}}{2\cdot 5}$

Line (5) $= \dfrac{\sqrt{15}}{10}$

Example 2: Simplify $(6\sqrt[3]{5}) \div (2\sqrt[3]{3})$

Solution:

Line (1) $(6\sqrt[3]{5}) \div (2\sqrt[3]{3}) = \dfrac{6\sqrt[3]{5}}{2\sqrt[3]{3}}$

Line (2) $= \dfrac{3\sqrt[3]{5}}{\sqrt[3]{3}}\cdot\dfrac{\sqrt[3]{9}}{\sqrt[3]{9}}$

Line (3) $= \dfrac{3\sqrt[3]{45}}{\sqrt[3]{27}}$

Line (4) $= \dfrac{3\sqrt[3]{45}}{3}$

Line (5) $= \sqrt[3]{45}$

(28)

Study Exercise Four

Perform the following multiplications and express the results in simplest form.

1. $(\sqrt{5} + 2\sqrt{3})(\sqrt{5} - 2\sqrt{3})$

2. $(2\sqrt{3} + 1)^2$

3. $\sqrt{3x^2y^3}\cdot\sqrt{12x^5y}$

4. $(3\sqrt{2} - 2\sqrt{3})^2$

5. $(2\sqrt{6} + 3\sqrt{14}) \div \sqrt{2}$

(29)

Addition and Subtraction of Radicals

Radicals with the same index and the same radicand are called *like radicals*. Like radicals may be "added" by means of the distributive principle. However, care should be exercised

$$\text{since } \sqrt{x} + \sqrt{y} \neq \sqrt{x + y}$$

(30)

Example 1: Simplify $3\sqrt{3} + 2\sqrt{3}$

Solution: $3\sqrt{3} + 2\sqrt{3} = (3 + 2)\sqrt{3}$

$= 5\sqrt{3}$

Example 2: Simplify $2\sqrt[3]{2} - \sqrt[3]{2}$

Solution: $2\sqrt[3]{2} - \sqrt[3]{2} = (2 - 1)\sqrt[3]{2}$

$= \sqrt[3]{2}$

(Frame 31, contd.)

Example 3: Simplify $\sqrt{8} - 3\sqrt{2}$

Solution:

$$\begin{aligned}\sqrt{8} - 3\sqrt{2} &= \sqrt{4}\sqrt{2} - 3\sqrt{2} \\ &= 2\sqrt{2} - 3\sqrt{2} \\ &= (2-3)\sqrt{2} \\ &= -\sqrt{2}\end{aligned}$$

(31)

Study Exercise Five

Express in simplest radical form:

1. $5\sqrt{12} + 3\sqrt{3}$
2. $3\sqrt{\dfrac{1}{2}} + \dfrac{5}{2}\sqrt{18}$
3. $\sqrt{72} - 3\sqrt{12} + \sqrt{32}$
4. $7\sqrt[3]{54} - 3\sqrt[3]{16}$
5. $\sqrt{xy^6} - y\sqrt{x^3} + x\sqrt{x}$ where $x \geqslant 0$ and $y \geqslant 0$

(32)

In the process of changing a radical to simplest form, many times the denominator has to be cleared of radicals.

The process of changing a fraction to an equivalent one where the denominator is free of radicals is called **rationalizing the denominator.**

(33)

To simplify $\dfrac{5}{\sqrt{2}}$ by rationalizing the denominator, we proceed as follows:

$$\begin{aligned}\frac{5}{\sqrt{2}} &= \frac{5\sqrt{2}}{\sqrt{2}\sqrt{2}} \\ &= \frac{5\sqrt{2}}{\sqrt{4}} \\ &= \frac{5\sqrt{2}}{2}\end{aligned}$$

(34)

But if the denominator is a binomial, what factor can be used to multiply numerator and denominator to clear the denominator of radicals?

What multiplier would you use to rationalize the denominator of $\dfrac{1}{\sqrt{2}+1}$?

(35)

Since $(\sqrt{2} + 1)(\sqrt{2} - 1) = \sqrt{4} - 1$
$= 2 - 1$
$= 1$

we will use $(\sqrt{2} - 1)$ as the rationalizing factor.

$$\frac{1}{\sqrt{2} + 1} = \frac{1(\sqrt{2} - 1)}{(\sqrt{2} + 1)(\sqrt{2} - 1)} = \frac{\sqrt{2} - 1}{\sqrt{4} - 1}$$

$$= \frac{\sqrt{2} - 1}{2 - 1}$$

$$= \sqrt{2} - 1$$

(36)

What is the rationalizing factor for $\dfrac{6}{2 + \sqrt{5}}$?

Answer: $2 - \sqrt{5}$

What is the rationalizing factor for $\dfrac{\sqrt{3}}{\sqrt{3} - \sqrt{2}}$?

Answer: $\sqrt{3} + \sqrt{2}$

(37)

Example: Simplify $\dfrac{\sqrt{5} + 1}{\sqrt{5} + \sqrt{2}}$

Solution:

Line (1) $$\frac{\sqrt{5} + 1}{\sqrt{5} + \sqrt{2}} = \frac{(\sqrt{5} + 1)(\sqrt{5} - \sqrt{2})}{(\sqrt{5} + \sqrt{2})(\sqrt{5} - \sqrt{2})}$$

Line (2) $$= \frac{\sqrt{25} + \sqrt{5} - \sqrt{10} - \sqrt{2}}{\sqrt{25} - \sqrt{4}}$$

Line (3) $$= \frac{5 + \sqrt{5} - \sqrt{10} - \sqrt{2}}{5 - 2}$$

Line (4) $$= \frac{5 + \sqrt{5} - \sqrt{10} - \sqrt{2}}{3}$$

(38)

The rationalizing factor is sometimes called the **conjugate** when the expression is of the form $\sqrt{a} + \sqrt{b}$ and either a or b are not perfect squares.

expression	*conjugate of expression*
$1 + \sqrt{3}$	$1 - \sqrt{3}$
$3 - \sqrt{5}$	$3 + \sqrt{5}$
$\sqrt{2} - 1$	$\sqrt{2} + 1$
$\sqrt{7} + \sqrt{2}$	$\sqrt{7} - \sqrt{2}$

(39)

Study Exercise Six

Simplify:

1. $\dfrac{3}{2+\sqrt{3}}$

2. $\dfrac{\sqrt{3}+\sqrt{2}}{\sqrt{3}-\sqrt{2}}$

3. $\dfrac{x}{x+\sqrt{y}}, \quad y > 0$

4. $\dfrac{\sqrt{x}-\sqrt{y}}{\sqrt{x}+\sqrt{y}}, \quad x > 0, \quad y > 0$

(40)

Example: Simplify $\dfrac{20+\sqrt{20}}{4}$

Solution:

Line (1) $\dfrac{20+\sqrt{20}}{4} = \dfrac{20+\sqrt{4\cdot 5}}{4}$

Line (2) $= \dfrac{20+2\sqrt{5}}{4}$

Line(3) $= \dfrac{2(10+\sqrt{5})}{4}$

Line(4) $= \dfrac{10+\sqrt{5}}{2}$

(41)

Study Exercise Seven

Simplify:

1. $\dfrac{10+\sqrt{48}}{2}$

2. $\dfrac{-16+\sqrt{32}}{4}$

3. $\dfrac{7+\sqrt{49-4(10)}}{2}$

(42)

One of the properties of radicals states that $\sqrt[n]{a}\cdot\sqrt[n]{b} = \sqrt[n]{a\cdot b}$.

However, is it possible to multiply radicals of different indices?

$$\sqrt{2}\cdot\sqrt{2} = \sqrt{4}$$
$$= 2$$
$$\sqrt{2}\cdot\sqrt[3]{2} = \underline{\quad ? \quad}$$

(43)

Example: Find $\sqrt{2}\cdot\sqrt[3]{2}$

Solution:

Line (1) $\sqrt{2}\cdot\sqrt[3]{2} = (2)^{1/2}\cdot(2)^{1/3}$

Line (2) $= (2)^{3/6}\cdot(2)^{2/6}$

Line (3) $= (2)^{3(1/6)}\cdot(2)^{2(1/6)}$

Line (4) $= (2^3\cdot 2^2)^{1/6}$

Line (5) $= \sqrt[6]{2^3\cdot 2^2}$

Line (6) $= \sqrt[6]{32}$

(44)

Study Exercise Eight

Multiply as indicated and simplify:

1. $\sqrt[3]{2} \cdot \sqrt{3}$ **2.** $\sqrt{2} \cdot \sqrt[4]{2}$

3. $\sqrt{2} \cdot \sqrt{\sqrt{4}}$

(45)

REVIEW EXERCISES

A. True or False

1. $\sqrt{25} = 5$ and -5

2. $\sqrt{9 + 16} = 3 + 4$

3. $\sqrt{x^2} = x$

4. $\sqrt{x^2 + y^2} = x + y$

5. $\sqrt[3]{-27}$ does not exist

B. Express in simplest form:

6. $\sqrt{72x^2}$

7. $\sqrt[3]{-54}$

8. $\sqrt{\frac{2}{3}}$

9. $\sqrt{\frac{49x^4}{2}}$

10. $\sqrt{\frac{1}{4} + \frac{1}{16}}$

11. $4\sqrt{7} - \sqrt{7}$

12. $\sqrt[3]{16} + 8\sqrt[3]{\frac{1}{4}}$

13. $\frac{\sqrt{10}}{\sqrt{2}}$

14. $\frac{\sqrt{6} + \sqrt{5}}{\sqrt{2}}$

15. $\frac{4 + \sqrt{3}}{2 + \sqrt{3}}$

16. $\sqrt{5}\sqrt[3]{2}$

17. $(5\sqrt{2} + 1)^2$

18. $(\sqrt{x})^2$

19. $\sqrt{x^2}$

(46)

SOLUTIONS TO REVIEW EXERCISES

A. **1.** false **2.** false **3.** false

4. false **5.** false

B. **6.** $\sqrt{36 \cdot 2 \cdot x^2} = 6\,|x|\,\sqrt{2}$

7. $\sqrt[3]{(-27(2)} = -3\sqrt[3]{2}$

8. $\frac{\sqrt{2}}{\sqrt{3}} \cdot \frac{\sqrt{3}}{\sqrt{3}} = \frac{\sqrt{6}}{3}$

9. $\frac{\sqrt{49x^4} \cdot \sqrt{2}}{\sqrt{2} \cdot \sqrt{2}} = \frac{7x^2\sqrt{2}}{2}$

10. $\sqrt{\frac{4}{16} + \frac{1}{16}} = \frac{\sqrt{5}}{4}$

11. $4\sqrt{7} - \sqrt{7} = 3\sqrt{7}$

12. $\sqrt[3]{8 \cdot 2} + 8 \cdot \frac{1}{\sqrt[3]{4}} \cdot \frac{\sqrt[3]{2}}{\sqrt[3]{2}} = 2\sqrt[3]{2} + 4\sqrt[3]{2}$

$= 6\sqrt[3]{2}$

SOLUTIONS TO REVIEW EXERCISES, CONTD.

(Frame 47, contd.)

13. $\frac{\sqrt{10}}{\sqrt{2}} \cdot \frac{\sqrt{2}}{\sqrt{2}} = \frac{\sqrt{20}}{2}$

$= \frac{\sqrt{4}\sqrt{5}}{2}$

$= \sqrt{5}$

14. $\frac{(\sqrt{6} + \sqrt{5})\sqrt{2}}{(\sqrt{2})\sqrt{2}} = \frac{\sqrt{12} + \sqrt{10}}{\sqrt{4}} = \frac{2\sqrt{3} + \sqrt{10}}{2}$

15. $\frac{(4 + \sqrt{3})(2 - \sqrt{3})}{(2 + \sqrt{3})(2 - \sqrt{3})} = \frac{8 - 2\sqrt{3} - \sqrt{9}}{4 - \sqrt{9}}$

$= 5 - 2\sqrt{3}$

16. $(5)^{1/2} \cdot 2^{1/3} = 5^{3/6} \cdot 2^{2/6}$

$= \sqrt[6]{5^3 \cdot 2^2}$

$= \sqrt[6]{500}$

17. $(5\sqrt{2} + 1)^2 = 25\sqrt{4} + 10\sqrt{2} + 1$

$= 50 + 10\sqrt{2} + 1$

$= 51 + 10\sqrt{2}$

18. $(\sqrt{x})^2 = x, x \geqslant 0$

19. $\sqrt{x^2} = |x|$

(47)

SOLUTIONS TO STUDY EXERCISES

Study Exercise One (Frame 7)

A. **1.** $\sqrt{5}$ **2.** $-\sqrt{46}$ **3.** $\sqrt{x^2 + y^2}$ **4.** $\sqrt{1 + 4x}$

B. **5.** 8 **6.** -5 **7.** does not exist

(7A)

Study Exercise Two (Frame 15)

1. 5 **2.** -3 **3.** 4 **4.** $|x|$ **5.** x

(15A)

Study Exercise Three (Frame 26)

1. $\sqrt{27x^4} = \sqrt{9 \cdot x^4 \cdot 3}$

$= 3x^2\sqrt{3}$

2. $\sqrt{98} = \sqrt{49 \cdot 2}$

$= 7\sqrt{2}$

3. $\sqrt[3]{-125} = -5$

4. $\sqrt[4]{32} = \sqrt[4]{16 \cdot 2}$

$= 2\sqrt[4]{2}$

SOLUTIONS TO STUDY EXERCISES, CONTD.

Study Exercise Three (Frame 26, contd.)

5. $\sqrt{\frac{3}{7}} = \frac{\sqrt{3}}{\sqrt{7}}$

$= \frac{\sqrt{3}\sqrt{7}}{\sqrt{7}\sqrt{7}}$

$= \frac{\sqrt{21}}{7}$

6. $\sqrt[3]{\frac{3}{8}} = \frac{\sqrt[3]{3}}{\sqrt[3]{8}}$

$= \frac{\sqrt[3]{3}}{2}$

7. $\sqrt{\frac{r}{4\pi}} = \frac{\sqrt{r}\sqrt{\pi}}{\sqrt{4\pi}\sqrt{\pi}}$

$= \frac{\sqrt{r\pi}}{\sqrt{4\pi^2}}$

$= \frac{\sqrt{r\pi}}{2\pi}, r \geqslant 0$

8. $5\sqrt{\frac{9}{20}} = \frac{5\sqrt{9}}{\sqrt{20}}$

$= \frac{5\sqrt{9}\sqrt{5}}{\sqrt{20}\sqrt{5}}$

$= \frac{15\sqrt{5}}{10}$

$= \frac{3\sqrt{5}}{2}$

9. $\sqrt{\frac{x}{y^2}} = \frac{\sqrt{x}}{\sqrt{y^2}}$

$= \frac{\sqrt{x}}{|y|}, x \geqslant 0, y \neq 0$

10. $\sqrt{20x^2} = \sqrt{4 \cdot x^2 \cdot 5}$

$= 2\,|x|\,\sqrt{5}$

11. $\frac{\sqrt[3]{27}}{\sqrt[3]{4}} = \frac{\sqrt[3]{27}\sqrt[3]{2}}{\sqrt[3]{4}\sqrt[3]{2}}$

$= \frac{3\sqrt[3]{2}}{\sqrt[3]{8}}$

$= \frac{3\sqrt[3]{2}}{2}$

12. $\frac{2}{\sqrt{3}} = \frac{2\sqrt{3}}{\sqrt{3}\sqrt{3}}$

$= \frac{2\sqrt{3}}{\sqrt{9}}$

$= \frac{2\sqrt{3}}{3}$

26A

Study Exercise Four (Frame 29)

1. $(\sqrt{5} + 2\sqrt{3})(\sqrt{5} - 2\sqrt{3}) = \sqrt{25} - 4\sqrt{9}$

$= 5 - 4 \cdot 3$

$= -7$

2. $(2\sqrt{3} + 1)^2 = 4\sqrt{9} + 4\sqrt{3} + 1$

$= 12 + 4\sqrt{3} + 1$

$= 13 + 4\sqrt{3}$

3. $\sqrt{3x^2y^3} \cdot \sqrt{12x^5y} = \sqrt{36x^7y^4}$

$= \sqrt{36x^6y^4} \cdot \sqrt{x}$

$= 6x^3y^2\sqrt{x}, x \geqslant 0, y \geqslant 0$

Study Exercise Four (Frame 29, contd.)

4. $(3\sqrt{2} - 2\sqrt{3})^2 = 9\sqrt{4} - 12\sqrt{6} + 4\sqrt{9}$

$= 18 - 12\sqrt{6} + 12$

$= 30 - 12\sqrt{6}$

5. $\dfrac{2\sqrt{6} + 3\sqrt{14}}{\sqrt{2}} = \dfrac{(2\sqrt{6} + 3\sqrt{14})\sqrt{2}}{\sqrt{2}\sqrt{2}}$

$= \dfrac{2\sqrt{12} + 3\sqrt{28}}{\sqrt{4}}$

$= \dfrac{2\sqrt{4 \cdot 3} + 3\sqrt{4 \cdot 7}}{2}$

$= \dfrac{4\sqrt{3} + 6\sqrt{7}}{2}$

$= \dfrac{2(2\sqrt{3} + 3\sqrt{7})}{2}$

$= 2\sqrt{3} + 3\sqrt{7}$

29A

Study Exercise Five (Frame 32)

1. $5\sqrt{12} + 3\sqrt{3} = 5\sqrt{4 \cdot 3} + 3\sqrt{3}$

$= 10\sqrt{3} + 3\sqrt{3}$

$= 13\sqrt{3}$

2. $3\sqrt{\dfrac{1}{2}} + \dfrac{5}{2}\sqrt{18} = \dfrac{3\sqrt{1}\sqrt{2}}{\sqrt{2}\sqrt{2}} + \dfrac{5\sqrt{9 \cdot 2}}{2}$

$= \dfrac{3\sqrt{2}}{2} + \dfrac{15\sqrt{2}}{2}$

$= \dfrac{18\sqrt{2}}{2}$

$= 9\sqrt{2}$

3. $\sqrt{72} - 3\sqrt{12} + \sqrt{32} = \sqrt{36 \cdot 2} - 3\sqrt{4 \cdot 3} + \sqrt{16 \cdot 2}$

$= 6\sqrt{2} - 6\sqrt{3} + 4\sqrt{2}$

$= 10\sqrt{2} - 6\sqrt{3}$

4. $7\sqrt[3]{54} - 3\sqrt[3]{16} = 7\sqrt[3]{27 \cdot 2} - 3\sqrt[3]{8 \cdot 2}$

$= 7 \cdot 3\sqrt[3]{2} - 3 \cdot 2\sqrt[3]{2}$

$= 21\sqrt[3]{2} - 6\sqrt[3]{2}$

$= 15\sqrt[3]{2}$

SOLUTIONS TO STUDY EXERCISES, CONTD.

Study Exercise Five (Frame 32, contd.)

5. $\sqrt{xy^6} - y\sqrt{x^3} + x\sqrt{x} = y^3\sqrt{x} - y\sqrt{x^2 \cdot x} + x\sqrt{x}$

$$= y^3\sqrt{x} - xy\sqrt{x} + x\sqrt{x}$$

$$= (y^3 - xy + x)\sqrt{x}$$

32A

Study Exercise Six (Frame 40)

1. $\dfrac{3}{2 + \sqrt{3}} = \dfrac{3(2 - \sqrt{3})}{(2 + \sqrt{3})(2 - \sqrt{3})}$

$$= \frac{3(2 - \sqrt{3})}{4 - \sqrt{9}}$$

$$= 3(2 - \sqrt{3})$$

2. $\dfrac{(\sqrt{3} + \sqrt{2})}{(\sqrt{3} - \sqrt{2})} = \dfrac{(\sqrt{3} + \sqrt{2})(\sqrt{3} + \sqrt{2})}{(\sqrt{3} - \sqrt{2})(\sqrt{3} + \sqrt{2})}$

$$= \frac{\sqrt{9} + 2\sqrt{6} + \sqrt{4}}{\sqrt{9} - \sqrt{4}}$$

$$= 5 + 2\sqrt{6}$$

3. $\dfrac{x}{x + \sqrt{y}} = \dfrac{x(x - \sqrt{y})}{(x + \sqrt{y})(x - \sqrt{y})}$

$$= \frac{x(x - \sqrt{y})}{x^2 - y}, \quad x^2 - y \neq 0$$

4. $\dfrac{\sqrt{x} - \sqrt{y}}{\sqrt{x} + \sqrt{y}} = \dfrac{(\sqrt{x} - \sqrt{y})(\sqrt{x} - \sqrt{y})}{(\sqrt{x} + \sqrt{y})(\sqrt{x} - \sqrt{y})}$

$$= \frac{\sqrt{x^2} - 2\sqrt{xy} + \sqrt{y^2}}{\sqrt{x^2} - \sqrt{y^2}}$$

$$= \frac{x + y - 2\sqrt{xy}}{x - y}, \quad x \neq y$$

40A

Study Exercise Seven (Frame 42)

1. $\dfrac{10 + \sqrt{48}}{2} = \dfrac{10 + \sqrt{16 \cdot 3}}{2}$

$$= \frac{10 + 4\sqrt{3}}{2}$$

$$= \frac{2(5 + 2\sqrt{3})}{2}$$

$$= 5 + 2\sqrt{3}$$

2. $\dfrac{-16 + \sqrt{32}}{4} = \dfrac{-16 + 4\sqrt{2}}{4}$

$$= \frac{4(-4 + \sqrt{2})}{4}$$

$$= -4 + \sqrt{2}$$

3. $\dfrac{7 + \sqrt{49 - 40}}{2} = \dfrac{7 + \sqrt{9}}{2}$

$$= \frac{10}{2} \text{ or } 5$$

42A

SOLUTIONS TO STUDY EXERCISES, CONTD.

Study Exercise Eight (Frame 45)

1. $\sqrt[3]{2} \cdot \sqrt{3} = 2^{1/3} \cdot 3^{1/2}$
$= 2^{2/6} \cdot 3^{3/6}$
$= (2^2 \cdot 3^3)^{1/6}$
$= \sqrt[6]{4 \cdot 27}$
$= \sqrt[6]{108}$

2. $\sqrt{2} \cdot \sqrt[4]{2} = 2^{1/2} \cdot 2^{1/4}$
$= 2^{2/4} \cdot 2^{1/4}$
$= (2^2 \cdot 2^1)^{1/4}$
$= \sqrt[4]{4 \cdot 2}$
$= \sqrt[4]{8}$

3. $\sqrt{2} \cdot \sqrt{\sqrt{4}} = \sqrt{2} \cdot \sqrt{2}$
$= 2$

45A

UNIT 8—SUPPLEMENTARY PROBLEMS

A. Write without a radical sign:

1. $\sqrt{100}$

2. $\sqrt{(2x)^2}$

3. $\sqrt{(-3)^2}$

4. $-\sqrt{(-4)^2}$

B. Perform the indicated operations and/or simplify:

5. $\sqrt{72}$

6. $\sqrt[3]{16}$

7. $\sqrt{\frac{3}{20}}$

8. $\sqrt{25 - 16}$

9. $\sqrt{\frac{1}{2} + \frac{1}{3}}$

10. $\sqrt{2} \cdot \sqrt{6}$

11. $\sqrt{\frac{2}{3}} \cdot \sqrt{\frac{1}{6}}$

12. $\sqrt{12} + 3\sqrt{8} - \sqrt{2x}$

13. $\sqrt{2}(\sqrt{18} - \sqrt{6})$

14. $\sqrt{24} + \sqrt{18} - \sqrt{54}$

15. $\sqrt{\frac{a}{b}} \cdot \sqrt{ab}, a > 0, b > 0$

16. $\sqrt{\frac{3}{2}}\sqrt{\frac{4}{3}}\sqrt{\frac{5}{4}}\sqrt{\frac{6}{5}}$

17. $\sqrt{\frac{a}{b}} \cdot \sqrt{\frac{b}{a}}, a > 0, b > 0$

18. $\sqrt{27x} - \sqrt{125x} + \sqrt{12x}$

19. $\frac{5\sqrt{2}}{2\sqrt{3} - \sqrt{2}}$

20. $3\sqrt[3]{4}$

21. $(2\sqrt{3} + 1)^2$

22. $\sqrt[4]{3x^5y^5}$

23. $(\sqrt{3})(\sqrt[3]{5})$

24. $(1 - \sqrt{2})(2 + \sqrt{2})$

25. $\frac{8 - 2\sqrt{12}}{4}$

26. $\frac{x}{\sqrt{x - 3}}$

27. $\sqrt[6]{8x^3y^6}$

28. $\sqrt[5]{\frac{1}{5}}$

29. $\frac{x}{\sqrt{xy}}$, $x > 0, y > 0$

30. $\sqrt[3]{54} + 2\sqrt[3]{128}$

31. $-\sqrt{160}$

32. $\sqrt{x^2 + 2x + 1}$

33. $(2 + \sqrt{3})(2 - \sqrt{3})$

34. $\frac{1}{1 + \sqrt{3}}$

C. Miscellaneous:

35. If $\sqrt{x} = 4$, what is x?

36. When is $\sqrt{x^2} = x$ a true statement?

37. Combine $5\sqrt{2} - \sqrt[4]{64} + 2\sqrt{32}$ and simplify the result.

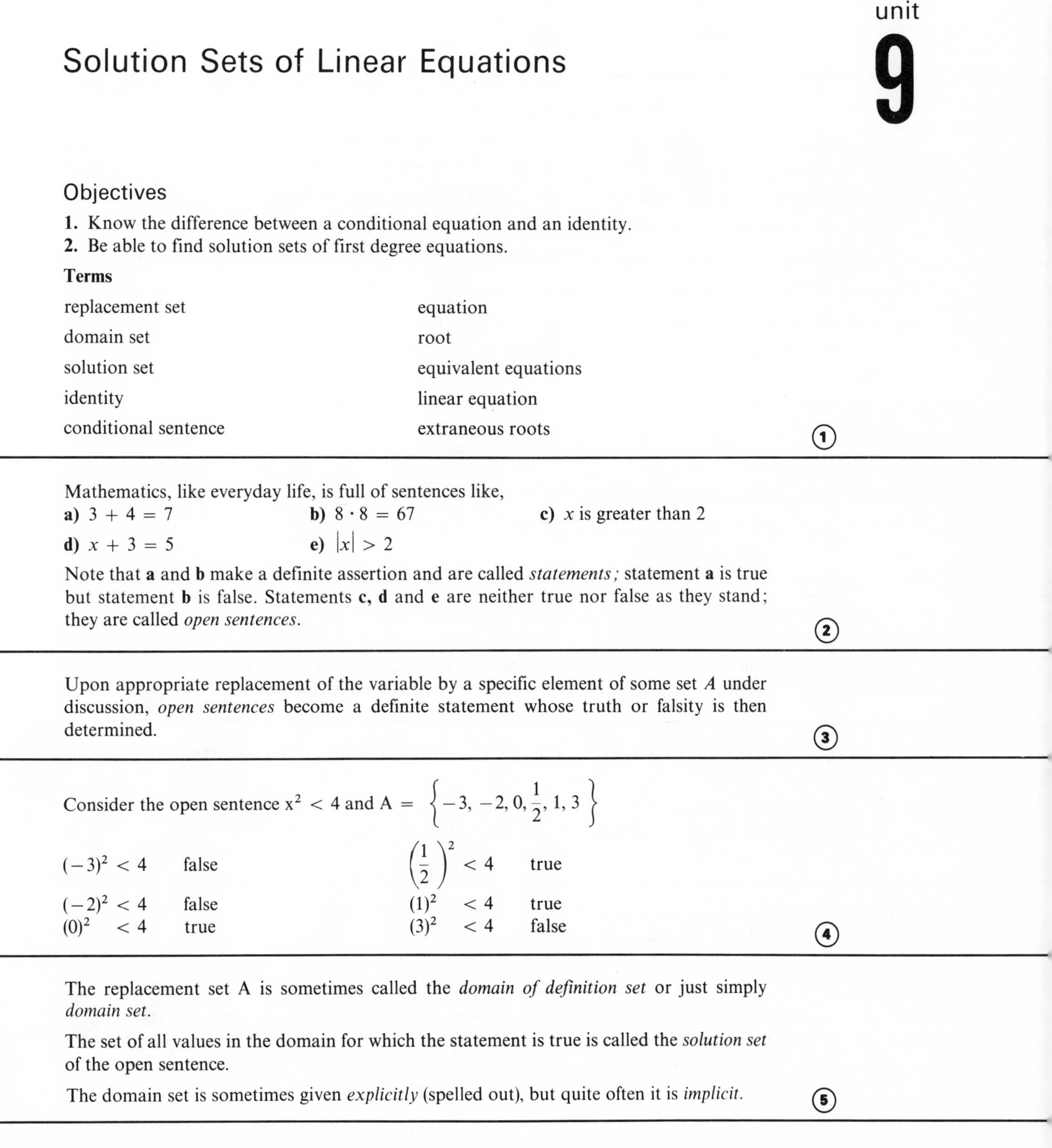

unit **9**

Solution Sets of Linear Equations

Objectives

1. Know the difference between a conditional equation and an identity.
2. Be able to find solution sets of first degree equations.

Terms

replacement set	equation
domain set	root
solution set	equivalent equations
identity	linear equation
conditional sentence	extraneous roots

(1)

Mathematics, like everyday life, is full of sentences like,

a) $3 + 4 = 7$ **b)** $8 \cdot 8 = 67$ **c)** x is greater than 2

d) $x + 3 = 5$ **e)** $|x| > 2$

Note that **a** and **b** make a definite assertion and are called *statements;* statement **a** is true but statement **b** is false. Statements **c, d** and **e** are neither true nor false as they stand; they are called *open sentences.*

(2)

Upon appropriate replacement of the variable by a specific element of some set A under discussion, *open sentences* become a definite statement whose truth or falsity is then determined.

(3)

Consider the open sentence $x^2 < 4$ and $A = \left\{-3, -2, 0, \frac{1}{2}, 1, 3\right\}$

$(-3)^2 < 4$	false	$\left(\frac{1}{2}\right)^2 < 4$	true
$(-2)^2 < 4$	false	$(1)^2 < 4$	true
$(0)^2 < 4$	true	$(3)^2 < 4$	false

(4)

The replacement set A is sometimes called the *domain of definition set* or just simply *domain set.*

The set of all values in the domain for which the statement is true is called the *solution set* of the open sentence.

The domain set is sometimes given *explicitly* (spelled out), but quite often it is *implicit.*

(5)

In our previous example where $x^2 < 4$ and $A = \left\{-3, -2, 0, \frac{1}{2}, 1, 3\right\}$,

The open sentence is $x^2 < 4$.

The domain set is $\left\{-3, -2, 0, \frac{1}{2}, 1, 3\right\}$.

The solution set is $\left\{0, \frac{1}{2}, 1\right\}$.

(6)

Study Exercise One

Find the solution sets of the following sentences over the indicated domain set.

1. $x^2 > 4$, domain set: $\{-2, -1, 2, 3, 5\}$

2. $|x| = 5$, domain set: integers

3. $x < 5$, domain set: natural numbers

4. $x + 2 = 5$, domain set: integers

5. $x + 2 = -2$, domain set: natural numbers

(7)

Sentences can be classed as *conditional sentences* or *identities.*

An *identity* is a sentence that is true for every value of the variable for which it is defined.

A *conditional sentence* is false for at least one value of the replacement set for which it is defined.

(8)

$\frac{2x}{2} + 3 = x + 3$ with domain set the real numbers is an identity.

$x + 1 = 5$ with domain set the real numbers is a conditional sentence.

The identity sentence is true for every real number; the conditional sentence is true only for the real number four.

(9)

A number from the domain set that makes a statement true is said to be a *root* or *solution* of the sentence.

Symbolic sentences involving equality relationships are called *equations*. A root of an equation is said to satisfy the equation. To *solve an equation* means to find its solution set.

(10)

Equivalent equations are equations with the same domain set that have identical solution sets.

If the domain of an equation is not specified, it is taken to be the *largest possible set of numbers for which the equation is defined.* The domain will be the set of real numbers in most cases.

(11)

Example 1: $x + 2 = 4$ with domain set N has solution set $\{2\}$.
$2x = 4$ with domain set N has solution set $\{2\}$.
Thus $x + 2 = 4$ and $2x = 4$ are equivalent equations.

Example 2: $x - 3 = 6$ and $x = 9$ are equivalent equations since the solution set is $\{9\}$ and each has the set of reals as the domain set.

(12)

Study Exercise Two

Determine if the following pairs of equations are equivalent.

1. $3x + 2 = 5$ with domain set N
$3x = 3$ with domain set N

2. $3x = 3$
$x = 1$

3. $x^2 - 1 = 0$
$x = 1$

(13)

To solve an equation we either determine the solution set by inspection or else generate equivalent equations and arrive at one with an obvious solution set. But how are equivalent equations generated?

(14)

We will let $P(x)$, $Q(x)$, and $R(x)$ be symbols that stand for expressions in x.

Theorem. For all values of x for which $P(x)$, $Q(x)$, and $R(x)$ are real numbers, $P(x) = Q(x)$ is equivalent to each of the following:

I. $P(x) + R(x) = Q(x) + R(x)$
II. $P(x) \cdot R(x) = Q(x) \cdot R(x)$ where $R(x) \neq 0$

(15)

Restatement of the theorem.

I. The addition of the same expression representing a real number to each member of an equation produces an equivalent equation.

II. The multiplication of each member of an equation by the same expression representing a non zero real number produces an equivalent equation.

(16)

A linear equation is an equation which is equivalent to an equation of the form $ax = b$, $a, b \in R, a \neq 0$.
Examples of linear equations include $3x = 5$, $7x - 2 = 8$, $x = 0$, and $5x - 1 = x + 3$.

(17)

Example: Find the solution set over R of $6x - 1 = 5$.

Solution:

Line (1) $\quad 6x - 1 + 1 = 5 + 1$
Line (2) $\quad 6x = 6$
Line (3) $\quad \frac{1}{6}(6x) = \frac{1}{6}(6)$
Line (4) $\quad x = 1$

the solution set is $\{1\}$

(18)

Example: Find the solution set over R of $3x + 3 = 3$.

Solution:

Line (1) $3x + 3 - 3 = 3 - 3$

Line (2) $3x = 0$

Line (3) $\frac{1}{3}(3x) = \frac{1}{3}(0)$

Line (4) $x = 0$

the solution set is $\{0\}$

(19)

Study Exercise Three

Find the solution sets over R.

1. $2x + 4 = 13$ **2.** $19 - 2x = 13$ **3.** $4(2x - 3) = 3(2x + 4)$

4. $\frac{2}{3}(x + 1) = -\frac{4}{5}$ **5.** $5x = 4x + 3 + x$ **6.** $\frac{2x - 4}{x - 2} = 2$

(20)

The following operations may lead to equations that are not equivalent to the original equation:

1. Multiplying or dividing both members by the same expression containing the unknown.
2. Raising both members to the same power.

(21)

Example: Solve $\frac{x}{x - 3} = \frac{3}{x - 3}$

Solution:

Line (1) $\frac{x}{(x - 3)} \cdot (x - 3) = \frac{3}{(x - 3)} \cdot (x - 3)$

Line (2) $x = 3$

Check. $x = 3$: $\frac{3}{3 - 3} = \frac{3}{3 - 3}$; does not check. Note also that 3 is not in the domain set of the original equation.

The solution set is: $\{\quad\}$ or $\varnothing$

(22)

Consider $x = 2$ and $x^2 = 4$. Solution sets are $\{2\}$ and $\{2, -2\}$ respectively. The two equations are not equivalent.

Consider also $x = 1$ and $x^2 = 1$. Solution sets are $\{1\}$ and $\{1, -1\}$ respectively. The two equations are not equivalent.

Whenever the new equation has roots that are not roots of the original, these roots are called extraneous roots.

(23)

Example: Solve $\frac{5}{x+2} - \frac{3}{x-2} = \frac{2}{(x+2)(x-2)}$

Line (1) $\frac{5}{(x + 2)} \cdot (x + 2)(x - 2) + \frac{-3}{(x - 2)} \cdot (x + 2)(x - 2)$

$= \frac{2}{(x + 2)(x - 2)} \cdot (x + 2)(x - 2)$

Line (2) $5(x - 2) + -3(x + 2) = 2$

Line (3) $5x - 10 - 3x - 6 = 2$

(Frame 24. contd.)

Line (4) $\quad 2x - 16 = 2$

Line (5) $\quad 2x = 18$

Line (6) $\quad x = 9$

The solution set is $\{9\}$.

Check.
$$\frac{5}{9+2} - \frac{3}{9-2} = \frac{2}{(9+2)(9-2)}$$
$$\frac{35-33}{77} = \frac{2}{77}$$
$$\frac{2}{77} = \frac{2}{77}$$

(24)

Study Exercise Four

Find the solution sets over R.

1. $x = 2 + \dfrac{x^2 - 4}{x}$

2. $\dfrac{3}{x} = \dfrac{5}{x}$

3. $\dfrac{2x+3}{2} + \dfrac{3x}{x-1} = x$

4. $\dfrac{x-7}{x^2 - 2x} = \dfrac{x}{x-2} - \dfrac{x+4}{x}$

(25)

Absolute Value

We know that $|3| = 3$, $|0| = 0$, $|-2| = 2$. The absolute value of a real number is always non negative. If we ask for $|x|$, we can't say it equals x because x may be negative.

Definition. $|x| = \begin{cases} x \text{ if } x > 0 \\ 0 \text{ if } x = 0 \\ -x \text{ if } x < 0 \end{cases}$

(26)

Suppose $|x - 2| = 5$. This means that $(x - 2)$ is either the number 5 or the number -5 since $|5| = 5$ and $|-5| = 5$.

$$x - 2 = 5 \quad \text{or} \quad x - 2 = -5$$
$$x = 7 \quad \text{or} \quad x = -3$$

The solution set is $\{-3, 7\}$.

Check. $x = -3$
$$|(-3) - 2| = 5$$
$$|-5| = 5$$
$$5 = 5$$

$x = 7$
$$|7 - 2| = 5$$
$$|5| = 5$$
$$5 = 5$$

(27)

Study Exercise Five

Find the solution set over R.

1. $|x| = 6$

2. $|2x + 1| = 5$

3. $|3 + 2x| = 3$

4. $|2x| = -2$

5. $\left|2x + \dfrac{1}{2}\right| = \dfrac{1}{4}$

(28)

Solving Equations For Specified Symbols

An equation containing more than one variable or containing symbols representing constants can often be solved for one of the symbols in terms of the remaining symbols by applying the transformations we have for equations to get equivalent equations. The desired symbol is to be obtained by itself as one member of an equation.

(29)

Example 1: Solve $v = Kt + g$ for K

Line (1) $v - g = Kt$

Line (2) $\frac{v - g}{t} = K$

Line (3) $K = \frac{v - g}{t}, t \neq 0$

Example 2: Solve $K^2a = 2a + 2c$ for a

Line (1) $K^2a - 2a = 2c$

Line (2) $(K^2 - 2)a = 2c$

Line (3) $a = \frac{2c}{K^2 - 2}, K^2 \neq 2$

(30)

Study Exercise Six

1. Solve $A = \frac{h}{2}(b + c)$ for c

2. Solve $t = a + (n - 1)d$ for n

3. Solve $\frac{1}{r} = \frac{1}{r_1} + \frac{1}{r_2}$ for r

4. Solve $s = \frac{1}{2}gt^2$ for g

5. Solve $P = 2l + 2w$ for w

(31)

Verbal Problems

One of the most important applications of algebra is to provide a means of solving verbal problems by representing the conditions of a problem as an equation and then solving it.

The most difficult part of solving word problems is to express the quantitative ideas symbolically as an equation.

(32)

If n represents the number, write the following expressions in algebraic form.

Example 1: Six less than four times the number.

Solution: $4n - 6$

Example 2: Five more than three times the number.

Solution: $3n + 5$

Example 3: Two less than $\frac{2}{3}$ of the number.

Solution: $\frac{2}{3}n - 2$

(33)

You must read a verbal problem carefully. First determine the quantities asked for and represent them by symbols. If possible, draw a sketch and label all known quantities and label the unknown in terms of the symbols. Then write an equation and solve it. Finally check your results against the original problem.

(34)

Example: Joe wants to buy a tablet that costs 11 cents more than twice as much money as he now has. How much does he have if the price of the tablet is 75 cents?

Solution:

Line (1) Let x represent amount of money Joe has

Line (2) $2x + 11$ is cost of the tablet.

Line (3) $2x + 11 = 75$

Line (4) $2x = 64$

Line (5) $x = 32$

Joe has 32 cents.

(35)

Example: The width of a garden is 2/3 of its length. If the perimeter is 80 feet, find its dimensions.

Solution:

Line (1) Let $x =$ length

Line (2) $\frac{2}{3}x =$ width

Line (3) $2x + 2\left(\frac{2}{3}x\right) = 80$

Line (4) $2x + \frac{4x}{3} = 80$

Line (5) $6x + 4x = 240$

Line (6) $10x = 240$

Line (7) $x = 24, \quad \frac{2}{3}x = 16$

The garden is 24 by 16 feet.

x

$2/3\,x$

(36)

Example: Tom sold his car for $1,320 which was 20% less than what he paid for it. What did he pay for the car?

Solution:

Line (1) Let $x =$ amount Tom paid for the car

Line (2) $1320 = x - .20x$

Line (3) $1320 = .80x$

Line (4) $13200 = 8x$

Line (5) $1650 = x$

Tom paid $1,650

(37)

Study Exercise Seven

1. Find two consecutive even integers whose sum is 46.
2. Mr. Doe sold a house for $14,950 which was 30% more than he paid for it. What was his purchase price?
3. What number will give a quotient of 24 with a remainder of 4 when it is divided by 124?
4. A rectangle is 3 times as long as it is wide and its perimeter is 144 inches. Find its dimensions.

(38)

REVIEW EXERCISES

A. Classify as a conditional equation or as an identity:

1. $3 + x = 2 + x$ **2.** $2(x + 1) = 2x + 2$

3. $3(x + 2) = x + 1$

B. Find the solution sets:

4. $4(2x - 3) = 3(2x + 4)$ **5.** $\frac{x}{x-1} = \frac{x+1}{x}$

6. $2x - 5 = 2x$ **7.** $|2x - 1| = 5$

C. Solve for the indicated letter:

8. Solve $C = 2\pi r$ for r **9.** Solve $C = \frac{5}{9}(F - 32)$ for F

D. Verbal problems:

10. One number is ten more than another. When the larger is subtracted from twelve times the smaller, the remainder is 45. Find the numbers.

11. Find three consecutive integers whose sum is sixty.

(39)

SOLUTIONS TO REVIEW EXERCISES

A. **1.** conditional **2.** identity **3.** conditional

B. **4.** $4(2x - 3) = 3(2x + 4)$

$8x - 12 = 6x + 12$

$2x = 24$

$x = 12$

the solution set is $\{12\}$

5. $\frac{x}{x-1} = \frac{x+1}{x}$

$x^2 = x^2 - 1$

the solution set is $\{\ \}$ or $\varnothing$

6. $2x - 5 = 2x$

the solution set is $\{\ \}$ or $\varnothing$

7. $2x - 1 = 5$ or $2x - 1 = -5$

$2x = 6$ or $2x = -4$

the solution set is $\{3, -2\}$

C. **8.** $r = \frac{C}{2\pi}$

9. $C = 5/9\,(F - 32)$

$9C = 5F - 160$

$\frac{9C + 160}{5} = F$

D. **10.** Let x = one number.

$x + 10$ = the other number

$12x - (x + 10) = 45$

$11x - 10 = 45$

$11x = 55$

$x = 5, \quad x + 10 = 15$

the two numbers are 5 and 15

11. Let x = first consecutive integer

$x + 1$ = second consecutive integer

$x + 2$ = third consecutive integer

$x + (x + 1) + (x + 2) = 60$

$3x + 3 = 60$

$x = 19, \quad x + 1 = 20, \quad x + 2 = 21$

the three numbers are 19, 20, 21.

(40)

SOLUTIONS TO STUDY EXERCISES

Study Exercise One (Frame 7)

1. $\{3, 5\}$ **2.** $\{5, -5\}$ **3.** $\{1, 2, 3, 4\}$

4. $\{3\}$ **5.** $\{\ \}$ or $\varnothing$

7A

Study Exercise Two (Frame 13)

1. equivalent; both have solution set $\{1\}$
2. equivalent; both have solution set $\{1\}$
3. not equivalent; the first equation has solution set $\{1, -1\}$ and the second equation has solution set $\{1\}$

13A

Study Exercise Three (Frame 20)

1.
$$2x + 4 = 13$$
$$2x = 9$$
$$x = \frac{9}{2}$$
the solution set is $\left\{\frac{9}{2}\right\}$

2.
$$19 - 2x = 13$$
$$-2x = -6$$
$$x = 3$$
the solution set is $\{3\}$

3.
$$4(2x - 3) = 3(2x + 4)$$
$$8x - 12 = 6x + 12$$
$$2x - 12 = 12$$
$$2x = 24$$
$$x = 12$$
the solution set is $\{12\}$

4.
$$\frac{2}{3}(x + 1) = -\frac{4}{5}$$
$$15 \cdot \frac{2}{3}(x + 1) = \frac{-4}{5} \cdot 15$$
$$10(x + 1) = -12$$
$$10x + 10 = -12$$
$$10x = -22$$
$$x = -\frac{11}{5}$$
the solution set is $\left\{-\frac{11}{5}\right\}$

5.
$$5x = 4x + 3 + x$$
$$5x = 5x + 3$$
the solution set is $\{\ \}$

6.
$$\frac{2x - 4}{x - 2} = 2$$
$$2x - 4 = 2x - 4, \quad x \neq 2$$
the solution set is the set of all real numbers except 2.

20A

Study Exercise Four (Frame 25)

1. $x = 2 + \frac{x^2 - 4}{x}$

$x^2 = 2x + x^2 - 4$
$0 = 2x - 4$
$4 = 2x$
$2 = x$
the solution set is $\{2\}$

2. $\frac{3}{x} = \frac{5}{x}$

the solution set is $\{\ \}$

3. $\frac{2x + 3}{2} + \frac{3x}{x - 1} = x$

$(2x + 3)(x - 1) + 3x(2) = x(2)(x - 1)$
$2x^2 + x - 3 + 6x = 2x^2 - 2x$
$7x - 3 = -2x$
$9x = 3$
$x = \frac{1}{3}$

the solution set is $\left\{\frac{1}{3}\right\}$

4. $\frac{x - 7}{x(x - 2)} = \frac{x}{x - 2} - \frac{(x + 4)}{x}$

$x - 7 = x^2 - [(x + 4)(x - 2)]$
$x - 7 = x^2 - x^2 - 2x + 8$
$3x = 15$
$x = 5$
the solution set is $\{5\}$

25A

Study Exercise Five (Frame 28)

1. the solution set is $\{6, -6\}$

2. $2x + 1 = 5$ or $2x + 1 = -5$
$2x = 4$ or $2x = -6$
$x = 2$ or $x = -3$
the solution set is $\{2, -3\}$

3. $3 + 2x = 3$ or $3 + 2x = -3$
$x = 0$ or $2x = -6$
the solution set is $\{0, -3\}$

4. since $|x|$ is non negative, the solution set is $\{\ \}$.

5. $2x + \frac{1}{2} = \frac{1}{4}$ or $2x + \frac{1}{2} = -\frac{1}{4}$

$8x + 2 = 1$ or $8x + 2 = -1$
$8x = -1$ or $8x = -3$
$x = -\frac{1}{8}$ or $x = -\frac{3}{8}$

the solution set is $\left\{-\frac{1}{8}, -\frac{3}{8}\right\}$.

28A

Study Exercise Six (Frame 31)

1. $A = \frac{h}{2}(b + c)$

$2A = bh + hc$
$2A - bh = hc$
$\frac{2A - bh}{h} = c, \quad h \neq 0$

2. $t = a + nd - d$
$t + d - a = nd$
$\frac{t + d - a}{d} = n, \quad d \neq 0$

Study Exercise Six (Frame 31)

3. $\frac{1}{r} = \frac{1}{r_1} + \frac{1}{r_2}$

$r_1r_2 = rr_2 + rr_1$

$r_1r_2 = (r_2 + r_1)r$

$r = \frac{r_1r_2}{r_1 + r_2}, \quad r_1 \neq -r_2, \quad r \neq 0, \quad r_1 \neq 0, \quad r_2 \neq 0$

4. $s = \frac{1}{2}gt^2$

$2s = gt^2$

$\frac{2s}{t^2} = g, \quad t \neq 0$

5. $p - 2l = 2w$

$\frac{p - 2l}{2} = w$

31A

Study Exercise Seven (Frame 38)

1. Let x = first consecutive even integer;

$x + 2$ = second consecutive even integer;

$x + x + 2 = 46$

$2x = 44$

$2x = 22, \quad x + 2 = 24$

The two integers are 22 and 24.

2. Let x = purchase price;

$x + .30x = 14{,}950$

$1.30x = 14{,}950$

$x = \frac{14{,}950}{1.30}$

$x = 11500$

The purchase price was $11,500.

3. Let x = the number:

$\frac{x}{124} = 24 + \frac{4}{124}$

$x = 24(124) + 4$

$x = 2976 + 4$

$x = 2980$

4. Let x = width

$3x$ = length

$x + 3x + x + 3x = 144$

$8x = 144$

$x = 18, \quad 3x = 54$

The rectangle is 18 by 54 inches.

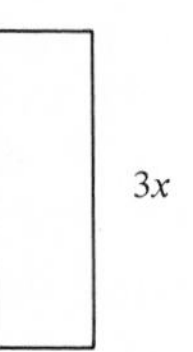

38A

UNIT 9—SUPPLEMENTARY PROBLEMS

A. Classify as conditional equations or identities:

1. $x = 4$

2. $2(x - 3) + 5 = 3(x - 2) + 5$

3. $2(3x + 2) - 4 = 2(3x - 2) + 4$

4. $6x + 4 \div 2 = (6x + 4) \div 2$

B. Find the solution sets over R:

5. $3x - 5 = 2x + 5$

6. $3x + 2 = 3x$

7. $5x = 0$

8. $\frac{3x}{4} + \frac{3}{2} = \frac{x}{3} - \frac{1}{6}$

9. $\frac{5}{x + 1} = \frac{2}{11}$

10. $2x - 4 = 6x + 11$

11. $\frac{x + 2}{3} - \frac{x - 3}{2} = x - 2$

12. $\frac{x^2 + 2x}{3} - x = \frac{x(x - 1)}{3}$

13. $-\frac{2x - 11}{x - 2} = \frac{7}{x - 2}$

14. $\frac{x + 4}{3} - 7 = 3 - \frac{x + 2}{4}$

15. $|x - 1| = 5$

16. $|x + 1| = 6$

17. $|3x + 2| = 1$

18. $|5x - 1| = 4$

C. Solve for the indicated letter.

19. $V = \frac{1}{3}\pi r^2 h$, solve for h

20. $I = prt$, solve for p

21. $y = mx + b$, solve for x

22. $\frac{t - a}{b} + \frac{t + b}{a} = 1$, solve for t

23. $\frac{a}{b - x} - \frac{b}{a - x} = 0$, solve for x

D. Find the solution sets and give the roots in simplest radical form.

24. $x\sqrt{2} - x = 4$

25. $3x - 6 = x\sqrt{3}$

E. Verbal Problems:

26. Find three consecutive integers whose sum is 75.

27. Find three consecutive odd integers such that 3 times the sum of the first two exceeds the third by 47.

28. The length of a rectangle exceeds twice its width by 3 inches. If the perimeter is 90 inches, find the dimensions.

29. The length of a rectangle exceeds twice its width by 2 feet. If the width is increased by 3 ft., and the length is decreased by 3 ft., the area remains unchanged. Find the dimensions of the original rectangle.

30. A boy has 24 coins consisting of nickels and quarters. They amount to \$2.00 in value. How many of each kind are there?

31. An office spent \$14.40 in postage to send out 310 pieces of mail, some of which carried 5¢ stamps and some 4¢ stamps. How many stamps of each denomination were there?

32. The selling price of an article is \$78. This represents a profit of 30% on the cost. Find the cost.

33. A sum of money is invested at 5% simple interest. A second sum, which is \$350 greater than the first, is invested at 6%. If the total annual income from both investments is \$87, find the amount invested at each rate.

34. Two planes take off from an airport at the same time and head in opposite directions at 260 mph and 275 mph respectively. In how many hours will they be 1605 miles apart? (Hint: distance equals rate times time.)

35. The speeds of two planes differ by 35 mph. They take off from the same airport at the same time and travel in opposite directions. After 3 hours, they are 1365 miles apart. What are their respective speeds?

36. A family drove into the country at an average rate of 35 mph, spent an hour for a picnic lunch, and then returned home at 28 mph. If the entire outing took $5\frac{1}{2}$ hr., how far into the country did they go?

unit

10

Solution Sets of Quadratic Equations

Objectives

1. Learn the quadratic formula and the discriminant.
2. Be able to find solution sets of quadratic equations by factoring, completing the square, and formula.

(1)

Equations that can be put in the form $ax^2 + bx + c = 0$ where $a, b, c \in R$, $a \neq 0$, are called quadratic equations.

The standard form of the quadratic equation will be $ax^2 + bx + c = 0$, $a, b, c \in R$, $a \neq 0$.

(2)

Examples of Quadratic Equations

Standard form is $ax^2 + bx + c = 0$, $\quad a, b, c \in R, a \neq 0$.

1. $3x^2 - 2x + 1 = 0$	$a = 3,$	$b = -2,$	$c = 1$
2. $x^2 + 2x = 0$	$a = 1,$	$b = 2,$	$c = 0$
3. $-5x^2 + 7 = 0$	$a = -5,$	$b = 0,$	$c = 7$
4. $3x^2 = 2x + 1$	$a = 3,$	$b = -2,$	$c = -1$

(3)

Theorem. $a, b \in R$. If $a = 0$ or $b = 0$, then $a \cdot b = 0$.

Proof:

1. $a, b \in R$, $a = 0$ or $b = 0$	1. **given**
2. assume $a = 0$ $ab = 0 \cdot b$ $= 0$	2. **mult. by zero**
3. assume $b = 0$ $ab = a \cdot 0$ $= 0$	3. **mult. by zero**
4. assume $a = 0, b = 0$ $a \cdot b = 0 \cdot 0$ $= 0$	4. **mult. by zero**

(4)

Theorem. $a, b \in R$. If $a \cdot b = 0$, then $a = 0$ or $b = 0$.

Proof:

1. $a, b \in R$, $a \cdot b = 0$	1. **given**
2. assume $a \neq 0$, thus $1/a$ exists	2. **mult. inv.**
3. $(a \cdot b) \cdot 1/a = 0 \cdot 1/a$	3. **multiplication law of equality**
4. $(a \cdot b) \cdot 1/a = 0$	4. **mult. by zero**
5. $(b \cdot a) \cdot 1/a = 0$	5. **cpm**

(Frame 5, contd.)

6. $b(a \cdot 1/a) = 0$ — 6. **apm**

7. $b = 0$ — 7. **mult. inv. and mult. ident.**

(5)

The product of two factors equals zero if and only if one or both of the factors equals zero.

$a \cdot b = 0$ if and only if $a = 0$ or $b = 0$.

The word "or" is used in the inclusive sense.

(6)

Solution of Quadratic Equations by Factoring

Line (1) $x^2 + 2x - 15 = 0$

Line (2) $[(x + 5)] [(x - 3)] = 0$

If the product of two numbers is zero, one or both of the numbers is zero.

Line (3) $(x + 5) = 0$ or $(x - 3) = 0$

Line (4) $x = -5$ or $x = 3$

Line (5) the solution set is $\{3, -5\}$

(7)

To solve quadratic equations by the method of factoring:

1. write the equation in the form $ax^2 + bx + c = 0$;
2. factor the left hand side;
3. set the factors equal to zero.

(8)

Example: Solve $x^2 - 3x = -2$ by factoring.

Solution: $x^2 - 3x = -2$

$x^2 - 3x + 2 = 0$

$(x - 2)(x - 1) = 0$

$x - 2 = 0$ or $x - 1 = 0$

$x = 2$ or $x = 1$

the solution set is $\{1, 2\}$

(9)

Example: Solve $x^2 = x$ by factoring.

Solution: $x^2 = x$

$x^2 - x = 0$

$x(x - 1) = 0$

$x = 0$ or $x - 1 = 0$

the solution set is $\{0, 1\}$

(10)

Sometimes students incorrectly solve equations like $x^2 = x$ by dividing through by x.

$$x^2 = x$$
$$x = 1$$

The solution set would be $\{1\}$, but we know from the previous frame the solution set is $\{0, 1\}$.

What is wrong?

(11)

Study Exercise One

Find the solution sets by factoring:

1. $x^2 - 6x + 8 = 0$ **2.** $x^2 + 3x = 0$ **3.** $x(2x - 3) = -1$

4. $(x - 2)(x + 1) = 4$ **5.** $x^2 = 2x$ **6.** $3 = \frac{10}{x^2} - \frac{7}{x}$

(12)

Now let's try the process in reverse. Given that the solution set of a quadratic equation is $\{2, 4\}$, can you find a possible quadratic equation with this solution set?

Line (1) the solution set is $\{2, 4\}$
Line (2) $x = 2$ or $x = 4$
Line (3) $x - 2 = 0$ or $x - 4 = 0$
Line (4) $(x - 2)(x - 4) = 0$
Line (5) $x^2 - 6x + 8 = 0$ is one such equation

(13)

Example: Find a possible quadratic equation with integral coefficients with solution set $\{-1/3, 2\}$.

Solution:

Line (1) the solution set is $\{-1/3, 2\}$
Line (2) $x = -1/3$ or $x = 2$
Line (3) $3x = -1$ or $x = 2$
Line (4) $3x + 1 = 0$ or $x - 2 = 0$
Line (5) $(3x + 1)(x - 2) = 0$
Line (6) $3x^2 - 5x - 2 = 0$

(14)

Study Exercise Two

Write a possible quadratic equation in standard form with integral coefficients with the following solution sets.

1. $\{3, 4\}$ **2.** $\left\{-\frac{1}{2}, 2\right\}$ **3.** $\left\{\frac{1}{3}, -1\right\}$ **4.** $\{4, -4\}$

(15)

We continue with more types of quadratic equations to be solved by factoring.

Example: Solve $5x^2 = 5$

Solution:

Line (1) $5x^2 = 5$
Line (2) $5x^2 - 5 = 0$
Line (3) $5(x - 1)(x + 1) = 0$
Line (4) $x - 1 = 0$ or $x + 1 = 0$
Line (5) $x = 1$ or $x = -1$
Line (6) The solution set is $\{1, -1\}$.

(16)

Example: Solve $x^2 - 10x + 25 = 0$.

Solution:

Line (1) $x^2 - 10x + 25 = 0$

Line (2) $(x - 5)(x - 5) = 0$

Line (3) $x - 5 = 0$ or $x - 5 = 0$

Line (4) $x = 5$ or $x = 5$

Line (5) the solution set is $\{5\}$

(17)

Study Exercise Three

Solve by factoring.

1. $3x^2 = 3$
2. $x^2 - 4x + 4 = 0$
3. $x^2 = 2ax + 8a^2$ (solve for x)
4. $\dfrac{2}{x-3} - \dfrac{6}{x-8} = -1$

(18)

Equations of the Form $x^2 = a$

Quadratic equations of the form $x^2 = a$ can be solved by a form known as the *extraction of roots method.*

If $x^2 = a$, then from the definition of a square root, x must be a square root of a. Since each positive real number has two square roots, the solution set, if $a > 0$, will be:

$$\{+\sqrt{a}, -\sqrt{a}\}.\text{ If } a < 0, \text{ the solution set is empty.}$$

(19)

To find the solution set of $x^2 = 25$, we use the extraction of roots method and set x equal to each square root of 25.

$x^2 = 25$

$x = +5, x = -5$

the solution set is $\{5, -5\}$

(20)

Example: Solve $9x^2 = 100$.

Solution:

Line (1) $9x^2 = 100$

Line (2) $(3x)^2 = 100$

Line (3) $3x = 10, \quad 3x = -10$

Line (4) $x = \dfrac{10}{3}, \quad x = -\dfrac{10}{3}$

Line (5) the solution set is $\left\{\dfrac{10}{3}, -\dfrac{10}{3}\right\}$

(21)

Example: Solve $(x - 1)^2 = 9$.

Solution:

Line (1) $(x - 1)^2 = 9$

Line (2) $(x - 1) = +3, \quad (x - 1) = -3$

Line (3) $x = 4, x = -2$

Line (4) the solution set is $\{4, -2\}$

(22)

Example: Solve $(x + 3)^2 = 15$

Solution:

Line (1) $(x + 3)^2 = 15$

Line (2) $x + 3 = +\sqrt{15}, \quad x + 3 = -\sqrt{15}$

Line (3) $x = -3 + \sqrt{15}, \quad x = -3 - \sqrt{15}$

Line (4) the solution set is $\{-3 + \sqrt{15}, -3 - \sqrt{15}\}$

(23)

Study Exercise Four

Find the solution sets by the extraction of roots method.

1. $x^2 = 36$ **2.** $(x - 1)^2 = 16$ **3.** $(x + 2)^2 = 5$ **4.** $(x - 3)^2 = 10$

(24)

Some quadratic polynomials are *perfect square trinomials.*

For example, $x^2 + 10x + 25 = (x + 5)^2$.

What do you think is needed as the third term of the expression $x^2 - 8x +$______ to make it a perfect square trinomial?

(25)

Since $(x + a)^2 = x^2 + 2ax + a^2$, the last term a^2 is the square of one half the coefficient of the x term. That is, $a^2 = [\frac{1}{2}(2a)]^2$.

(26)

A quadratic polynomial of the form $x^2 + k \cdot x$ can be transformed into a perfect square trinomial by adding the square of one half the coefficient of x. The process is known as *completing the square.*

(27)

Example: Complete the square on $x^2 + 6x$.

Solution:

$$[\tfrac{1}{2}(6)]^2 = 3^2 = 9$$

$$x^2 + 6x + 9 = (x + 3)^2$$

(28)

Example: Complete the square on $x^2 - 7x$

Solution:

$$[\tfrac{1}{2}(-7)]^2 = \left[\frac{-7}{2}\right]^2 = \frac{49}{4}$$

$$x^2 - 7x + \frac{49}{4} = \left(x - \frac{7}{2}\right)^2$$

(29)

Study Exercise Five

Find the term that must be added to make the expression a perfect square. Write the expression as the square of a binomial.

1. $x^2 - 6x$ **2.** $x^2 + 3x$ **3.** $x^2 + \frac{1}{2}x$ **4.** $x^2 - \frac{1}{4}x$

(30)

Method of Completing the Square

Example: Solve $x^2 - 4x - 6 = 0$

Line (1) $x^2 - 4x = 6$

Line (2) $x^2 - 4x + 4 = 6 + 4$

Line (3) $(x - 2)^2 = 10$

Line (4) $x - 2 = +\sqrt{10}, x - 2 = -\sqrt{10}$

Line (5) $x = 2 + \sqrt{10}, x = 2 - \sqrt{10}$

Line (6) The solution set is $\{2 + \sqrt{10}, 2 - \sqrt{10}\}$.

(31)

To solve a quadratic equation by completing the square:

1. If the coefficient of x^2 is not one, divide the equation by the appropriate number so that the coefficient of x^2 becomes one.
2. Keep the terms containing x^2 and x in the left member.
3. Add to both members the square of one half the coefficient of x.
4. Proceed as in the extraction of roots method.

(32)

Example: Solve by completing the square $2x^2 + x - 2 = 0$.

Solution:

Line (1) $2x^2 + x = 2$

Line (2) $x^2 + \frac{1}{2}x = 1$

Line (3) $x^2 + \frac{1}{2}x + \frac{1}{16} = 1 + \frac{1}{16}$

Line (4) $\left(x + \frac{1}{4}\right)^2 = \frac{17}{16}$

Line (5) $x + \frac{1}{4} = +\sqrt{\frac{17}{16}}, \quad x + \frac{1}{4} = -\sqrt{\frac{17}{16}}$

Line (6) $x = -\frac{1}{4} + \frac{\sqrt{17}}{4}, \quad x = -\frac{1}{4} - \frac{\sqrt{17}}{4}$

The solution set is $\left\{\frac{-1 + \sqrt{17}}{4}, \frac{-1 - \sqrt{17}}{4}\right\}$.

(33)

Study Exercise Six

Solve by completing the square:

1. $x^2 + 2x = 3$
2. $x^2 - 6x - 7 = 0$
3. $2x^2 = 4 - 3x$
4. $2x^2 + 3x - 2 = 0$

(34)

The Quadratic Formula

We now will complete the square on $ax^2 + bx + c = 0, \quad a \neq 0.$

Line (1) $\quad ax^2 + bx + c = 0$

Line (2) $\quad x^2 + \frac{b}{a}x = -\frac{c}{a}$

Line (3) $\quad x^2 + \frac{b}{a}x + \frac{b^2}{4a^2} = \frac{b^2}{4a^2} - \frac{c}{a}, \quad \left[\frac{1}{2}\cdot\frac{b}{a}\right]^2 = \frac{b^2}{4a^2}$

Line (4) $\quad x^2 + \frac{b}{a}x + \frac{b^2}{4a^2} = \frac{b^2}{4a^2} - \frac{4ac}{4a^2}$

Line (5) $\quad \left(x + \frac{b}{2a}\right)^2 = \frac{b^2 - 4ac}{4a^2}$

Line (6) $\quad x + \frac{b}{2a} = +\sqrt{\frac{b^2 - 4ac}{4a^2}}, \quad x + \frac{b}{2a} = -\sqrt{\frac{b^2 - 4ac}{4a^2}}$

Line (7) $\quad x = -\frac{b}{2a} + \sqrt{\frac{b^2 - 4ac}{4a^2}}, \quad x = -\frac{b}{2a} - \sqrt{\frac{b^2 - 4ac}{4a^2}}$

Case 1. $\quad a > 0$ (then $\sqrt{a^2} = a$)

Line (8) $\quad x = \frac{-b}{2a} + \frac{\sqrt{b^2 - 4ac}}{2a}, \quad x = \frac{-b}{2a} - \frac{\sqrt{b^2 - 4ac}}{2a}$

Line (9) $\quad$ The solution set is $\left\{\frac{-b + \sqrt{b^2 - 4ac}}{2a}, \frac{-b - \sqrt{b^2 - 4ac}}{2a}\right\}$

Case 2. $\quad a < 0$ (then $\sqrt{a^2} = -a$)

Line (10) $\quad x = \frac{-b}{2a} + \frac{\sqrt{b^2 - 4ac}}{-2a}, \quad x = \frac{-b}{2a} - \frac{\sqrt{b^2 - 4ac}}{-2a}$

The solution set is $\left\{\frac{-b - \sqrt{b^2 - 4ac}}{2a}, \frac{-b + \sqrt{b^2 - 4ac}}{2a}\right\}$

In either case, $a > 0$ or $a < 0$, the solution set is

$$\left\{\frac{-b + \sqrt{b^2 - 4ac}}{2a}, \frac{-b - \sqrt{b^2 - 4ac}}{2a}\right\}$$

This result is considered a formula for the roots of a quadratic equation expressed in terms of the coefficients.

(35)

Sometimes the symbol $\pm$ is used to condense the writing of the solution set $\left\{\frac{-b + \sqrt{b^2 - 4ac}}{2a}, \frac{-b - \sqrt{b^2 - 4ac}}{2a}\right\}$ into $\left\{\frac{-b \pm \sqrt{b^2 - 4ac}}{2a}\right\}$.

The solution set of $ax^2 + bx + c = 0$, $a, b, c \in R$, $a \neq 0$ is $\left\{\frac{-b \pm \sqrt{b^2 - 4ac}}{2a}\right\}$.

(36)

Example: Find the solution set of $x^2 - x - 12 = 0$ by the formula.

Solution: In $x^2 - x - 12 = 0,\quad a = 1,\quad b = -1,\quad c = -12$

$$x = \frac{-b \pm \sqrt{b^2 - 4ac}}{2a}$$

$$x = \frac{-(-1) \pm \sqrt{(-1)^2 - 4(1)(-12)}}{2(1)}$$

$$x = \frac{1 \pm \sqrt{1 + 48}}{2}$$

$$x = \frac{1 \pm \sqrt{49}}{2}$$

$$x = \frac{1 \pm 7}{2} \quad \text{or} \quad \frac{1+7}{2}, \frac{1-7}{2}$$

the solution set is $\{4, -3\}$

(37)

Example: Solve $x^2 - 3x + 1 = 0$

Solution: In $x^2 - 3x + 1 = 0,\quad a = 1,\quad b = -3,\quad c = 1$

$$x = \frac{-b \pm \sqrt{b^2 - 4ac}}{2a}$$

$$x = \frac{-(-3) \pm \sqrt{(-3)^2 - 4(1)(1)}}{2(1)}$$

$$x = \frac{3 \pm \sqrt{9 - 4}}{2}$$

$$x = \frac{3 \pm \sqrt{5}}{2}$$

the solution set is $\left\{\frac{3 + \sqrt{5}}{2}, \frac{3 - \sqrt{5}}{2}\right\}$

(38)

Study Exercise Seven

Find the solution sets by the formula:

1. $x^2 - 5x - 6 = 0$ **2.** $x^2 - 3x + 1 = 0$ **3.** $3x^2 - 5x = 1$

4. $x^2 + 4x = -4$ **5.** $x^2 - x + 2 = 0$

(39)

The solution set of a quadratic equation is sometimes empty.

The solution set of $ax^2 + bx + c = 0, a \neq 0$ is $\left\{\frac{-b + \sqrt{b^2 - 4ac}}{2a}, \frac{-b - \sqrt{b^2 - 4ac}}{2a}\right\}$.

The solution set will be empty if $\sqrt{b^2 - 4ac}$ does not exist. $\sqrt{b^2 - 4ac}$ will not exist if $b^2 - 4ac$ is a negative number.

(40)

The solution set of a quadratic equation will be non empty if $\sqrt{b^2 - 4ac}$ exists. $\sqrt{b^2 - 4ac}$ exists if $b^2 - 4ac$ is greater than or equal to zero.

Since $b^2 - 4ac$ controls the type of solution set, it is called the *discriminant* (symbolized by D).

Thus $D = b^2 - 4ac.$

(41)

In the equation $ax^2 + bx + c = 0, \quad a \neq 0$, the discriminant is $b^2 - 4ac$.

If $b^2 - 4ac = 0$, then there is one real solution.

If $b^2 - 4ac < 0$, then there are no real solutions.

If $b^2 - 4ac > 0$, then there are two real solutions.

(42)

Example 1: Without solving, classify the solutions of $2x^2 - x + 2 = 0$.

Solution:

$$2x^2 - x + 2 = 0$$
$$\begin{aligned} D &= b^2 - 4ac \\ &= (-1)^2 - 4(2)(2) \\ &= 1 - 16 \\ &= -15 \end{aligned}$$

Since $D < 0$, there are no real solutions.

Example 2: Classify the solutions of $x^2 - 3x + 1 = 0$

Solution:

$$x^2 - 3x + 1 = 0$$
$$\begin{aligned} D &= b^2 - 4ac \\ &= (-3)^2 - 4(1)(1) \\ &= 9 - 4 \\ &= 5 \end{aligned}$$

Since $D > 0$, there are two real solutions.

(43)

Study Exercise Eight

A. Use the discriminant to classify the solutions:

1. $x^2 - 2x + 9 = 0$ **2.** $x^2 - 2x + 1 = 0$ **3.** $x^2 + 2x - 5 = 0$

B. Use the discriminant to determine k so that there is one real solution.

4. $x^2 + kx + 16 = 0$ **5.** $x^2 - 5x + k = 0$

(44)

We will now see how quadratic equations may be used in problem solving.

Example: The sum of the length and width of a rectangle is 13 inches and the area is 40 square inches. Find the dimensions.

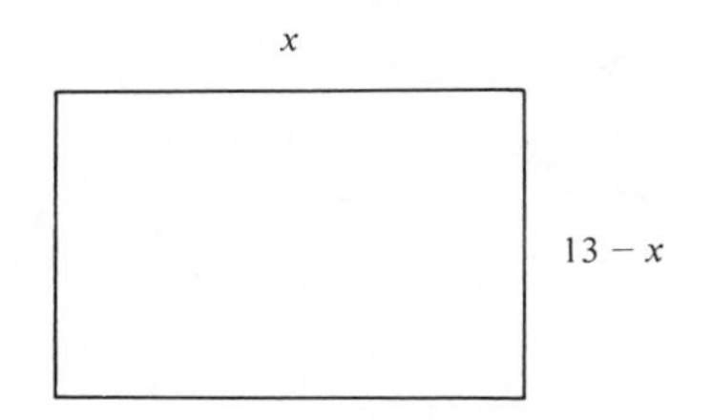

(Frame 45, contd.)

Solution: Let x = length in inches

$13 - x$ = width

area = (length) (width)

$40 = x(13 - x)$

$40 = 13x - x^2$

$x^2 - 13x + 40 = 0$

$(x - 5)(x - 8) = 0$

$x = 5$ or $x = 8$

$x = 5$ inches length | $x = 8$ inches length

$13 - x = 8$ inches width | $13 - x = 5$ inches width

The dimensions are 5 inches long by 8 inches wide or 8 inches long by 5 inches wide.

(45)

Example: Find two numbers such that each is equal to 5 more than 6 times its reciprocal.

Solution: Let x = one of the numbers.

$\frac{1}{x}$ = the reciprocal.

$6\left(\frac{1}{x}\right) + 5$ = 5 more than 6 times the reciprocal

$x = \frac{6}{x} + 5$

$x^2 = 6 + 5x$

$x^2 - 5x - 6 = 0$

$(x - 6)(x + 1) = 0$

$x - 6 = 0$ or $x + 1 = 0$

$x = 6$ or $x = -1$

The two numbers are -1 and 6.

(46)

Study Exercise Nine

1. The product of two consecutive even integers is 80. Find them.
2. If $(6 - x)$, $(13 - x)$, and $(14 - x)$ are the lengths of a right triangle, find the value of x.
3. A rectangle is 3 inches longer than its width and its area exceeds four times the number of inches in its width by twelve. Find its dimensions.

(47)

REVIEW EXERCISES

A. **1.** What is the standard form of the quadratic equation?

2. What is the solution set of the standard form of the quadratic equation?

3. What is the discriminant?

B. Find the solution sets by factoring

4. $6x^2 - x - 1 = 0$ **5.** $x^2 = 4x$

C. Find the solution sets by completing the square

6. $x^2 - 8x = 5$ **7.** $x^2 - 2\sqrt{2}x + 3 = 0$

D. Find the solution sets by the formula

8. $3x^2 - 2x = 2$ **9.** $3x^2 + \sqrt{3}x - 2 = 0$

E. Without solving, classify the solutions

10. $2x^2 - x + 1 = 0$ **11.** $x^2 + 3x - 4 = 0$

F. Write a quadratic equation with the following solution sets.

12. $\left\{\frac{1}{2}, -8\right\}$ **13.** $\{4\}$

(48)

SOLUTIONS TO REVIEW EXERCISES

A. **1.** $ax^2 + bx + c = 0, \quad a, b, c \in R, a \neq 0$

2. $\left\{\frac{-b + \sqrt{b^2 - 4ac}}{2a}, \frac{-b - \sqrt{b^2 - 4ac}}{2a}\right\}$

3. $b^2 - 4ac$

B. **4.** $(3x + 1)(2x - 1) = 0$

$3x + 1 = 0$ or $2x - 1 = 0$

$3x = -1$ or $2x = 1$

$x = -\frac{1}{3}$ or $x = \frac{1}{2}$

the solution set is $\left\{\frac{1}{2}, -\frac{1}{3}\right\}$

5. $x^2 - 4x = 0$

$x(x - 4) = 0$

$x = 0$ or $x = 4$

the solution set is $\{0, 4\}$

C. **6.** $x^2 - 8x + 16 = 5 + 16$

$(x - 4)^2 = 21$

$x - 4 = \sqrt{21}$ or $x - 4 = -\sqrt{21}$

the solution set is $\{4 + \sqrt{21}, 4 - \sqrt{21}\}$

7. $x^2 - 2\sqrt{2}x = -3$

$x^2 - 2\sqrt{2}x + 2 = -3 + 2$

$(x - \sqrt{2})^2 = -1$

the solution set is empty

(Frame 49, contd.)

D. **8.** $3x^2 - 2x - 2 = 0$

$$x = \frac{-(-2) \pm \sqrt{(-2)^2 - 4(3)(-2)}}{2(3)}$$

$$= \frac{2 \pm \sqrt{4 + 24}}{6}$$

$$= \frac{2 \pm \sqrt{28}}{6}$$

$$= \frac{2 \pm 2\sqrt{7}}{6}$$

$$= \frac{2(1 \pm \sqrt{7})}{6}$$

the solution set is $\left\{\frac{1 + \sqrt{7}}{3}, \frac{1 - \sqrt{7}}{3}\right\}$

9. $3x^2 + \sqrt{3}x - 2 = 0$

$$x = \frac{-\sqrt{3} \pm \sqrt{(\sqrt{3})^2 - 4(3)(-2)}}{2(3)}$$

$$= \frac{-\sqrt{3} \pm \sqrt{3 + 24}}{6}$$

$$= \frac{-\sqrt{3} \pm 3\sqrt{3}}{6}$$

the solution set is $\left\{\frac{\sqrt{3}}{3}, \frac{-2\sqrt{3}}{3}\right\}$

E. **10.** $D = (-1)^2 - 4(2)(1)$
$= 1 - 8$
$= -7$
there is no real solution

11. $D = (3)^2 - 4(1)(-4)$
$= 9 + 16$
$= 25$
there are two real solutions

F. **12.** $x = \frac{1}{2}$ or $x = -8$

$x - \frac{1}{2} = 0$ or $x + 8 = 0$

$2x - 1 = 0$ or $x + 8 = 0$

$(2x - 1)(x + 8) = 0$

$2x^2 + 15x - 8 = 0$

13. $x = 4$ or $x = 4$

$(x - 4) = 0$ or $(x - 4) = 0$

$(x - 4)(x - 4) = 0$

$x^2 - 8x + 16 = 0$

(49)

SOLUTIONS TO STUDY EXERCISES

Study Exercise One (Frame 12)

1. $(x - 4)(x - 2) = 0$
$x - 4 = 0$ or $x - 2 = 0$
the solution set is $\{4, 2\}$

2. $x(x + 3) = 0$
$x = 0$ or $x + 3 = 0$
the solution set is $\{0, -3\}$

3. $2x^2 - 3x + 1 = 0$
$(2x - 1)(x - 1) = 0$
$2x - 1 = 0$ or $x - 1 = 0$
the solution set is $\left\{\frac{1}{2}, 1\right\}$

4. $x^2 - x - 2 = 4$
$x^2 - x - 6 = 0$
$(x - 3)(x + 2) = 0$
$x - 3 = 0$ or $x + 2 = 0$
the solution set is $\{3, -2\}$

5. $x^2 - 2x = 0$
$x(x - 2) = 0$
$x = 0$ or $x - 2 = 0$
the solution set is $\{0, 2\}$

6. $3 = \frac{10}{x^2} - \frac{7}{x}$
$3x^2 = 10 - 7x$
$3x^2 + 7x - 10 = 0$
$(3x + 10)(x - 1) = 0$
$3x + 10 = 0$ or $x - 1 = 0$
the solution set is $\left\{\frac{-10}{3}, 1\right\}$

(12A)

Study Exercise Two (Frame 15)

1. $x = 3, x = 4$
$(x - 3) = 0$ or $x - 4 = 0$
$(x - 3)(x - 4) = 0$
$x^2 - 7x + 12 = 0$

2. $x = -\frac{1}{2}, x = 2$
$2x = -1, x - 2 = 0$
$(2x + 1)(x - 2) = 0$
$2x^2 - 3x - 2 = 0$

3. $x = \frac{1}{3}, x = -1$
$3x = 1, x + 1 = 0$
$(3x - 1)(x + 1) = 0$
$3x^2 + 2x - 1 = 0$

4. $x = 4, x = -4$
$x - 4 = 0, x + 4 = 0$
$(x - 4)(x + 4) = 0$
$x^2 - 16 = 0$

(15A)

Study Exercise Three (Frame 18)

1. $3x^2 - 3 = 0$
$3(x + 1)(x - 1) = 0$
$x + 1 = 0$ or $x - 1 = 0$
the solution set is $\{-1, 1\}$

2. $(x - 2)(x - 2) = 0$
$x - 2 = 0$ or $x - 2 = 0$
the solution set is $\{2\}$

3. $x^2 - 2ax - 8a^2 = 0$
$(x - 4a)(x + 2a) = 0$
$x - 4a = 0, \quad x + 2a = 0$
$x = 4a, \quad x = -2a$
the solution set is $\{-2a, 4a\}$

4. $\frac{2}{x - 3} - \frac{6}{x - 8} = -1$
$2(x - 8) - 6(x - 3) = -1(x - 3)(x - 8)$
$2x - 16 - 6x + 18 = -x^2 + 11x - 24$
$x^2 - 15x + 26 = 0$
$(x - 13)(x - 2) = 0$
$x - 13 = 0$ or $x - 2 = 0$
the solution set is $\{2, 13\}$

(18A)

Study Exercise Four (Frame 24)

1. the solution set is $\{6, -6\}$

2. $(x - 1)^2 = 16$

$x - 1 = 4, \quad x - 1 = -4$

$x = 5, \quad x = -3$

the solution set is $\{5, -3\}$

3. $(x + 2)^2 = 5$

$x + 2 = \sqrt{5}, \quad x + 2 = -\sqrt{5}$

the solution set is $\{-2 + \sqrt{5}, \quad -2 - \sqrt{5}\}$

4. $(x - 3)^2 = 10$

$x - 3 = \sqrt{10}, x - 3 = -\sqrt{10}$

the solution set is $\{3 + \sqrt{10}, \quad 3 - \sqrt{10}\}$

24A

Study Exercise Five (Frame 30)

1. $\left[\frac{1}{2}(-6)\right]^2 = 9; \quad x^2 - 6x + 9 = (x - 3)^2$

2. $\left[\frac{1}{2}(3)\right]^2 = \frac{9}{4}; \quad x^2 + 3x + \frac{9}{4} = \left(x + \frac{3}{2}\right)^2$

3. $\left[\frac{1}{2}\left(\frac{1}{2}\right)\right]^2 = \frac{1}{16}; \quad x^2 + \frac{1}{2}x + \frac{1}{16} = \left(x + \frac{1}{4}\right)^2$

4. $\left[\frac{1}{2}\left(-\frac{1}{4}\right)\right]^2 = \frac{1}{64}; x^2 - \frac{1}{4}x + \frac{1}{64} = \left(x - \frac{1}{8}\right)^2$

30A

Study Exercise Six (Frame 34)

1. $x^2 + 2x = 3$

$(x + 1)^2 = 4$

$x + 1 = \sqrt{4}, \quad x + 1 = -\sqrt{4}$

solution set is $\{1, -3\}$

2. $x^2 - 6x = 7$

$x^2 - 6x + 9 = 16$

$(x - 3)^2 = 16$

$x - 3 = \sqrt{16}, \quad x - 3 = -\sqrt{16}$

$x - 3 = 4, \quad x - 3 = -4$

solution set is $\{7, -1\}$

SOLUTIONS TO STUDY EXERCISES, CONTD.

Study Exercise Six (Frame 34, contd.)

3. $2x^2 + 3x = 4$

$$x^2 + \frac{3}{2}x = 2$$

$$x^2 + \frac{3}{2}x + \frac{9}{16} = 2 + \frac{9}{16}$$

$$\left(x + \frac{3}{4}\right)^2 = \frac{41}{16}$$

$$x + \frac{3}{4} = \sqrt{\frac{41}{16}}, \quad x + \frac{3}{4} = -\sqrt{\frac{41}{16}}$$

$$x = -\frac{3}{4} + \frac{\sqrt{41}}{4}, \quad x = -\frac{3}{4} - \frac{\sqrt{41}}{4}$$

the solution set is $\left\{\frac{-3 + \sqrt{41}}{4}, \frac{-3 - \sqrt{41}}{4}\right\}$

4. $2x^2 + 3x = 2$

$$x^2 + \frac{3}{2}x = 1$$

$$x^2 + \frac{3}{2}x + \frac{9}{16} = 1 + \frac{9}{16}$$

$$\left(x + \frac{3}{4}\right)^2 = \frac{25}{16}$$

$$x + \frac{3}{4} = \sqrt{\frac{25}{16}}, \quad x + \frac{3}{4} = -\sqrt{\frac{25}{16}}$$

$$x = -\frac{3}{4} + \frac{5}{4}, \quad x = -\frac{3}{4} - \frac{5}{4}$$

the solution set is $\left\{\frac{1}{2}, -2\right\}$

34A

Study Exercise Seven (Frame 39)

1. $a = 1, \quad b = -5, \quad c = -6$

$$x = \frac{-(-5) \pm \sqrt{(-5)^2 - 4(1)(-6)}}{2(1)}$$

$$x = \frac{5 \pm \sqrt{49}}{2}$$

$$x = \frac{5 \pm 7}{2}$$

the solution set is $\{6, -1\}$

2. $a = 1, \quad b = -3, \quad c = 1$

$$x = \frac{-(-3) \pm \sqrt{(-3)^2 - 4(1)(1)}}{2(1)}$$

$$x = \frac{3 \pm \sqrt{5}}{2}$$

the solution set is $\left\{\frac{3 + \sqrt{5}}{2}, \frac{3 - \sqrt{5}}{2}\right\}$

3. $a = 3, \quad b = -5, \quad c = -1$

$$x = \frac{-(-5) \pm \sqrt{(-5)^2 - 4(3)(-1)}}{2(3)}$$

$$x = \frac{5 \pm \sqrt{37}}{6}$$

the solution set is $\left\{\frac{5 + \sqrt{37}}{6}, \frac{5 - \sqrt{37}}{6}\right\}$

4. $a = 1, \quad b = 4, \quad c = 4$

$$x = \frac{-4 \pm \sqrt{4^2 - 4(1)(4)}}{2(1)}$$

$$x = \frac{-4 \pm \sqrt{0}}{2}$$

the solution set is $\{-2\}$

5. $a = 1, \quad b = -1, \quad c = 2$

$$x = \frac{-(-1) \pm \sqrt{(-1)^2 - 4(1)(2)}}{2(1)}$$

$$x = \frac{1 \pm \sqrt{-7}}{2}, \quad \text{does not exist}$$

the solution set is $\{\quad\}$

39A

SOLUTIONS TO STUDY EXERCISES, CONTD.

Study Exercise Eight (Frame 44)

A. 1. $b^2 - 4ac = (-2)^2 - 4(1)(9)$
$= 4 - 36$
$= -32$
Since $D < 0$, there are no real solutions.

2. $b^2 - 4ac = (-2)^2 - 4(1)(1)$
$= 0$
Since $D = 0$, there is one real solution.

3. $b^2 - 4ac = (2)^2 - 4(1)(-5)$
$= 4 + 20$
$= 24$
Since $D > 0$, there are two real solutions.

B. 4. $b^2 - 4ac = 0$ if there is to be one real solution:
$k^2 - 4(16) = 0$
$k^2 = 64$
$k = 8 \quad \text{or} \quad k = -8$

5. $(-5)^2 - 4(1)(k) = 0$
$25 - 4k = 0$
$-4k = -25$
$k = \dfrac{25}{4}$

44A

Study Exercise Nine (Frame 47)

1. Let $x =$ first consecutive even integer:
$x + 2 =$ second consecutive even integer:
$(x)(x + 2) = 80$
$x^2 + 2x - 80 = 0$
$(x + 10)(x - 8) = 0$
$x = -10, \quad x = 8$
$x + 2 = -8, \quad x + 2 = 10$

The two consecutive even integers are -10 and -8 or 8 and 10.

2. $(14 - x)^2 = (6 - x)^2 + (13 - x)^2$ by the pythagorean theorem:
$196 - 28x + x^2 = 36 - 12x + x^2 + 169 - 26x + x^2$
$0 = x^2 - 10x + 9$
$0 = (x - 9)(x - 1)$
$x = 9, x = 1$
9 will not work since $6 - x$ would be negative.
thus $x = 1$;

3. $x =$ width in inches
$x + 3 =$ length in inches
$x(x + 3) =$ area

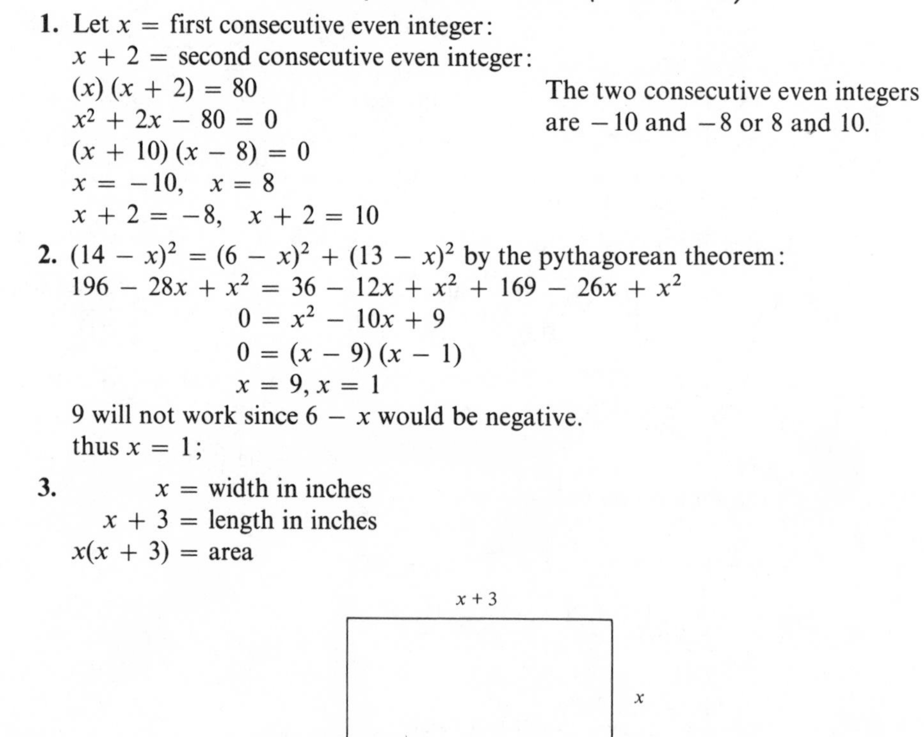

SOLUTIONS TO STUDY EXERCISES, CONTD.

Study Exercise Nine (Frame 47, contd.)

$x(x + 3) - 4x = 12$
$x^2 + 3x - 4x = 12$
$x^2 - x - 12 = 0$
$(x - 4)(x + 3) = 0$
$x = 4, x = -3$
$x = -3$ will not work since x is negative.
Thus $x = 4$ inches width
$x + 3 = 7$ inches length

47A

UNIT 10—SUPPLEMENTARY PROBLEMS

A. Find a, b, c according to the standard form $ax^2 + bx + c = 0$.
1. $x^2 + 2x - 3 = 0$ **2.** $4x^2 + x = 1$ **3.** $x^2 = 9$

B. **4.** If $(x + y)(y + 2) = 0$ and $y > 0$, what can be deduced about $(x + y)$?
5. What can be deduced about the values of x and y if $2xy = 0$?

C. Find the solution sets by factoring.
6. $3x^2 + 3x = 0$ **7.** $x(x - 1) = 12$ **8.** $x^2 = 4$ **9.** $4 + 9x^2 = 12x$
10. $10x = 3x^2$ **11.** $x^2 - 2x - 15 = 0$ **12.** $\dfrac{x - 4}{x + 1} - \dfrac{15}{4} = \dfrac{x + 1}{x - 4}$

D. Find the solution sets by completing the square.
13. $x^2 - 2x - 5 = 0$ **14.** $x^2 + 4x + 1 = 0$ **15.** $x^2 - 6x + 1 = 0$
16. $x^2 + x - 1 = 0$ **17.** $2x^2 - 4x = 7$

E. Find the solution sets by the quadratic formula.
18. $x^2 - 6x + 2 = 0$ **19.** $2x(x + 1) = 7$ **20.** $4x^2 + 7x - 1 = 0$
21. $x^2 + 3x + 4 = 0$ **22.** $x^2 = 2x$ **23.** $3x^2 - 2x = 2$

F. Find a quadratic equation with the given solution set.
24. $\{3, -2\}$ **25.** $\{6, -6\}$ **26.** $\left\{\dfrac{1}{4}, \dfrac{3}{2}\right\}$
27. $\{2 + \sqrt{2}, 2 - \sqrt{2}\}$ **28.** $\left\{\dfrac{\sqrt{2}}{2}, -\dfrac{\sqrt{2}}{2}\right\}$

G. Classify the types of solutions by using the discriminant.
29. $x^2 - 4x - 2 = 0$ **30.** $4x^2 - x + 1 = 0$
31. $4x^2 + 4x + 1 = 0$ **32.** $x^2 + 9 = 0$

H. Solve for the indicated letter.
33. $(x + a)^2 - b^2 = 0$, solve for x **34.** $4a^2r^2 - 5s^2 = 0$, solve for r
35. $x^5 - 5x^3 + 4x = 0$, solve for x

I. Miscellaneous:
36. Find two numbers whose sum is 8 and whose product is -33.
37. The area of a rectangle is 120 feet and its length is 7 feet more than its width. Find its dimensions.
38. Find the area of a right triangle with sides x inches, $(x + 7)$ inches, and $(x + 8)$ inches.
39. The perimeter of a rectangle is 20 feet and its area is 21 square feet. Find the length and width.
40. Prove that there is no real number such that the sum of it and its reciprocal is 1.

unit

11

Relations and Functions

Objectives

1. Understand functional notation.
2. Be able to graph relations and functions.

Terms

cross product	domain	ordinate
relation	range	components
function	abscissa	

①

A pair is a set containing two elements. Order thus far has been unimportant. Thus $\{a, b\} = \{b, a\}$.

If we wish to take a set of two elements and consider the elements in a specified order, we will use parentheses instead of braces.

②

An ordered pair consists of two elements, a and b, arranged to show that a is the first element and b is the second.

The notation for an ordered pair is (a, b) where a is the first element and b is the second. Note that $(a, b) \neq (b, a)$.

In the ordered pair (a, b), the a and b are called *components*.

③

Two ordered pairs (x, y) and (a, b) are equal if and only if their first components and their second components are equal; that is, if and only if $x = a$ and $y = b$.

Any set of ordered pairs will be called a *relation*.
Thus $\{(1, 3), (2, 5), (1, 4)\}$ is a relation.

④

For two sets A and B, the set of all ordered pairs (a, b) where $a \in A$ and $b \in B$ is called the *cross product* of A and B and is denoted by $A \text{ x } B$ (read A cross B).

Thus $A \text{ x } B = \{(a, b) | a \in A \text{ and } b \in B\}$

Example: If $A = \{1\}$, $B = \{2, 3\}$,
$A \text{ x } B = \{(1, 2), (1, 3)\}$
$B \text{ x } A = \{(2, 1), (3, 1)\}$
$B \text{ x } B = \{(2, 2), (2, 3), (3, 2), (3, 3)\}$

⑤

Study Exercise One

1. If $A = \{3, 5\}$, $B = \{1, 3, 6\}$, find $A \times B$
2. If $A = \{1, 5\}$, find $A \times A$
3. $A = \{l, o, v, e\}$, $B = \{m, a, t, h\}$. Find $A \times B$.
4. In general, does $A \times B = B \times A$?

(6)

If the elements of A are displayed on a horizontal line and the elements of B on a vertical line, then the network of points determined by vertical lines through $a \in A$ meeting horizontal lines through $b \in B$ is called the *graph* of $A \times B$.

Example: $A = \{r, t\}$, $B = \{c, d\}$.

$$A \times B = \{(r, c), (r, d), (t, c), (t, d)\}$$

(7)

In dealing with sets of real numbers, it is agreed that to display $A \times B$, we assign elements of the first set to the horizontal real line and elements of the second set to a vertical real line through the origin (the zero point). This point below represents the ordered pair (1, 2).

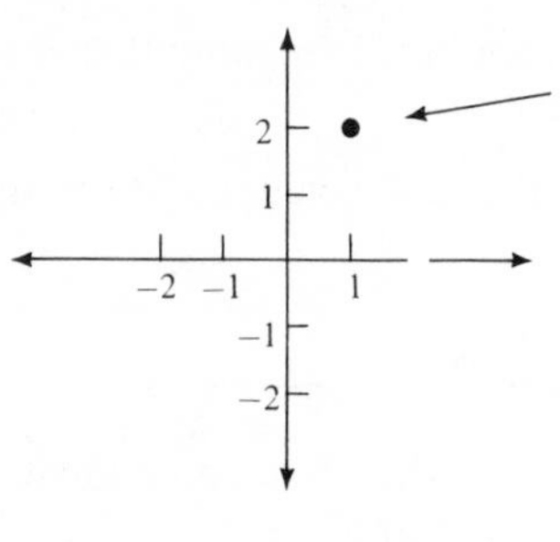

(8)

If $U = \{1, 2, 3, 4\}$, then $U \times U$ is $\{(1,1), (1,2), (1,3), (1,4), (2,1), (2,2), (2,3), (2,4), (3,1), (3,2), (3,3), (3,4), (4,1), (4,2), (4,3), (4,4)\}$

The graph of $U \times U$ is shown below:

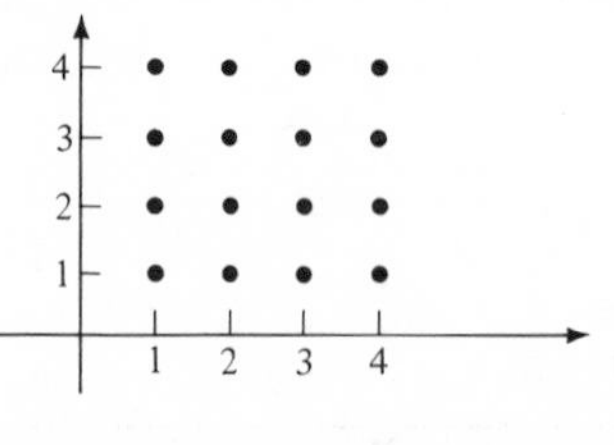

(9)

If x and y are variables with replacement set such that $x \in A$ and $y \in B$, then an open sentence in two variables x and y is said to have as replacement set the ordered pairs from $A \times B$.

The solution set of the open sentence will be all those ordered pairs of $A \times B$ that make the corresponding statement true.

(10)

Consider $\{(x, y) \in U \times U | y = x - 1\}$ where $U = \{1, 2, 3, 4\}$.

From the replacement set $U \times U$ we are to select those ordered pairs that make the statement true. The only ordered pairs that make the statement true are (2, 1), (3, 2), and (4, 3).

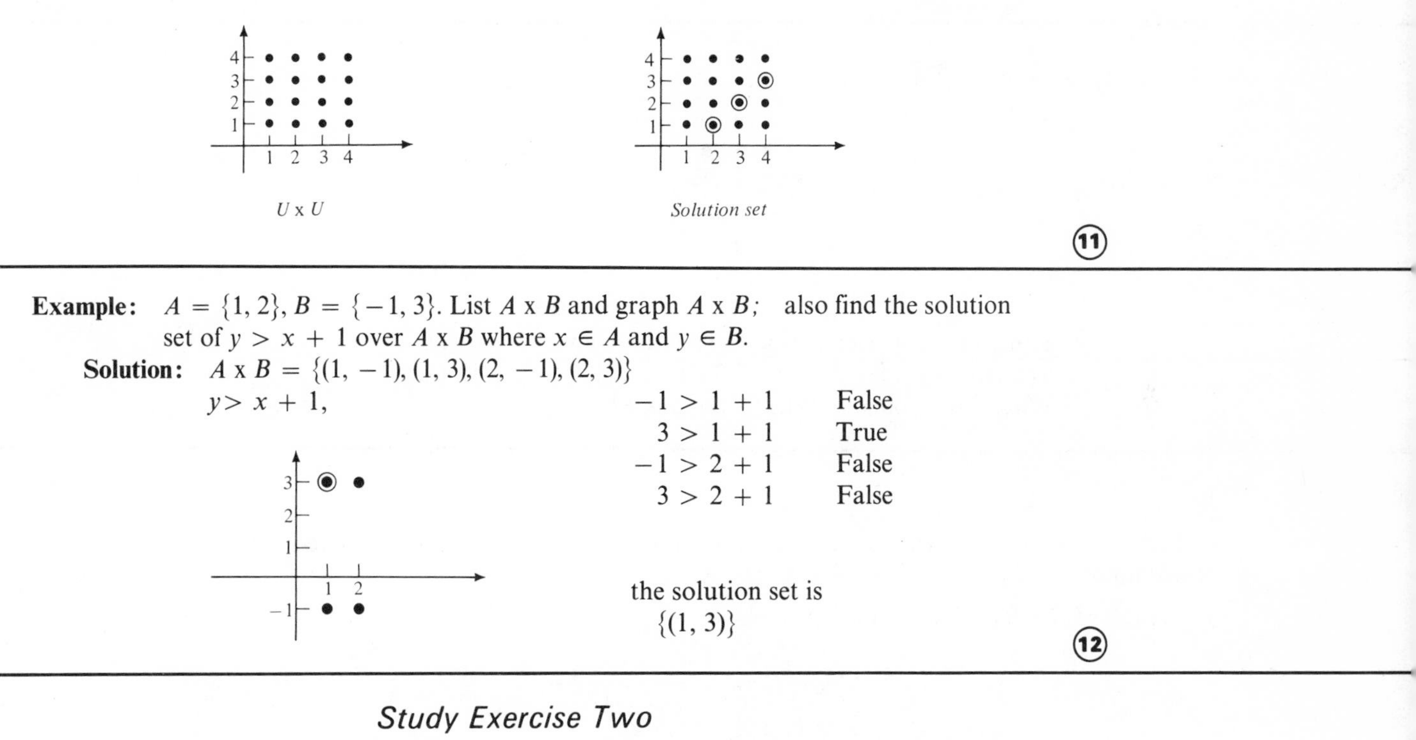

U x U *Solution set*

(11)

Example: $A = \{1, 2\}, B = \{-1, 3\}$. List $A \times B$ and graph $A \times B$; also find the solution set of $y > x + 1$ over $A \times B$ where $x \in A$ and $y \in B$.

Solution: $A \times B = \{(1, -1), (1, 3), (2, -1), (2, 3)\}$

$y > x + 1,$

$-1 > 1 + 1$	False
$3 > 1 + 1$	True
$-1 > 2 + 1$	False
$3 > 2 + 1$	False

the solution set is $\{(1, 3)\}$

(12)

Study Exercise Two

If $x \in A, y \in B, A = \{-2, -1, 0\}, B = \{0, 1, 2\}$, find the solution set of the following over $A \times B$.

1. $x \leqslant y$ **2.** $y = x + 1$ **3.** $x > y^2$

(13)

If we let x denote a variable belonging to A, and y a variable belonging to B so that a general element of $A \times B$ is denoted by (x, y), then the horizontal axis of the graph of $A \times B$ is called the *x axis* and the vertical axis the *y axis.*

x and y will also be referred to as coordinates of the point (x, y). In particular x will be called the *abscissa* and y the *ordinate.*

(14)

Relations

Any set of ordered pairs is called a relation. The set of first components in a relation is called the *domain set* and the set of all second components in a relation is called the *range set.*

(Frame 15, contd.)

Example: $\{(1, 2)\ (-1, 3)\ (2, 3)\ (0, 1)\}$ is a relation.
The domain is $\{-1, 0, 1, 2\}$
The range is $\{1, 2, 3\}$

(15)

A variable representing an element in the domain of a relation (first component) is often called the *independent variable*, while the variable representing an element in the range (second component) is often called the *dependent variable.*

Usually we will refer to (x, y) where x is the independent variable and y is the dependent variable.

(16)

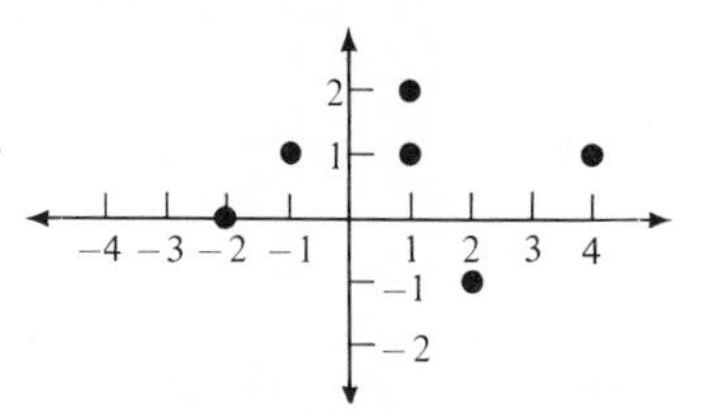

Can you list the elements in the relation above?
The relation is $\{(-2, 0), (-1, 1), (1, 1), (1, 2), (2, -1), (4, 1)\}$
What is the domain? What is the range?

(17)

$R \times R$

If R represents the set of real numbers, then $R \times R$ represents every possible ordered pair of real numbers.

The graph of $R \times R$ consists of the entire plane.

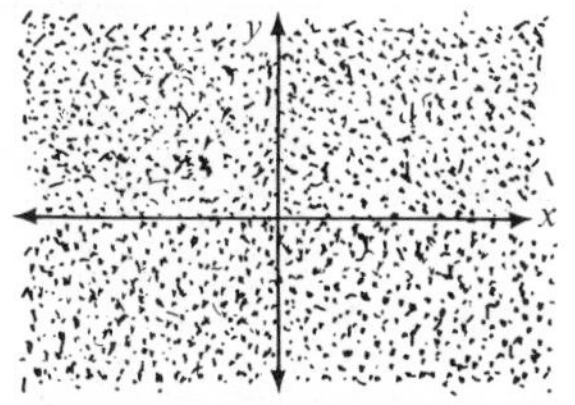

(18)

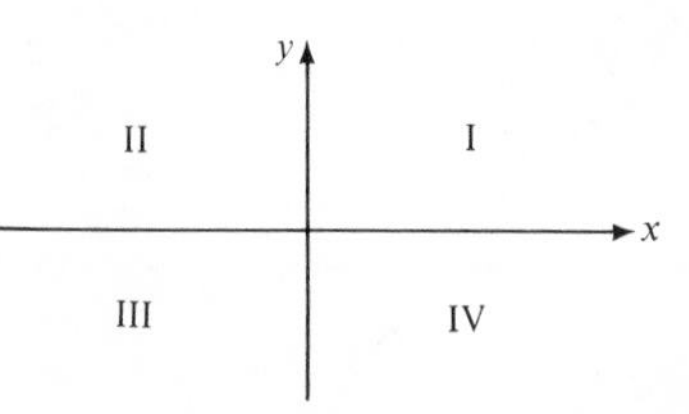

Ordered pairs that are in the region labeled I are said to be in *Quadrant I* (Quadrant I does not include the axes).

Similar statements can be made for the regions labeled II, III and IV.

(19)

Study Exercise Three

Fill in the blanks.

1. The ______ of the ordered pair (2, 3) is 2.
2. $(-6, 1)$ is an ordered pair found in Q ______. (Q is an abbreviation for quadrant)
3. $(-2, -1)$ is found in Q ______.
4. $(0, -3)$ is found on ______.
5. The abscissa of $(-3, 4)$ is ______.

(20)

Function

A *function* is a relation in which no two ordered pairs have the same first component and different second components.

As an alternate definition, we can say a function is a set of ordered pairs (x, y) such that for each x there corresponds one and only one y.

(21)

Example 1:
$\{(2, 3), (1, 2), (0, -2), (-1, 1)\}$

Example 2:
$\{(2, 5), (3, 5), (1, 6)\}$

Example 3:
$\{(1, 2), (2, 4), (3, 8), (4, 16), (5, 32)\}$

Example 4:
$\{(1, 3), (2, 1), (1, 4), (0, -2)\}$

Examples 1, 2 and 3 are functions. Example 4 is not a function.

(22)

It is not difficult to recognize from its graph a relation which is also a function. If the graph of a relation has two or more points on the same vertical line, then it is not a function; otherwise it is.

(23)

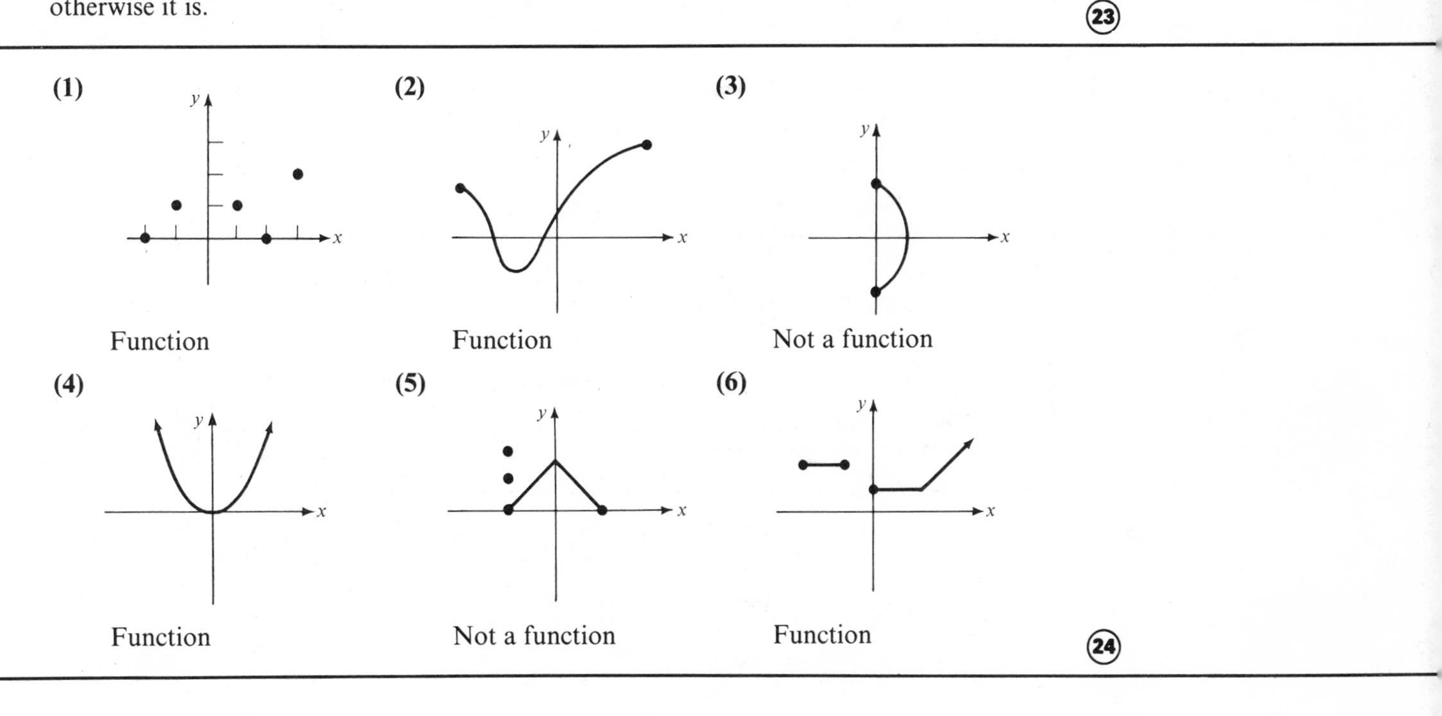

(24)

Study Exercise Four

Which of the following relations are also functions over $R \times R$?

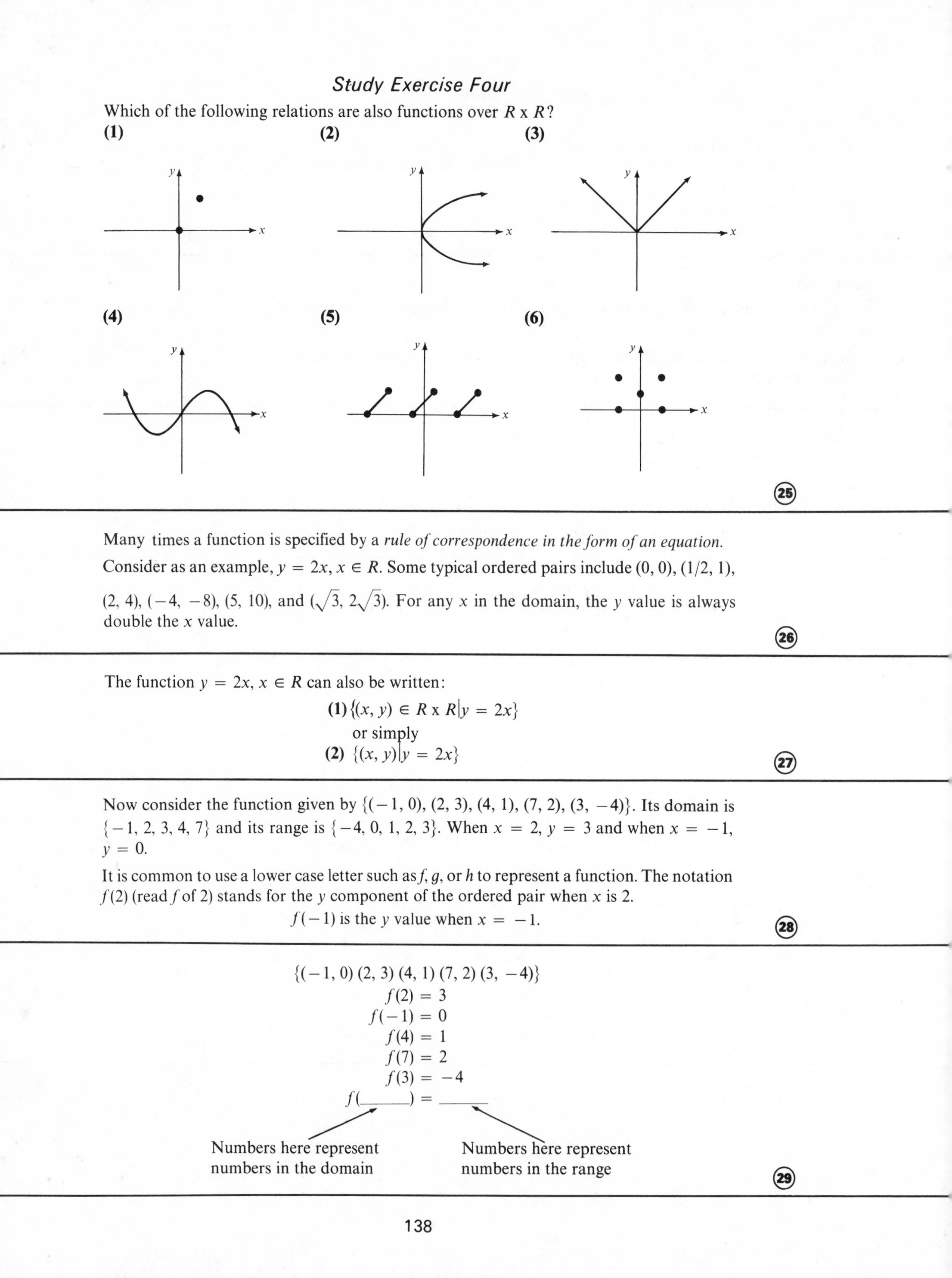

(25)

Many times a function is specified by a *rule of correspondence in the form of an equation.* Consider as an example, $y = 2x, x \in R$. Some typical ordered pairs include (0, 0), (1/2, 1), (2, 4), $(-4, -8)$, (5, 10), and $(\sqrt{3}, 2\sqrt{3})$. For any x in the domain, the y value is always double the x value.

(26)

The function $y = 2x, x \in R$ can also be written:

(1) $\{(x, y) \in R \times R | y = 2x\}$

or simply

(2) $\{(x, y) | y = 2x\}$

(27)

Now consider the function given by $\{(-1, 0), (2, 3), (4, 1), (7, 2), (3, -4)\}$. Its domain is $\{-1, 2, 3, 4, 7\}$ and its range is $\{-4, 0, 1, 2, 3\}$. When $x = 2$, $y = 3$ and when $x = -1$, $y = 0$.

It is common to use a lower case letter such as f, g, or h to represent a function. The notation $f(2)$ (read f of 2) stands for the y component of the ordered pair when x is 2.

$f(-1)$ is the y value when $x = -1$.

(28)

$\{(-1, 0)\ (2, 3)\ (4, 1)\ (7, 2)\ (3, -4)\}$

$f(2) = 3$

$f(-1) = 0$

$f(4) = 1$

$f(7) = 2$

$f(3) = -4$

$f(\_\_\_\_) = \_\_\_\_$

Numbers here represent numbers in the domain

Numbers here represent numbers in the range

(29)

$$\{(1, 2), (2, 4), (3, 6), (4, 8), (5, 10), \ldots\}$$
$$f(1) = 2$$
$$f(5) = 10$$
$$f(6) = 12$$
$$f(9) = 18$$
$$f(x) = 2x$$

For any x in the domain, y is double the x value; i, e. $y = 2x$.

(30)

Several ways of writing the function $y = 7x$ are indicated below:

(1) $\{(x, y) \in R \times R | y = 7x\}$

(2) $\{(x, y) \mid y = 7x\}$

(3) $y = 7x$

(4) $f(x) = 7x$

(31)

Study Exercise Five

If a function f is defined by $f(x) = \frac{2}{3}x - 4$, find:

1. $f(0)$

2. $f(3)$

3. $f(-2)$

4. $f(3/4)$

5. $f(3) - f(0)$

(32)

Example: Draw the graph of $f(x) = 2x - 1$ over $R \times R$

Solution:

x	$f(x)$
0	-1
1	1
2	3
3	5
-1	-3
-2	-5
$\frac{1}{2}$	0
$-\frac{1}{2}$	-2

(33)

Study Exercise Six

Draw the graph of the following over $R \times R$.

1. $\{(x, f(x)) \in R \times R | f(x) = 3x - 2\}$

2. $\{(x, y) \mid y = x^2 - 1\}$

3. $h(x) = 0 \cdot x + 3$

(34)

A function consists of three parts:

(1) a set called the domain
(2) a set called the range
(3) a rule that assigns to each element of the domain one and only one element of the range.

(35)

When a function is specified by an equation and no mention is made of its domain, it is agreed that it will be the largest possible set for which the expression is defined.

Example 1: $f(x) = 3x^2$, the domain is the set of reals.

(Frame 36, contd.)

Example 2: $g(x) = 1/x$, the domain is all reals except zero.
Example 3: $h(x) = \sqrt{x}$, the domain is $\{x|x \geqslant 0, x \in R\}$.

(36)

Study Exercise Seven

Give the domain of each function.

1. $f(x) = 6x$ **2.** $g(x) = \sqrt{x - 1}$

3. $h(x) = x^2 - 1$ **4.** $f(x) = \dfrac{1}{x - 2}$

(37)

Consider the function shown below.

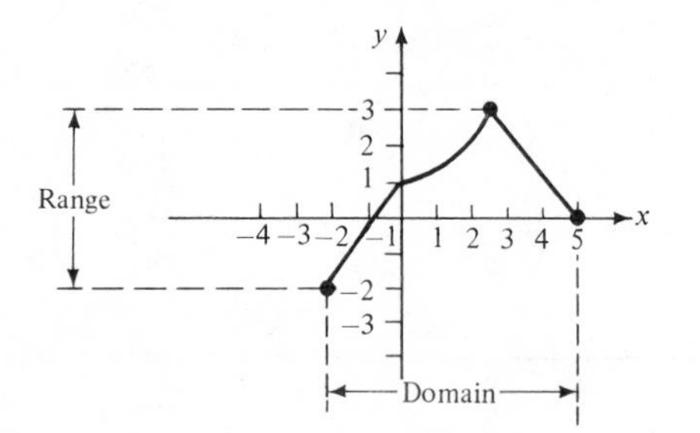

The domain of this function is $\{x \mid -2 \leqslant x \leqslant 5\}$.
The range is $\{y \mid -2 \leqslant y \leqslant 3\}$.

(38)

Zeros of a Function

In the ordered pair (x, y), x is called a *zero* of the function if y is zero.

Consider the function given by $\{(1, 3), (2, 0), (0, 3), (-1, 2), (4, 0)\}$. The zeros of this function are 2 and 4 since y is zero in the ordered pairs $(2, 0)$ and $(4, 0)$.

(39)

Example 1: Find the zeros of the function given by $y = 3x - 2$.

Solution:

Line (1) $0 = 3x - 2$
Line (2) $2 = 3x$
Line (3) $2/3 = x$
the zero is $2/3$

Example 2: Find the zeros of the function $y = x^2 + 4x - 21$.

Solution:

Line (1) $0 = x^2 + 4x - 21$
Line (2) $0 = (x + 7)(x - 3)$
Line (3) $x = -7, x = 3$
the zeros are -7 and 3

(40)

Study Exercise Eight

A. Find the zeros of following functions:

1. $y = 5x - 2$ **2.** $f(x) = x^2 - x - 6$ **3.** $g(x) = x^2 - 25$

(Frame 41, contd.)

B. Give the domain and range of the functions whose graph is given below.

4.

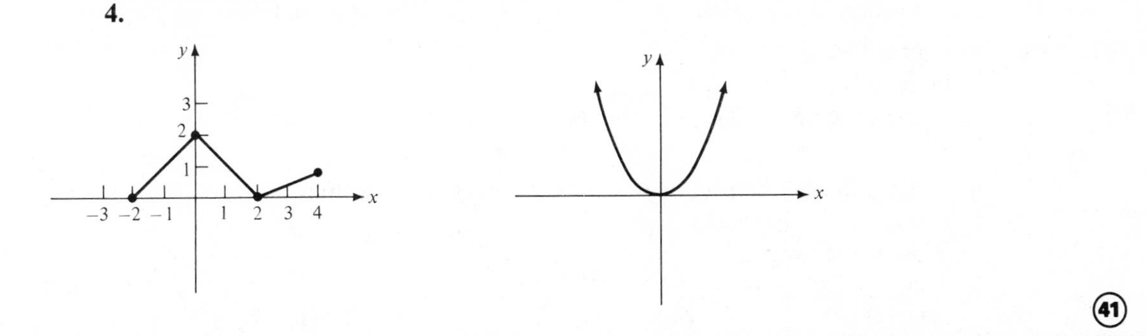

(41)

REVIEW EXERCISES

A. True or False.

1. A relation is a function. **2.** A function is a relation.

B. Fill in the blank.

3. $(-3, -2)$ is a point in Q______.

4. In the ordered pair $(7, -6)$, the second component, -6, is called the ______.

C. **5.** If $A = \{1, 3, 5\}$, $B = \{-1, 2\}$, find $A \times B$.

6. What is the solution set of $y > x$ over the set $A \times B$ from problem 5.

7. If $f(x) = x^2$, what is the domain?
What is the range?

8. If $g(x) = 2x^2 - x - 1$, find:
(a) $g(2)$ **(b)** $g(-1)$ **(c)** $g(1) - g(-1)$

9. What does $R \times R$ represent?

10. Draw the graph of $f(x) = x$ over $R \times R$.

11. Find the zeros of the function given by:
(a) $y = 2x + 5$ **(b)** $y = 2x^2 + 5x - 3$

42

SOLUTIONS TO REVIEW EXERCISES

A. **1.** false **2.** true

B. **3.** III **4.** ordinate

C. **5.** $A \times B = \{(1, -1), (1, 2), (3, -1), (3, 2), (5, -1), (5, 2)\}$

6. $\{(1, 2)\}$

7. domain is $\{x \mid x \in R\}$
range is $\{y \mid y \geqslant 0, y \in R\}$

8. **(a)** $g(2) = 2(2)^2 - 2 - 1$
$= 5$

(b) $g(-1) = 2(-1)^2 - (-1) - 1$
$= 2$

(c) $g(1) = 2(1)^2 - (1) - 1$
$= 0$
$g(1) - g(-1) = 0 - 2$
$= -2$

9. every possible ordered pair of real numbers

10.

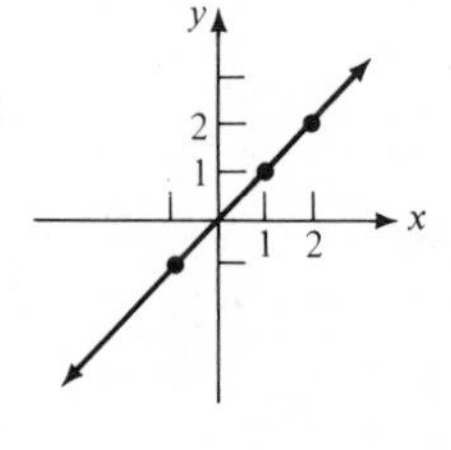

11. **(a)**
$0 = 2x + 5$
$-5 = 2x$
$-5/2 = x$
the zero is $-5/2$

(b) $0 = 2x^2 + 5x - 3$
$0 = (2x - 1)(x + 3)$
$2x - 1 = 0 \qquad x + 3 = 0$
$x = 1/2 \qquad x = -3$
the zeros are $1/2$ and -3

43

SOLUTIONS TO STUDY EXERCISES

Study Exercise One (Frame 6)

1. $\{(3, 1), (3, 3), (3, 6), (5, 1), (5, 3), (5, 6)\}$
2. $\{(1, 1), (1, 5), (5, 1), (5, 5)\}$
3. $\{(l, m), (l, a), (l, t), (l, h), (o, m), (o, a), (o, t), (o, h), (v, m), (v, a), (v, t), (v, h), (e, m), (e, a), (e, t), (e, h)\}$
4. No

6A

Study Exercise Two (Frame 13)

$A \times B = \{(-2, 0), (-2, 1), (-2, 2), (-1, 0), (-1, 1), (-1, 2), (0, 0), (0, 1), (0, 2)\}$

1. $\{(-2, 0), (-2, 1), (-2, 2), (-1, 0), (-1, 1), (-1, 2), (0, 0), (0, 1), (0, 2)\}$
2. $\{(-1, 0), (0, 1)\}$
3. $\{\ \}$ or $\varnothing$

13A

Study Exercise Three (Frame 20)

1. abscissa **2.** II **3.** III **4.** the y axis **5.** -3

20A

Study Exercise Four (Frame 25)

1. function **2.** not a function **3.** function
4. function **5.** function **6.** not a function

25A

Study Exercise Five (Frame 32)

1. $f(0) = 2/3(0) - 4$
$= -4$

2. $f(3) = 2/3(3) - 4$
$= -2$

3. $f(-2) = 2/3(-2) - 4$
$= -16/3$

4. $f(3/4) = 2/3(3/4) - 4$
$= \frac{1}{2} - 4$
$= -7/2$

5. $f(3) - f(0) = (2) - (-4)$
$= 2$

32A

Study Exercise Six (Frame 34)

1. **2.** **3.**

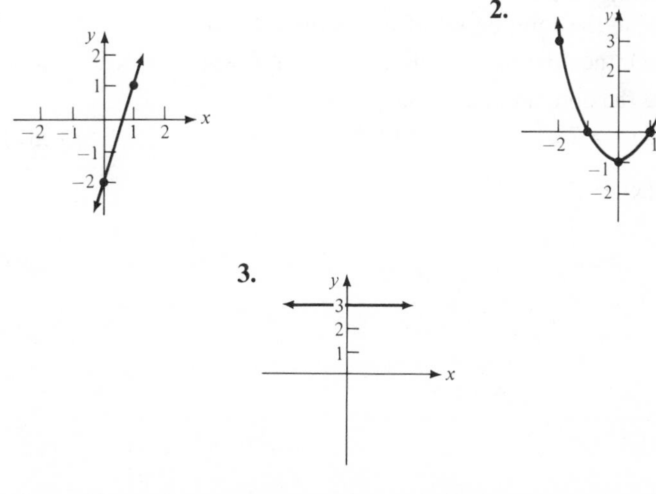

34A

Study Exercise Seven (Frame 37)

1. $\{x|x \in R\}$
2. $\{x|x \geqslant 1\}$
3. $\{x|x \in R\}$
4. $\{x|x \neq 2 \text{ and } x \in R\}$

37A

Study Exercise Eight (Frame 41)

A.

1. $0 = 5x - 2$
   $2 = 5x$
   $2/5 = x$
   the zero is 2/5
2. $0 = x^2 - x - 6$
   $0 = (x - 3)(x + 2)$
   $x = 3, x = -2$
   the zeros are 3 and -2
3. $0 = x^2 - 25$
   $0 = (x + 5)(x - 5)$
   $x = 5, x = -5$
   the zeros are 5 and -5

B.

4. domain is $\{x|-2 \leqslant x \leqslant 4\}$
   range is $\{y|0 \leqslant y \leqslant 2\}$
5. domain is $\{x|x \in R\}$
   range is $\{y|y \geqslant 0\}$

41A

UNIT 11—SUPPLEMENTARY PROBLEMS

1. If the coordinates of a point are such that the abscissa is zero and the ordinate is positive, where is the point located in the plane?
2. The abscissa of the ordered pair (a, b) is ______.
3. If $A = \{-1, 1\}$, find $A \text{ x } A$.
4. Which of the sets describes a function?
   a) $\{(1, 2), (2, 3), (3, 4)\}$ **b)** $\{(1, 2), (1, 3), (2, 3), (3, 3)\}$
5. **a)** Does $\{(x, y)|y^2 = x\}$ represent a relation?
   b) Does $\{(x, y)|y^2 = x\}$ represent a function? Hint: find x when $y = 1$ and -1.
6. If $f(x) = 2x - 5$, find:
   a) $f(-1)$ **b)** $f(2) - f(-2)$
7. If $f = \{(x, f(x))|f(x) = x^2 - x + 1\}$, find:
   a) $f(0)$ **b)** $f(1) + f(2)$ **c)** $f(t + 1)$ where $(t + 1)$ is in the domain of f
8. If $(x + h)$ is in the domain of f and $f(x) = 2x^2 - 1$, find $f(x + h)$.
9. $A = \{1, 2, 3\}$, $B = \{0, 1, 4\}$
   a) Graph $A \text{ x } B$
   b) Find the solution set of $y > x$ over $A \text{ x } B$
10. Sketch the relation $\{(x, y)|y = 2x, x \in I, -2 \leq x \leq 3\}$
11. Find the domain and range of:
    a) $f(x) = 2x$ **b)** $h(x) = x^2$ **c)** $g(x) = \sqrt{x}$
    d) $i(x) = \sqrt{x - 1}$

UNIT 11—SUPPLEMENTARY PROBLEMS

12. Which of the following graphs represent functions?

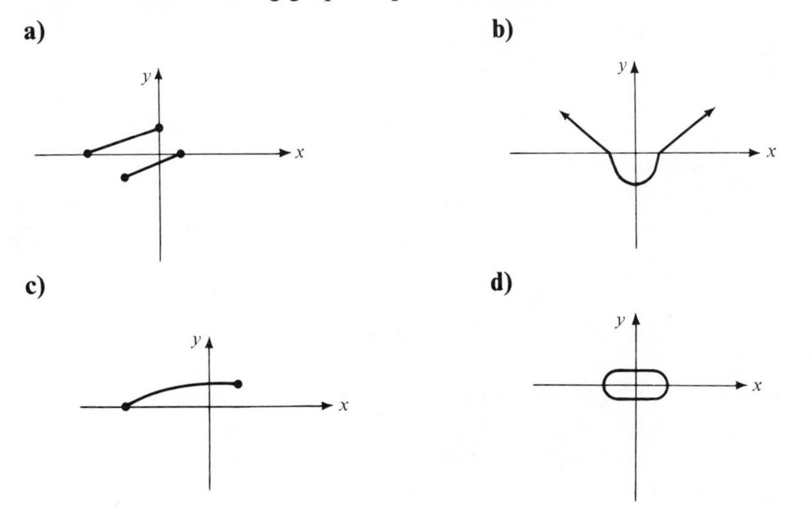

13. Sketch over $R \times R$, $h(x) = \begin{cases} x \text{ if } x \geq 0 \\ -x \text{ if } x < 0 \end{cases}$

14. Do you recognize the function in number 13? What is another way to write it? $h(x) =$ ______.

15. $f(x) = x^2 + 1$. Find $f(3) - f(2) + f(0)$.

16. **a)** Does $y = \pi$ represent a function? Why or why not?
b) Draw the graph of $y = \pi$ over $R \times R$.

17. Find the zeros of the function given by:
a) $y = 5x + 7$ **b)** $f(x) = x^2 - 49$ **c)** $y = x^2 + 3x - 2$

18. True or false: every function is a relation.

19. True or false: every relation is a function.

unit

The Linear Function

12

Objectives

1. Be able to find the slope and intercepts of a straight line.
2. Be able to draw the graph of a straight line.
3. Know the slope intercept and point slope form of a straight line.
4. Be able to compute distance between two points.

Terms

linear function   *y* intercept   *x* intercept   slope

①

Any first degree equation in two variables of the form $Ax + By + C = 0$, $A, B, C \in R$, $B \neq 0$ defines a function.

Examples: $3x - 2y + 1 = 0$ with $A = 1, B = -2, C = 1$;
$5x + y = 0$ with $A = 5, B = 1, C = 0$; and
$4y + 5 = 0$ with $A = 0, B = 4, C = 5$.

②

The function defined by $Ax + By + C = 0$ with $A, B, C \in R, B \neq 0$ is called a linear function since the graph of this function over $R \times R$ is always a straight line.

③

Not all lines, however, represent functions.

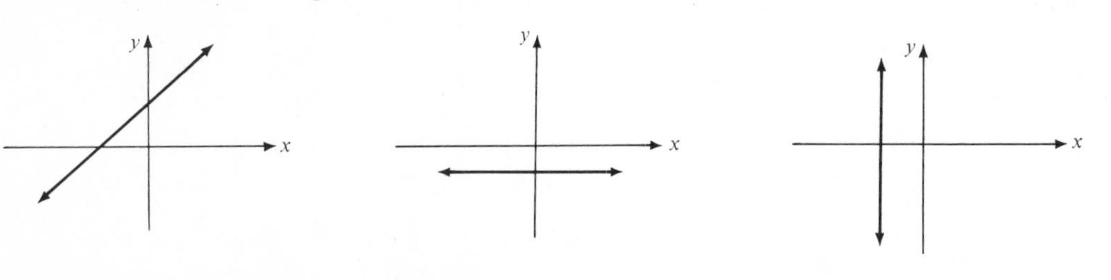

The first two graphs represent functions but the third does not.

④

Vertical lines do not represent functions. All non-vertical lines do represent functions.

In order to sketch the graph of a line, a minimum of two points is required.

⑤

If a line is neither vertical nor horizontal, it will have one point in common with the x-axis. and one point in common with the y-axis. These points will be called the *x intercept and y intercept* respectively.

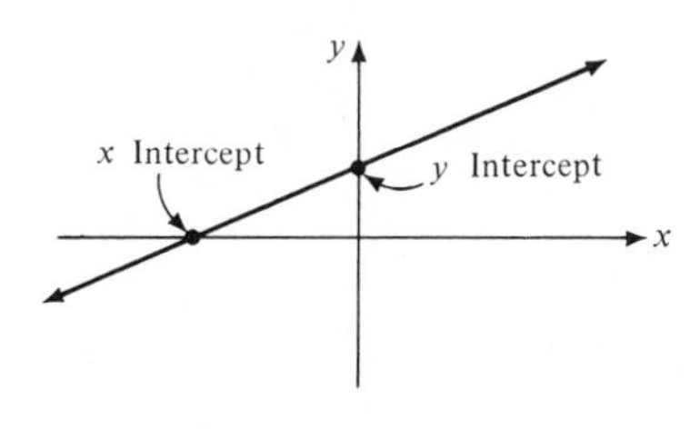

(6)

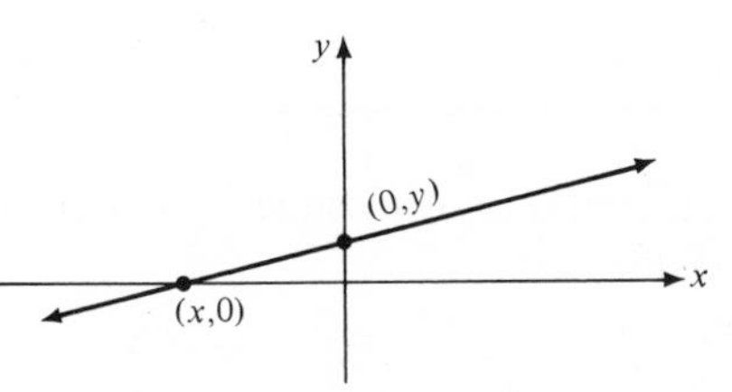

The x intercept point always has its second coordinate equal to zero.

The y intercept point always has its first coordinate equal to zero.

(7)

Thus, to find the x intercept, set $y = 0$ and solve the resulting equation for x. The point represented by $(x, 0)$ will be the x intercept.

To find the y intercept, set $x = 0$ and solve the resulting equation for y. The point $(0, y)$ will be the y intercept point.

(8)

Example: Find the x and y intercepts of $2x + y = 8$.

Solution: To find the x intercept, set $y = 0$;

$$2x + 0 = 8$$
$$2x = 8$$
$$x = 4$$

The x intercept is $(4, 0)$.
To find the y intercept, set $x = 0$

$$2(0) + y = 8$$
$$0 + y = 8$$
$$y = 8$$

The y intercept is $(0, 8)$.

(9)

Study Exercise One

Find the x and y intercepts of the following lines. Then draw the graphs by plotting the intercepts.

1. $x + y = 4$ **2.** $y = 3x + 2$ **3.** $y + 2x = 0$

(10)

The Constant Function

Consider $0 \cdot x + y = 4$ or $y = 4$.

For each x this equation assigns 4 as the value for y. Therefore, we have a function such that to each real x, y is 4.

That is, $f(x) = 4$. The graph is a horizontal line.

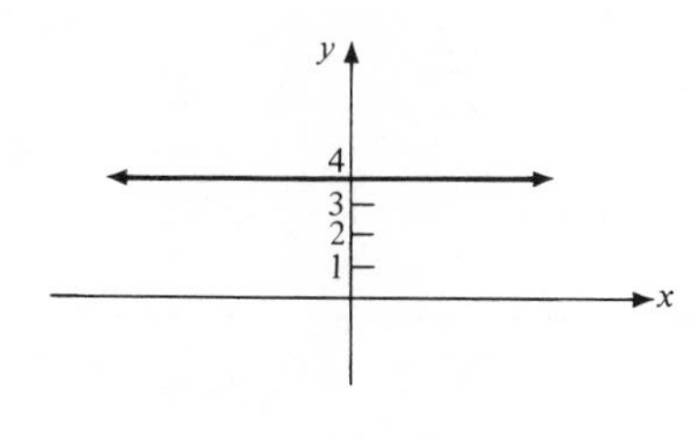

(11)

The function $f(x) = c$ where c is some fixed constant is known as the *constant function*. The graph of the constant function is a horizontal line.

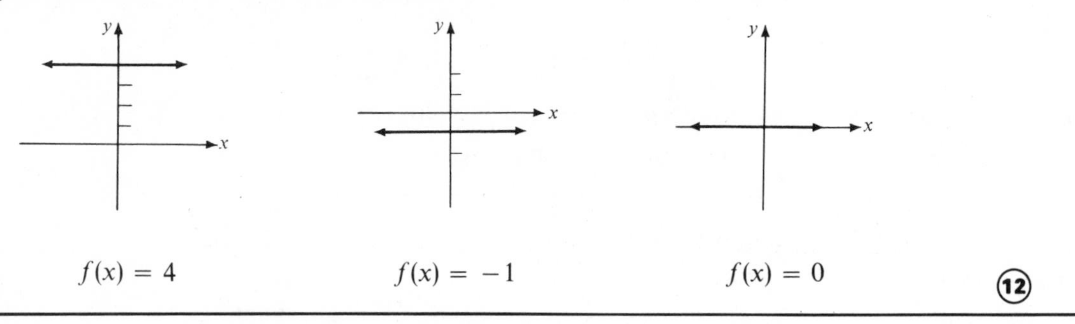

$f(x) = 4$ $f(x) = -1$ $f(x) = 0$

(12)

Another special type of linear equation is $x + 0 \cdot y = 3$ or $x = 3$.

Only one value is permissible for x (namely 3) while any value may be assigned to y. The graph of $x = 3$ is a vertical line and does not define a function.

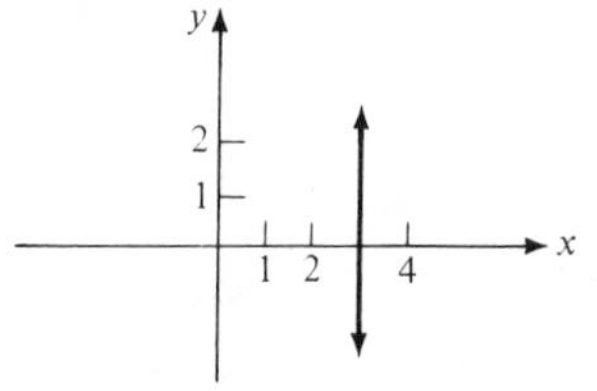

(13)

Study Exercise Two

Draw the graph:

1. $f(x) = -2$ **2.** $x = -2$ **3.** $y = 0$

(14)

Directed Distances

A, B, C, designate points on the real line with coordinates a, b, c, respectively.

The directed distance AB (from A to B) is the real number $b - a$.
The directed distance BA (from B to A) is the real number $a - b$.

(15)

---

c a b

C 0 A B

$AB = b - a \qquad BA = a - b$
If $a = 4, \quad b = 6, \quad c = -2$
$AB = b - a = 6 - 4$ or 2
$BA = a - b = 4 - 6$ or -2
$AC = c - a = (-2) - (4) = -6$
$CB = b - c = 6 - (-2) = 8$

(16)

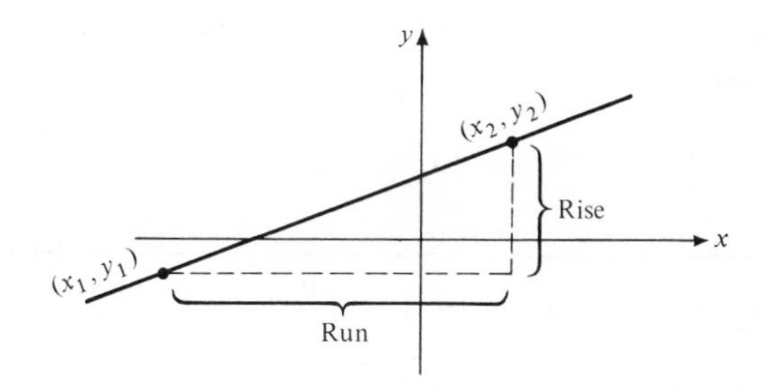

Given two points on a line, (x_1, y_1) and (x_2, y_2)

The directed distance $x_2 - x_1$ is called the *run*.
The directed distance $y_2 - y_1$ is called the *rise*.

The ratio of the rise to the run is the *slope*.

(17)

$$\text{Slope} = \frac{\text{rise}}{\text{run}}$$

$$= \frac{y_2 - y_1}{x_2 - x_1}$$

Notice that $x_2 - x_1$ cannot equal zero. That is, $x_2 - x_1 \neq 0$ or $x_2 \neq x_1$.

(18)

Slope is commonly indicated by the letter m.

$$m = \frac{\text{rise}}{\text{run}}$$

$$= \frac{y_2 - y_1}{x_2 - x_1}$$

$$= \frac{\Delta y}{\Delta x}$$

Δy is a symbol for the change in y. That is $y_2 - y_1$.
Δx is a symbol for the change in x. That is $x_2 - x_1$.

19

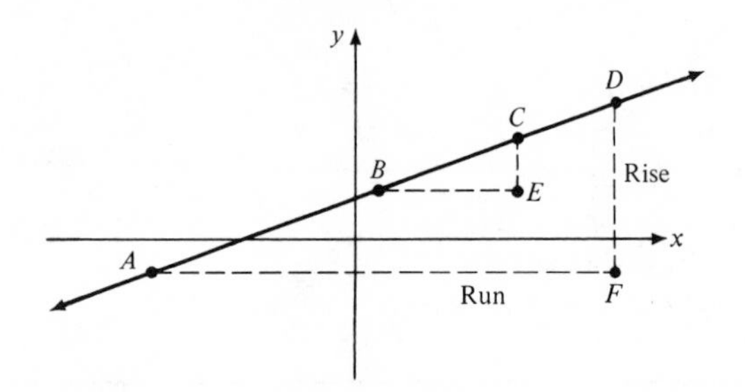

Since triangle AFD is similar to triangle BEC, the corresponding sides are in proportion. Thus any two points on a line will give the same slope.

$$m = \frac{\text{rise}}{\text{run}}$$

20

Example: (3, 2) and (5, 6) are points on a line. Find the slope of the line.

Solution:

$$m = \frac{\Delta y}{\Delta x}$$

$$= \frac{y_2 - y_1}{x_2 - x_1}$$

$$= \frac{6 - 2}{5 - 3}$$

$$= 4/2$$

$$= 2$$

21

Example: $(-2, -3)$ and $(-4, 1)$ are points on a line. Find the slope of the line.

Solution:

$$m = \frac{\Delta y}{\Delta x}$$

$$= \frac{y_2 - y_1}{x_2 - x_1}$$

$$= \frac{1 - (-3)}{-4 - (-2)}$$

$$= \frac{1 + 3}{-4 + 2}$$

$$= \frac{4}{-2}$$

$$= -2$$

22

Study Exercise Three

Given are two points on a line. Find the slope of the line.

1. (4, 1) and (6, 4) **2.** (1, 3) and (2, 4) **3.** (3, −1) *and* (−1, −4)

4. (2, 3) and (5, 3) **5.** (1, 2) and (1, 6)

(23)

The slope of a line is a measure of its steepness. With positive slopes, the larger the slope, the steeper the line.

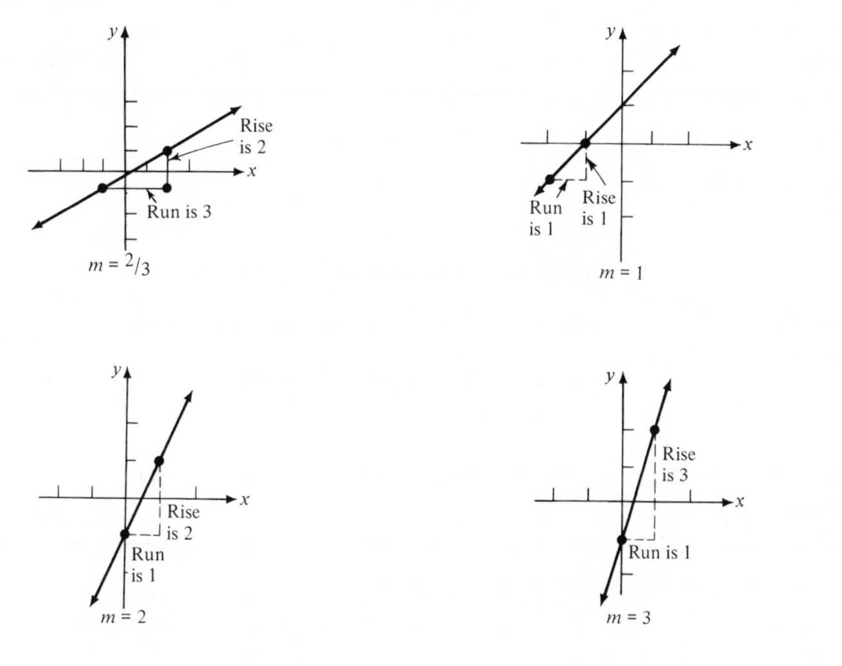

(24)

Below are some graphs of lines with negative slopes.

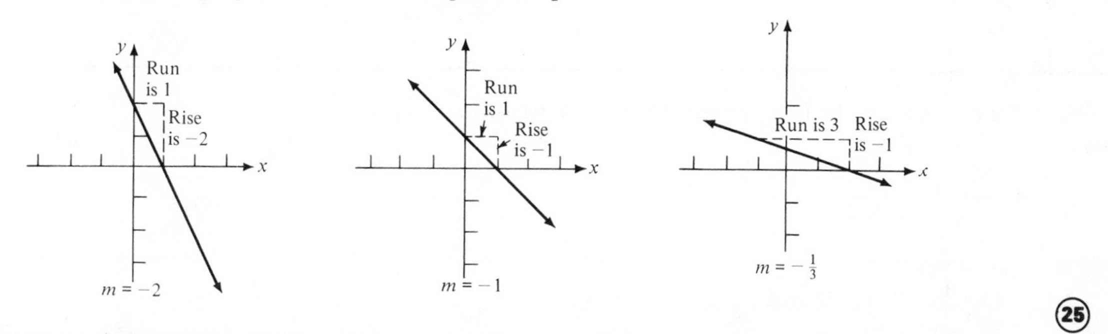

(25)

A horizontal line has slope zero.

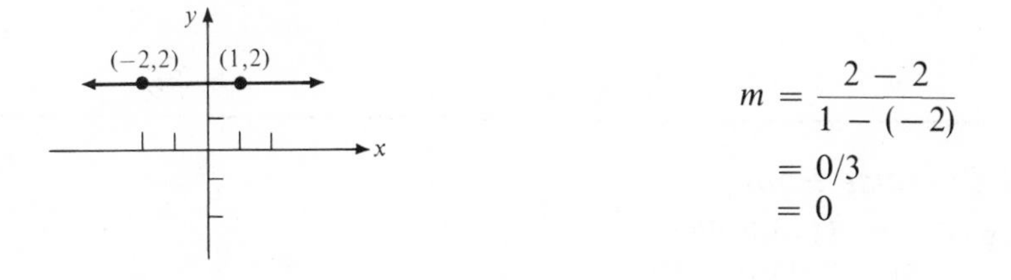

$$
\begin{aligned}
m &= \frac{2-2}{1-(-2)} \\
&= 0/3 \\
&= 0
\end{aligned}
$$

(Frame 26, contd.)

The slope of a vertical line is undefined.

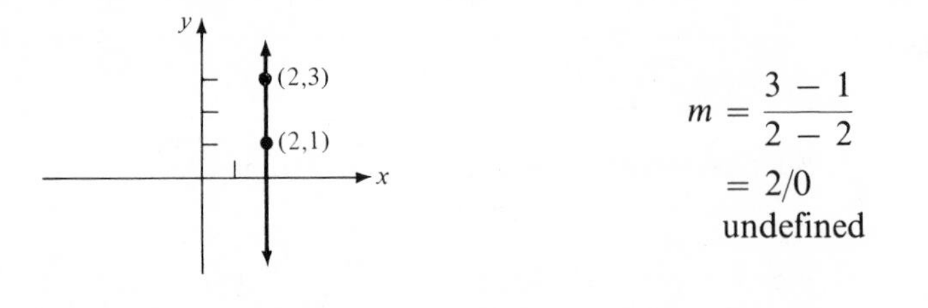

$$m = \frac{3-1}{2-2}$$
$$= 2/0$$
undefined

(26)

Point Slope Form of a Straight Line

Given a line l with slope m passing through (x_1, y_1)

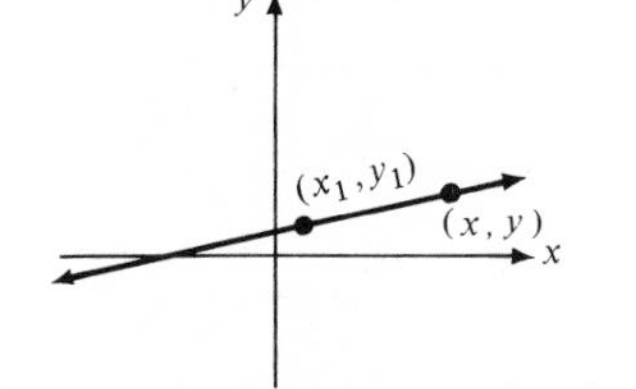

Let (x, y) be any other point on line l.

Then $m = \dfrac{y - y_1}{x - x_1}$. If we multiply both members by $(x - x_1)$ we obtain $m(x - x_1) = y - y_1$ or $y - y_1 = m(x - x_1)$.

Thus, the equation of a line with slope m passing through (x_1, y_1) is $y - y_1 = m(x - x_1)$. This form is known as the *point-slope form.*

(27)

Example: Write the equation of a line through (2, 4) with slope 3.

Solution:

Line (1) $m = 3, \quad (x_1, y_1) = (2, 4)$
Line (2) $y - y_1 = m(x - x_1)$
Line (3) $y - 4 = 3(x - 2)$
Line (4) $y - 4 = 3x - 6$
Line (5) $-3x + y + 2 = 0$

(28)

Example: Write the equation of the line passing through (4, 6) and 3, -1).

Solution:

Line (1) $m = \dfrac{\Delta y}{\Delta x}$
Line (2) $= \dfrac{6 - (-1)}{4 - 3}$
Line (3) $= 7$
Choose one of the points; say (4, 6).
Line (4) $y - y_1 = m(x - x_1)$
Line (5) $y - 6 = 7(x - 4)$
Line (6) $y - 6 = 7x - 28$
Line (7) $-7x + y + 22 = 0$

(29)

Study Exercise Four

1. Write the equation of the line through $(-3, -1)$ with slope 2.
2. Find the equation of the line through (5, 5) with slope $-2/3$.
3. Write the equation of the line through $(-4, 2)$ with slope zero.

Study Exercise Four (Frame 30, contd.)

4. Find the equation of the line through (2, 1) and (−1, 3).
5. Find the equation of the line parallel to the x axis and passing through (1, 3).

(30)

Slope-Intercept Form

Given a line with slope m and y intercept $(0, b)$

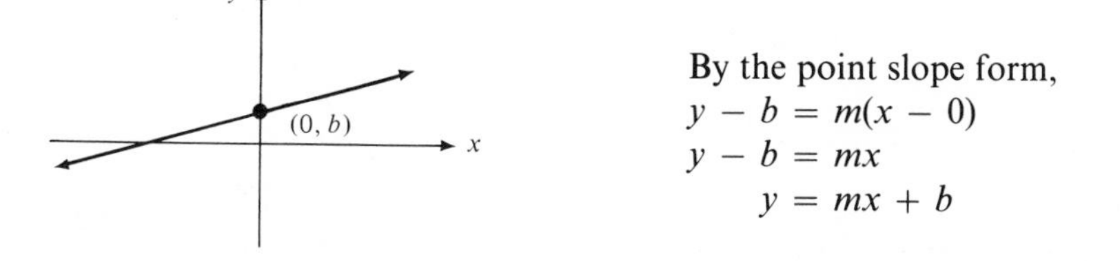

By the point slope form,

$$y - b = m(x - 0)$$
$$y - b = mx$$
$$y = mx + b$$

The form $y = mx + b$ is called the *slope-intercept form*.

Notice the final form still indicates the slope and y intercept of the line.

(31)

In $y = mx + b$, m is the slope and b represents the y intercept $(0, b)$.

Example 1: $y = 3x + 2$ — slope is 3; y intercept is (0, 2)

Example 2: $y = -2x + 1/2$ — slope is −2; y intercept is (0, 1/2)

Example 3: $y = x$ — slope is 1; y intercept is (0, 0)

(32)

Any first degree equation in x and y, when solved for y will be in the slope intercept form and then the slope and y intercept may be read directly.

Example:

$$3x + 2y - 5 = 0$$
$$3x + 2y = 5$$
$$2y = -3x + 5$$
$$y = -\tfrac{3}{2}x + 5/2$$

the slope is −3/2
the y intercept is (0, 5/2)

(33)

It is possible to draw the graph of a line if the slope and y intercept are known. Consider $y = \frac{3}{2}x - 1$.
The slope is 3/2 and y intercept is (0, −1).

1. First plot the y intercept (0, −1).
2. The slope indicates a run of 2 and a rise of 3.
3. From the y intercept, take a run of 2 and rise of 3 and plot the point.
4. Draw the line connecting the two points.

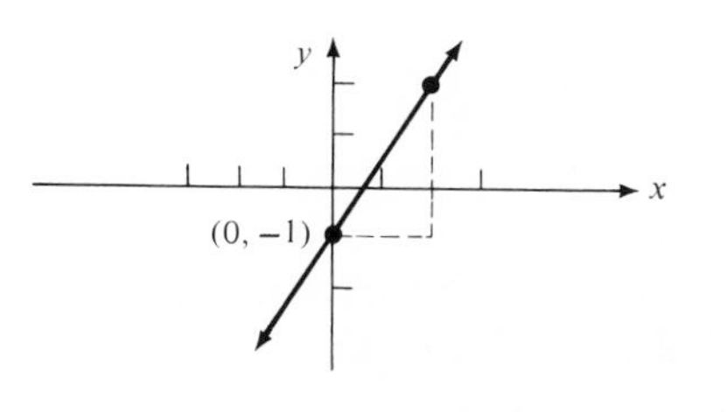

(34)

Example: Draw the graph of $y = -\frac{1}{2}x + 1$

Solution: First plot the y intercept (0, 1) then, from (0, 1) take a run of 2 and a rise of -1 (drop 1 unit down).

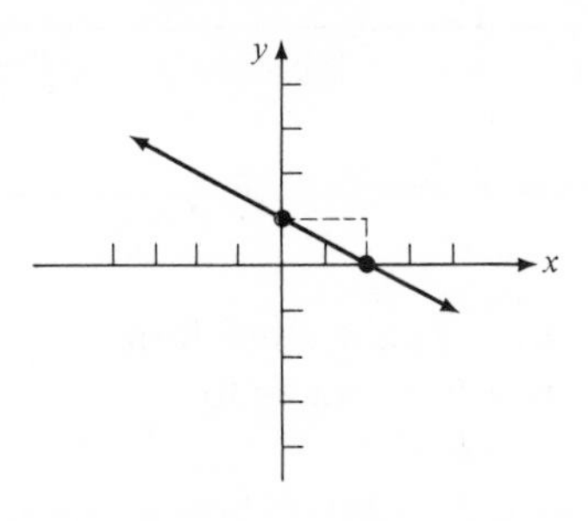

(35)

Study Exercise Five

Find the slope, y intercept, and draw the graph.

1. $x + y = 3$ **2.** $2x - 3y = 0$ **3.** $y = -2x - 1$

4. $y = -x$ **5.** $y = 3$

(36)

Parallel Lines

Theorem. Two non vertical lines are parallel if and only if their slopes are equal.

There are two parts to the theorem. When the slopes of two non vertical lines are equal, it means the lines are parallel and when the lines are parallel, their slopes are equal.

As a special case, two vertical lines are also parallel, but their slopes are undefined.

(37)

Perpendicular Lines

Theorem. Two lines are perpendicular if and only if the product of their slopes is -1

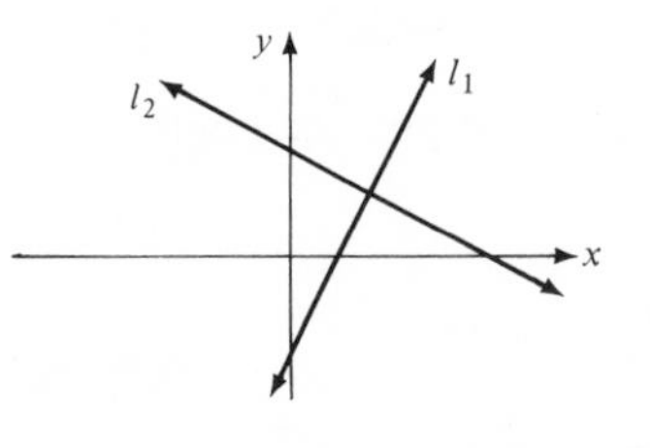

There are two parts to this theorem.

When two lines are perpendicular, it means the product of their slopes is -1 and when the product of the slopes of two lines is -1, it means the lines are perpendicular.

As a special case, a vertical and a horizontal line are also perpendicular even though one slope is undefined.

(38)

Example: A line l has slope 3/4. What is the slope of a line
(a) parallel to l
(b) perpendicular to l

Solution: **(a)** A line parallel to l has slope 3/4.
(b) A line perpendicular to l has slope equal to the negative reciprocal of the slope of l. Hence, the line perpendicular to l has slope $-4/3$.

(39)

Study Exercise Six

1. What is the slope of a line perpendicular to a line with slope 2?
2. Find the slope of a line parallel to the line containing (3, 2) and (5, 8).
3. Find the slope of a line perpendicular to the line containing (−2, 3) and (1, 7).
4. Show that the line through (3, −2) and (8, 3) is perpendicular to the line through (−2, 8) and (3, 3).

(40)

Pythagorean Theorem

For any right triangle, the square of the length of the *hypotenuse* equals the sum of the squares of the lengths of the other two sides.

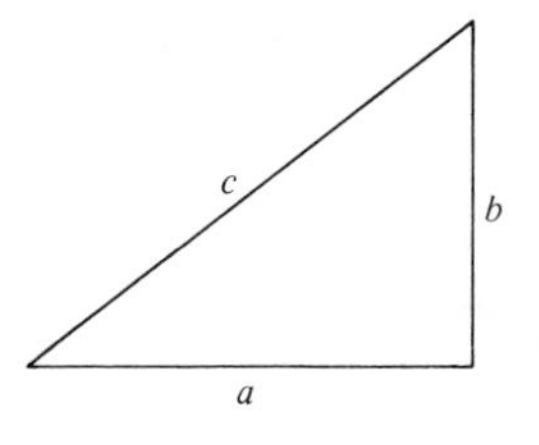

a, b, c are lengths of the sides and c is the length of the hypotenuse.

$$c^2 = a^2 + b^2$$

(41)

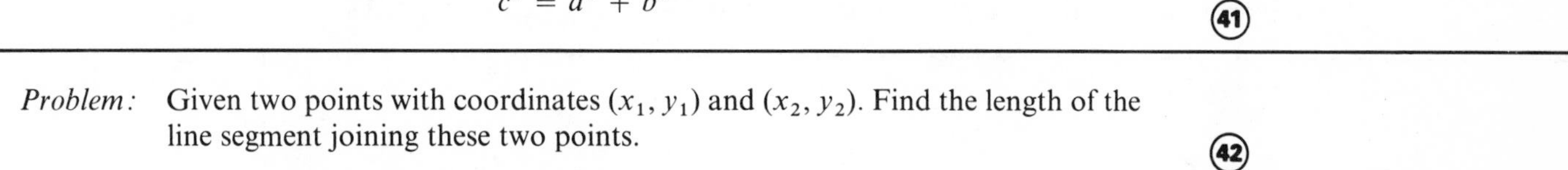

Problem: Given two points with coordinates (x_1, y_1) and (x_2, y_2). Find the length of the line segment joining these two points.

(42)

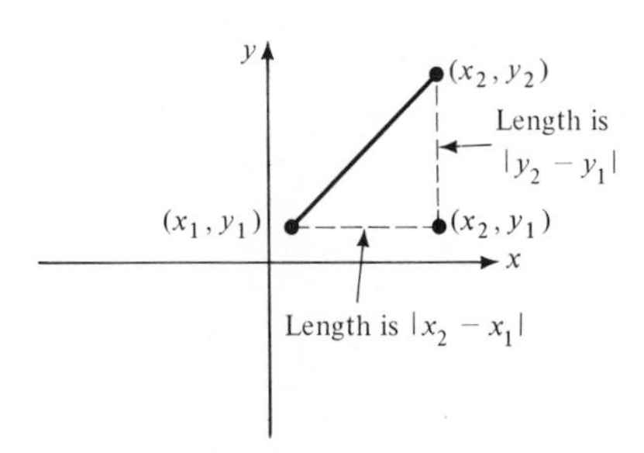

Let d be the length between (x_1, y_1) *and* (x_2, y_2).

Line (1) $d^2 = |x_2 - x_1|^2 + |y_2 - y_1|^2$

Line (2) $d^2 = (x_2 - x_1)^2 + (y_2 - y_1)^2$

Line (3) $d = \sqrt{(x_2 - x_1)^2 + (y_2 - y_1)^2}$

(43)

The Distance Formula

If d represents the distance between two points in the plane, (x_1, y_1) and (x_2, y_2), then;

$$d = \sqrt{(x_2 - x_1)^2 + (y_2 - y_1)^2}$$

(44)

Example 1: Find the distance between (4, 6) and (2, 1)

Solution: $d = \sqrt{(x_2 - x_1)^2 + (y_2 - y_1)^2}$

Line (1) $d = \sqrt{(4 - 2)^2 + (6 - 1)^2}$

Line (2) $d = \sqrt{2^2 + 5^2}$

Line (3) $d = \sqrt{4 + 25}$

Line (4) $d = \sqrt{29}$

Example 2: Find the distance between $(-3, 1)$ and $(-2, -7)$

Solution:

Line (1) $d = \sqrt{[-3 - (-2)]^2 + [1 - (-7)]^2}$

Line (2) $d = \sqrt{(-1)^2 + (8)^2}$

Line (3) $d = \sqrt{65}$

(45)

Study Exercise Seven

1. Find the length of the line segment joining $(-2, 3)$ and $(2, -6)$.
2. Find the length of the line segment joining $(6, -3)$ and $(2, -9)$.
3. Find the perimeter of the triangle formed by joining $(-2, -1)$, $(1, 6)$, and $(8, 1)$.

(46)

REVIEW EXERCISES

A. Fill in the blanks with the correct response:

1. The x intercept of $x + y = 6$ is ______________.

2. The slope of $y = 6x - 2$ is ______________.

3. The slope of a horizontal line is ______________.

4. The slope of a vertical line is ______________.

5. A given line has slope 2/3. If a line is to be perpendicular to this line, it must have slope ______________.

B. Miscellaneous:

6. Find the slope of the line containing the points $(6, -5)$ and $(-2, -3)$.

7. Given $2x + 3y = 1$

(a) Find the slope of the line. **(b)** Find the x and y intercepts.

(c) Draw the graph.

8. Write the equation of the line with slope -3 and y intercept of $(0, 5)$.

9. Write the equation of the line passing through $(-2, -2)$ and $(3, -1)$.

10. Find the length of the line segment joining $(-2, -2)$ and $(3, -1)$.

10. Write the equation of the line through $(0, 4)$ and perpendicular to the line in problem 9.

47

SOLUTIONS TO REVIEW EXERCISES

A **1.** (6, 0) **2.** 6 **3.** zero **4.** undefined **5.** $-3/2$

B. **6.**
$$m = \frac{\Delta y}{\Delta x} = \frac{(-5) - (-3)}{(6) - (2)} = \frac{-2}{8} = -\frac{1}{4}$$

7. (a) $3y = -2x + 1$

$y = -2/3x + 1/3$

the slope is $-2/3$

(b) the y intercept is $\left(0, \frac{1}{3}\right)$

the x intercept is $\left(\frac{1}{2}, 0\right)$

(c)

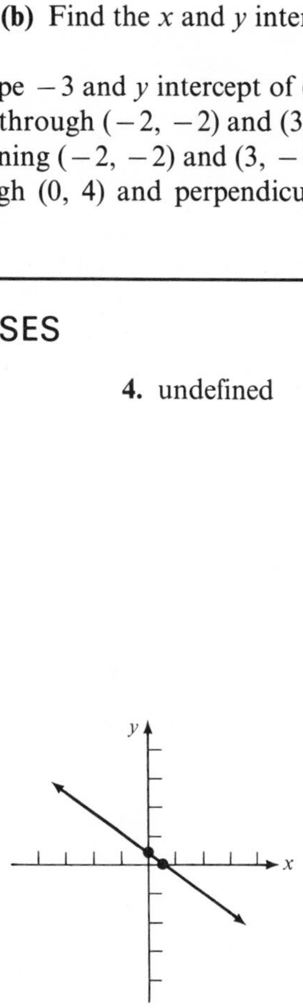

y, *x*

8. $y = -3x + 5$

9.
$$m = \frac{(-1) - (-2)}{(3) - (-2)} = 1/5$$

$y - (-1) = 1/5(x - 3)$

$y + 1 = 1/5(x - 3)$

$5y + 5 = x - 3$

$-x + 5y + 8 = 0$ or $x - 5y - 8 = 0$

SOLUTIONS TO REVIEW EXERCISES, CONTD.

(Frame 48, contd.)

10. $d = \sqrt{[3 - (-2)]^2 + [-1 - (-2)]^2}$

$d = \sqrt{5^2 + 1^2}$

$d = \sqrt{26}$

11. The line in problem 9 has slope 1/5.
A line perpendicular to this line will have slope -5.
$m = -5; (x_1, y_1) = (0, 4)$
$y - 4 = -5(x - 0)$
$y - 4 = -5x$ or $5x + y - 4 = 0$.

(48)

SOLUTIONS TO STUDY EXERCISES

Study Exercise One (Frame 10)

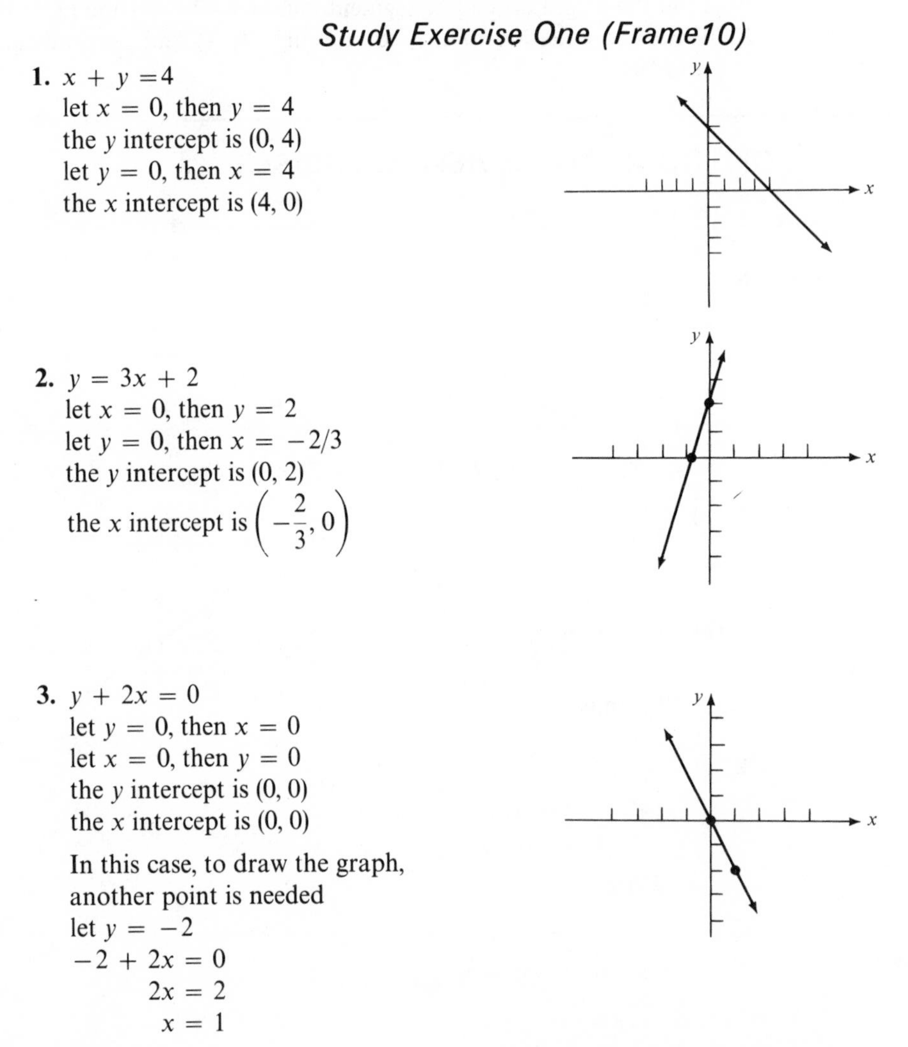

1. $x + y = 4$
let $x = 0$, then $y = 4$
the y intercept is $(0, 4)$
let $y = 0$, then $x = 4$
the x intercept is $(4, 0)$

2. $y = 3x + 2$
let $x = 0$, then $y = 2$
let $y = 0$, then $x = -2/3$
the y intercept is $(0, 2)$
the x intercept is $\left(-\frac{2}{3}, 0\right)$

3. $y + 2x = 0$
let $y = 0$, then $x = 0$
let $x = 0$, then $y = 0$
the y intercept is $(0, 0)$
the x intercept is $(0, 0)$

In this case, to draw the graph, another point is needed
let $y = -2$
$-2 + 2x = 0$
$2x = 2$
$x = 1$

(10A)

Study Exercise Two (Frame 14)

1.

2.

3.

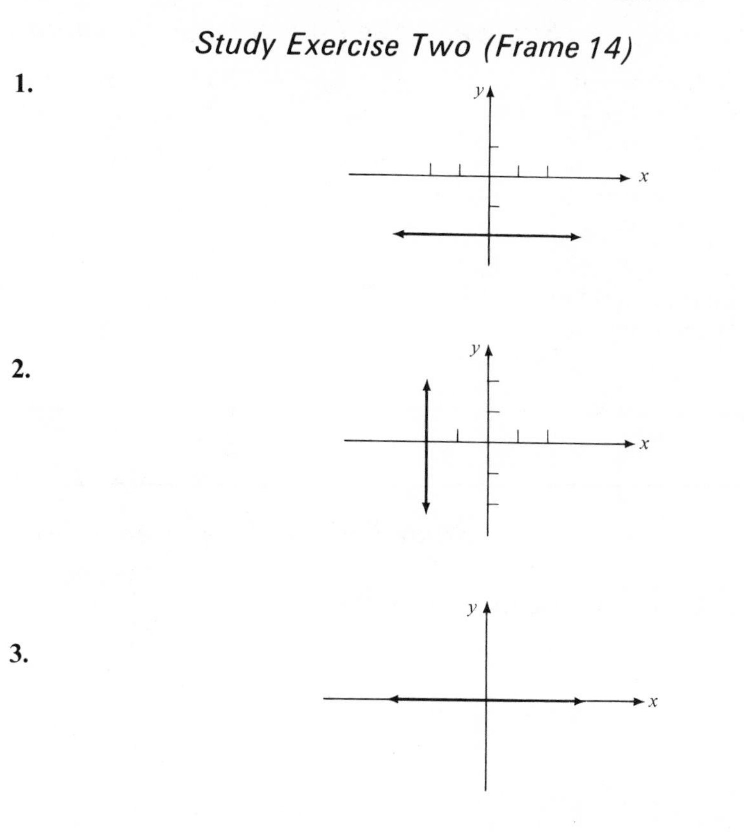

$y = 0$ is the x axis.

14A

Study Exercise Three (Frame 23)

1. $m = \dfrac{4 - 1}{6 - 4}$
$= 3/2$

2. $m = \dfrac{4 - 3}{2 - 1}$
$= 1$

3. $m = \dfrac{(-4) - (-1)}{(-1) - 3}$
$= -3/-4$
$= 3/4$

4. $m = \dfrac{3 - 3}{5 - 2}$
$= 0/3$
$= 0$

5. $m = \dfrac{6 - 2}{1 - 1}$
$= 4/0$
$=$ undefined

23A

Study Exercise Four (Frame 30)

1. $(-3, -1), m = 2$
$y - (-1) = 2(x - [-3])$
$y + 1 = 2x + 6$
$-2x + y - 5 = 0$
or $2x - y + 5 = 0$

2. $(5, 5), m = -2/3$
$y - 5 = \dfrac{-2}{3}(x - 5)$
$3y - 15 = -2x + 10$
$2x + 3y - 25 = 0$

SOLUTIONS TO STUDY EXERCISES, CONTD.

Study Exercise Four (Frame 30, contd.)

3. $(-4, 2), m = 0$

$y - 2 = 0(x - [-4])$

$y - 2 = 0$

5. $(1, 3), m = 0$

$y - 3 = 0(x - 1)$

$y - 3 = 0$

a line parallel to the x axis has slope zero

4. $(2, 1), (-1, 3)$

$$m = \frac{3 - 1}{-1 - 2}$$

$$= \frac{2}{-3}$$

$$(2, 1), m = -\frac{2}{3}$$

$$y - 1 = -\frac{2}{3}(x - 2)$$

$3y - 3 = -2x + 4$

$2x + 3y - 7 = 0$

30A

Study Exercise Five (Frame 36)

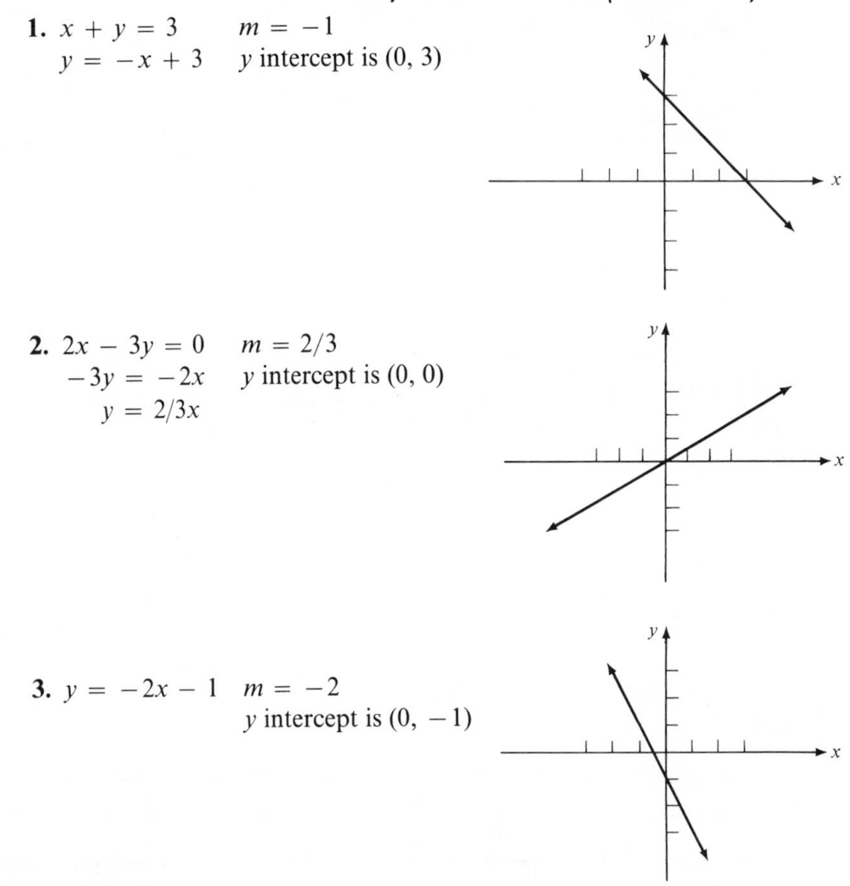

1. $x + y = 3$ $\quad m = -1$

$y = -x + 3$ $\quad y$ intercept is $(0, 3)$

2. $2x - 3y = 0$ $\quad m = 2/3$

$-3y = -2x$ $\quad y$ intercept is $(0, 0)$

$y = 2/3x$

3. $y = -2x - 1$ $\quad m = -2$

y intercept is $(0, -1)$

Study Exercise Five (Frame 36) contd.)

4. $y = -x$ $\quad m = -1$
   y intercept is $(0, 0)$

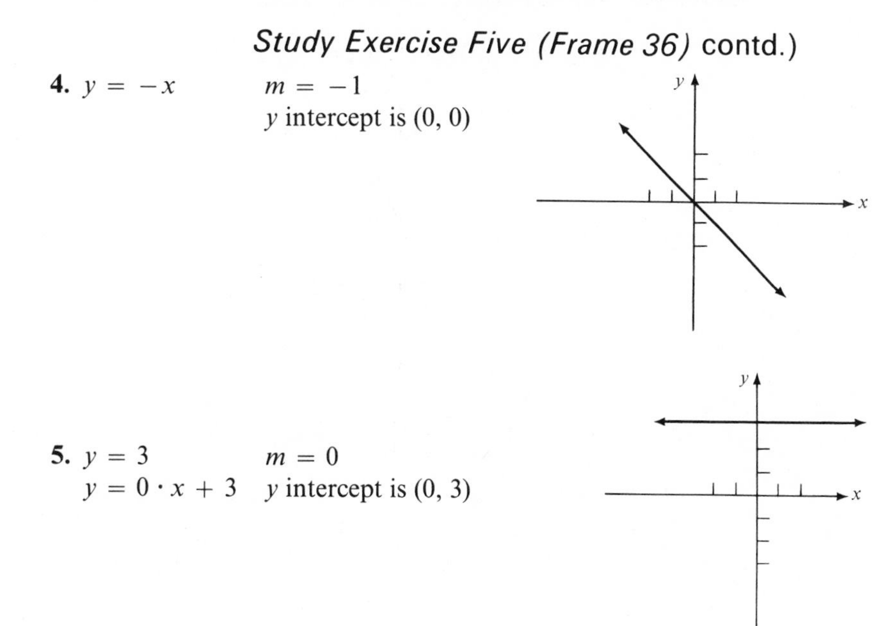

5. $y = 3$ $\quad m = 0$
   $y = 0 \cdot x + 3$ $\quad y$ intercept is $(0, 3)$

36A

Study Exercise Six (Frame 40)

1. $-1/2$
2. $m = \dfrac{8 - 2}{5 - 3}$
   $= 6/2$ or 3
   The slope of the parallel line will also be 3.
3. $m = \dfrac{7 - 3}{1 - (-2)}$
   $= \dfrac{4}{3}$
   The slope of the perpendicular line is $-3/4$.
4. $(3, -2), (8, 3)$ $\qquad (-2, 8), (3, 3)$
   $m = \dfrac{3 - (-2)}{8 - 3}$ $\qquad m = \dfrac{8 - 3}{-2 - 3}$
   $= 5/5$ or 1 $\qquad = 5/-5$ or -1
   Since the product of the slopes is -1, the lines are perpendicular.

40A

Study Exercise Seven (Frame 46)

1. $(-2, 3), (2, -6)$

$$d = \sqrt{[2 - (-2)]^2 + [-6 - (3)]^2}$$
$$= \sqrt{4^2 + [-9]^2}$$
$$= \sqrt{16 + 81}$$
$$= \sqrt{97}$$

2. $(6, -3), (2, -9)$

$$d = \sqrt{(6 - 2)^2 + [-3 - (-9)]^2}$$
$$= \sqrt{4^2 + 6^2}$$
$$= \sqrt{52}$$
$$= 2\sqrt{13}$$

SOLUTIONS TO STUDY EXERCISES, CONTD.

Study Exercise Seven (Frame 46) contd.)

3.

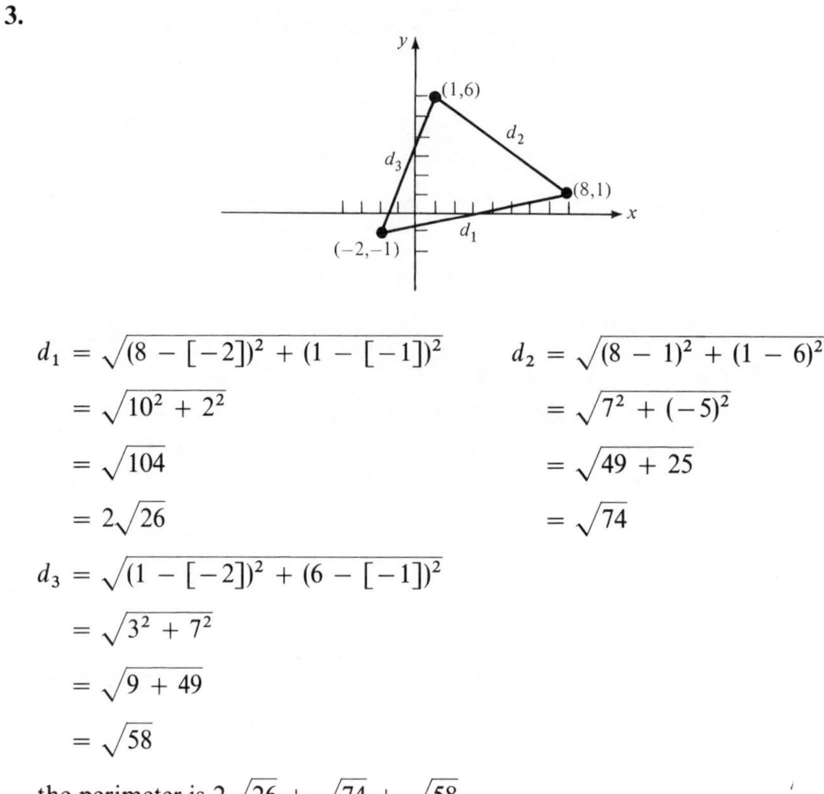

$$d_1 = \sqrt{(8 - [-2])^2 + (1 - [-1])^2}$$
$$= \sqrt{10^2 + 2^2}$$
$$= \sqrt{104}$$
$$= 2\sqrt{26}$$

$$d_2 = \sqrt{(8 - 1)^2 + (1 - 6)^2}$$
$$= \sqrt{7^2 + (-5)^2}$$
$$= \sqrt{49 + 25}$$
$$= \sqrt{74}$$

$$d_3 = \sqrt{(1 - [-2])^2 + (6 - [-1])^2}$$
$$= \sqrt{3^2 + 7^2}$$
$$= \sqrt{9 + 49}$$
$$= \sqrt{58}$$

the perimeter is $2\sqrt{26} + \sqrt{74} + \sqrt{58}$

46A

UNIT 12—SUPPLEMENTARY PROBLEMS

A. True or false:

1. The line $x = 5$ is a horizontal line.
2. The lines $y = 3x - 2$ and $2x - 3y = 6$ have the same x intercept.
3. The line through (3, 2) with slope zero is the line $x = 2$.
4. The line $y = x$ is perpendicular to the line $y = -x$.
5. The slope of a horizontal line is zero.
6. The distance between two points with the same ordinate is given by the greater abscissa minus the smaller abscissa.

B. Find the slope of the line joining the two given points.

7. (0, 0) (−3, 1) **8.** (4, 7) (−4, 7) **9.** (−2, −2) (3, −7)
10. (1/2, 1/3) (3, 2)

C. Write the equation of a line satisfying the given conditions. Write your answer in the form $Ax + By + C = 0$.

11. $m = 1/2$, passes through (4, 3).
12. passes through (4, −2) and (0, −3).
13. has x intercept (5, 0) and y intercept (0, 3).
14. passes through (2, −3) and is parallel to $x - 3y = 6$.
15. passes through (5, 0) and is perpendicular to the line $3x + 2y = 8$.

D. Find the slope and y intercept of the given line.

16. $y = 4x$ **17.** $2x - 3y = 12$ **18.** $x = 3$

19. $y = 5$ **20.** $2y = x + 4$

E. Find the length of the line segment joining the given points.

21. $(3, -1), (3, -5)$ **22.** $(2, 1), (6, 5)$ **23.** $(5, 6), (-6, 5)$

F. Miscellaneous:

24. Does $(0, 3)$ lie on the line determined by $(-1, 5)$ and $(4, -5)$?

25. Find k for which $(-5, k)$ is on the line $2x + 3y = 8$.

26. If $y = 3x - 10$, find Δy when $\Delta x = -3/2$.

27. A linear function is such that $f(-1) = 4$ and $f(1) = -2$. Find the rule of correspondence.

28. If $(8, 3)$ is on a circle whose center is $(5, -1)$, find the radius of the circle.

29. $ABCD$ is a rectangle having coordinates $(-5, -2)$, $(-7, 4)$, and $(2, 7)$ respectively. Find the coordinates of D.

30. If $f(x) = 5x + 2$, evaluate $\dfrac{f(a) - f(b)}{a - b}$ if a and b are numbers in the domain.

unit

13

The Quadratic Function

Objectives

1. Know the rule for a quadratic function.
2. Be able to graph a quadratic function.
3. Be able to find any zeros of a quadratic function.
4. Be able to find maximum or minimum value and coordinates of the vertex.

Terms

quadratic function
vertex of a parabola
fill and spill parabola
maximum and minimum

(1)

A function whose rule of correspondence is given by $y = ax^2 + bx + c, a, b, c \in R, a \neq 0$ is called a *quadratic function.*

The quadratic function may also be written:
$f(x) = ax^2 + bx + c, a, b, c \in R, a \neq 0.$

(2)

Recall that the zeros of a function are those values in the domain (x values) for which the range values (y values) are zero.
Thus, to find the zeros of the quadratic function, set y equal to zero and solve for x. This is the same as finding the solution set of the quadratic equation $ax^2 + bx + c = 0, a \neq 0$.

(3)

Example: Find the zeros of $f(x) = x^2 - 5x + 6$

Solution:

Line (1) set $f(x) = 0$
Line (2) $0 = x^2 - 5x + 6$
Line (3) $0 = (x - 3)(x - 2)$
Line (4) $x - 3 = 0$ or $x - 2 = 0$
Line (5) $x = 3$ or $x = 2$
The zeros are 2 and 3.

Therefore, (2, 0) and (3, 0) belong to the function $f(x)$.

(4)

Study Exercise One

Find the zeros:

1. $f(x) = x^2 - 2x - 3$ **2.** $y = 3x^2 - x$ **3.** $y = x^2 + x + 2$

(5)

Consider $y = x^2 - 3$. Ordered pairs that satisfy $y = x^2 - 3$ may be found by arbitrarily assigning values to x and computing related values for y.

For example, when $x = 2$, $y = 2^2 - 3$ or 1. When $x = -1$, $y = (-1)^2 - 3$ or -2.

(6)

---

x	y
-2	1
-1	-2
0	-3
1	-2
2	1
-3	6
3	6
$\frac{1}{2}$	$-2\frac{3}{4}$
$-\frac{1}{2}$	$-2\frac{3}{4}$
$\pm\sqrt{3}$	0

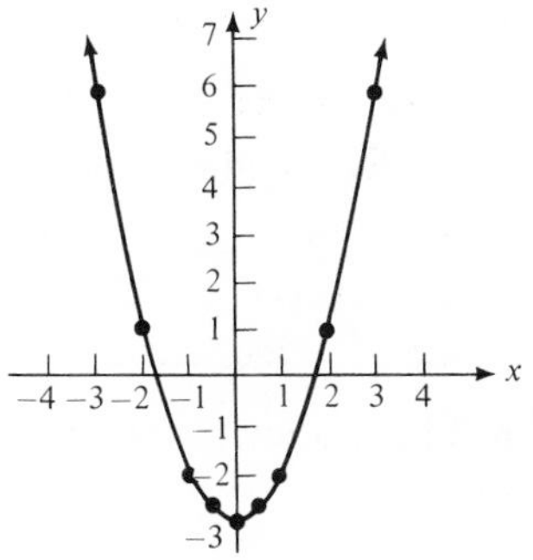

The curve is an example of a *parabola*.

(7)

---

x	y
-2	-3
-1	3
0	5
1	3
2	-3
$\frac{1}{2}$	$4\frac{1}{2}$
$-\frac{1}{2}$	$4\frac{1}{2}$
$\pm\sqrt{2.5}$	0

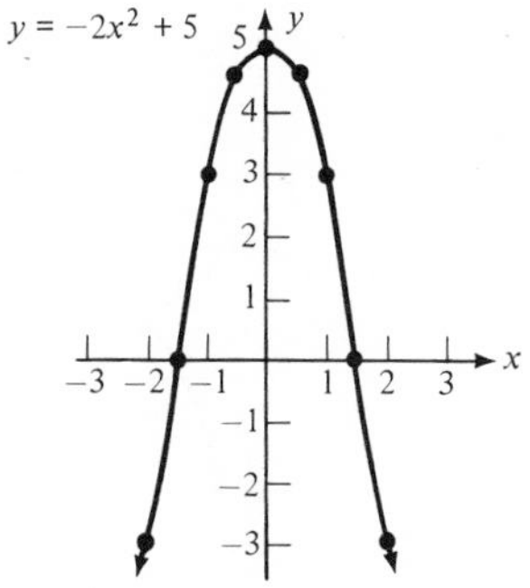

$y = -2x^2 + 5$

(8)

Study Exercise Two

Graph the following functions over $R \times R$ defined by the given equation.

1. $y = x^2 + 2$ **2.** $y = -x^2 + 4$ **3.** $y = x^2 + 2x + 1$ **4.** $y = -3x^2 - x$

(9)

We assume without proof:

1. The graph of the function given by $y = ax^2 + bx + c$, $a, b, c \in R$, $a \neq 0$ is a parabola.
2. The parabola opens up  if $a > 0$.
3. The parabola opens down  if $a < 0$.

(10)

Parabolas that open up will be called *fill* parabolas.
Parabolas that open down will be called *spill* parabolas.

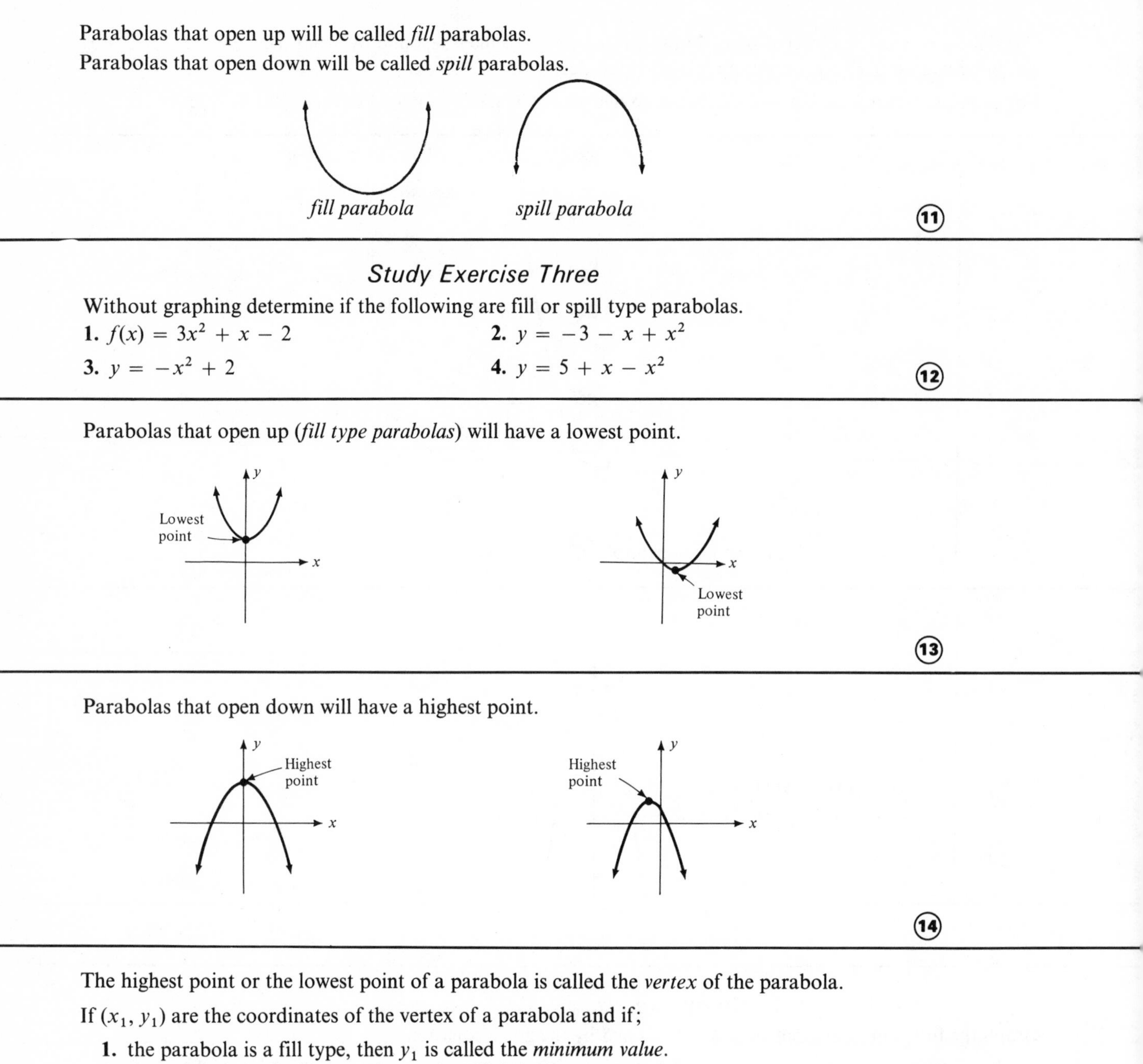

fill parabola *spill parabola*

(11)

Study Exercise Three

Without graphing determine if the following are fill or spill type parabolas.

1. $f(x) = 3x^2 + x - 2$ **2.** $y = -3 - x + x^2$

3. $y = -x^2 + 2$ **4.** $y = 5 + x - x^2$

(12)

Parabolas that open up (*fill type parabolas*) will have a lowest point.

(13)

Parabolas that open down will have a highest point.

(14)

The highest point or the lowest point of a parabola is called the *vertex* of the parabola.

If (x_1, y_1) are the coordinates of the vertex of a parabola and if;

1. the parabola is a fill type, then y_1 is called the *minimum value*.
2. the parabola is a spill type, then y_1 is called the *maximum value*.

(15)

Study Exercise Four

From the graph determine whether or not the parabola has a maximum or minimum value and calculate maximum or minimum value. Also state the coordinates of the vertex.

1. **2.** **3.**

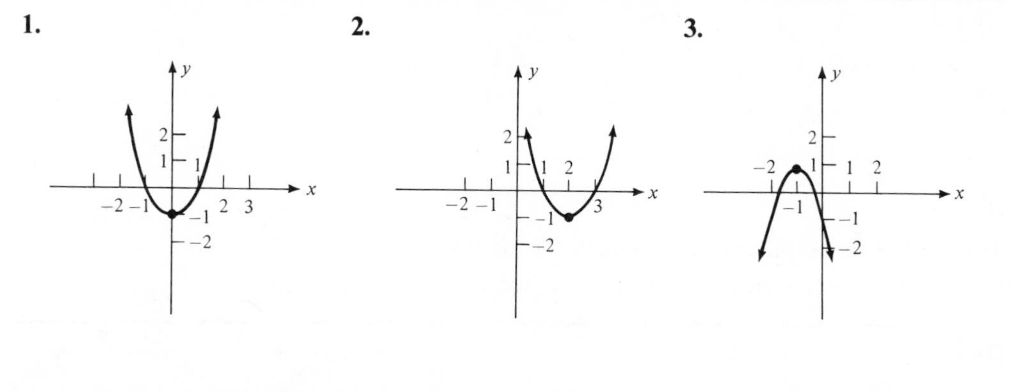

(16)

Intercepts of the Quadratic Function

A. To find the y intercept(s) of the graph set $x = 0$ in $y = ax^2 + bx + c$ and solve for y. Thus $y = c$. Therefore, in $y = ax^2 + bx + c$, the c represents the y intercept $(0, c)$.

B. To find the x intercept(s) of the graph set $y = 0$ in $y = ax^2 + bx + c$ and solve for x. Thus $0 = ax^2 + bx + c$. This is the same as the zeros of the function or the solution of the quadratic equation.

(17)

Example: Find the intercepts of $y = 3x^2 + 5x - 2$

Solution: In $y = 3x^2 + 5x - 2$, -2 represents the y intercept $(0, -2)$. To get the x intercept, set $y = 0$.

$0 = 3x^2 + 5x - 2$
$0 = (3x - 1)(x + 2)$
$3x - 1 = 0$ or $x + 2 = 0$
$x = 1/3$ or $x = -2$

The x intercepts are $(1/3, 0)$ and $(-2, 0)$.

(18)

Study Exercise Five

Find the x and y intercepts. Then, plot a few more points and draw the graph.

1. $y = x^2 - 6x + 9$ **2.** $y = -x^2 + 4$ **3.** $y = x^2 - 3x$

(19)

Since the zeros of $f(x) = ax^2 + bx + c$ and the roots of $ax^2 + bx + c = 0$ are the same, the intercepts of the quadratic function represent the roots of the quadratic equation.

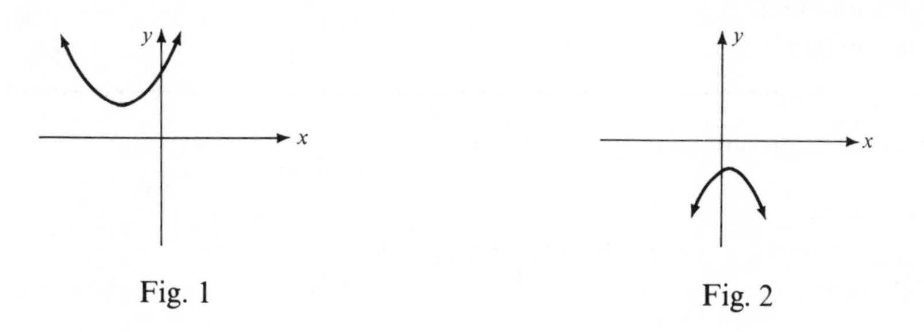

Fig. 1 Fig. 2

(Frame 20, contd.)

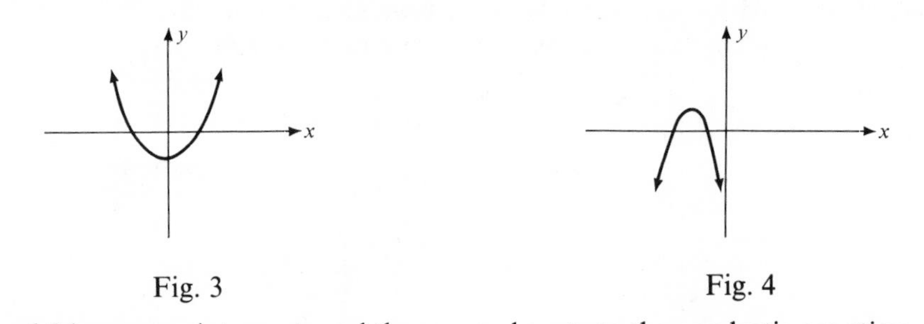

Fig. 3 Fig. 4

Fig. 1 and 2 have no x intercepts and thus no real roots to the quadratic equation.
Fig. 3 and 4 each have two x intercepts and thus two roots to the quadratic equation.

(20)

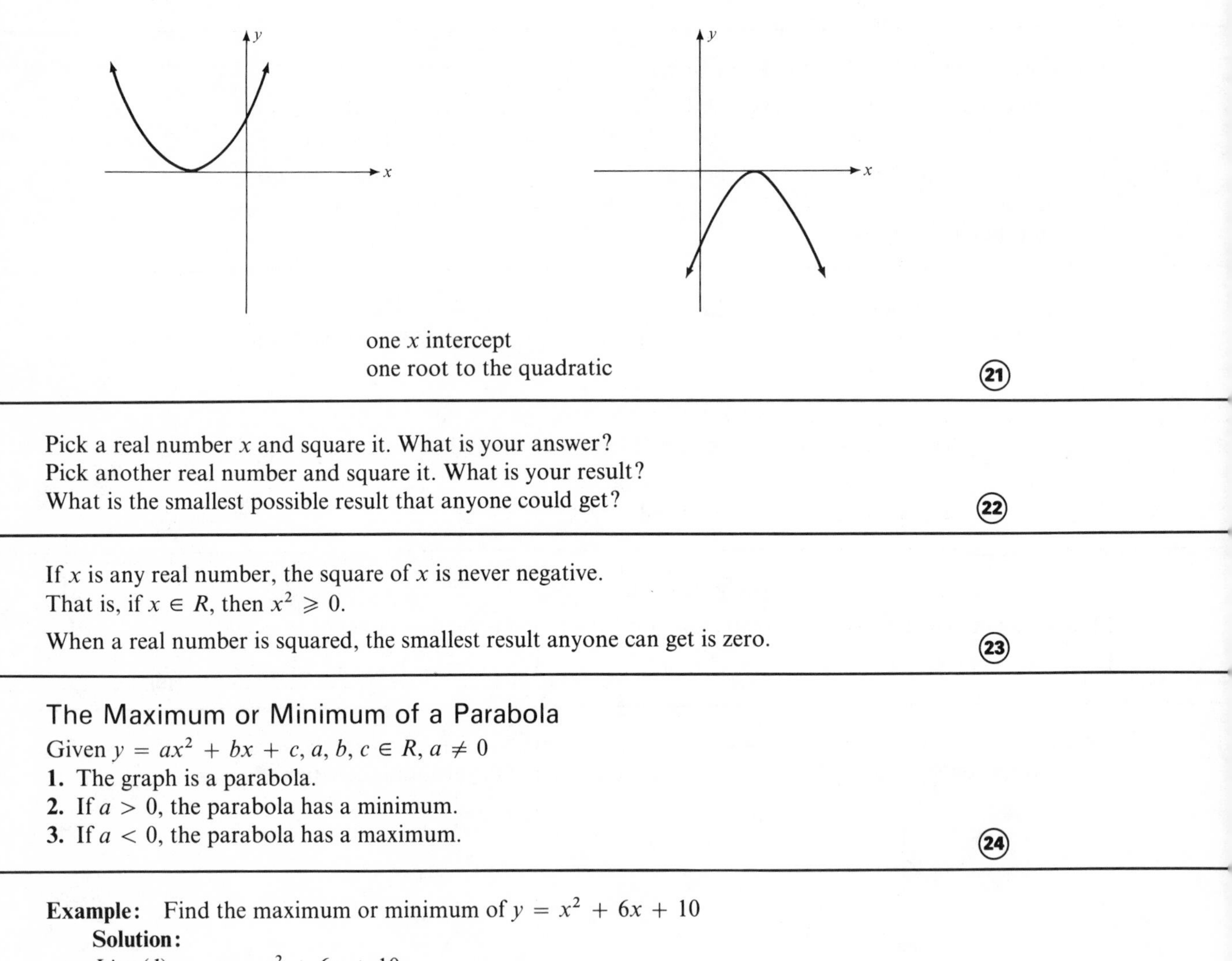

one x intercept
one root to the quadratic

(21)

Pick a real number x and square it. What is your answer?
Pick another real number and square it. What is your result?
What is the smallest possible result that anyone could get?

(22)

If x is any real number, the square of x is never negative.
That is, if $x \in R$, then $x^2 \geqslant 0$.

When a real number is squared, the smallest result anyone can get is zero.

(23)

The Maximum or Minimum of a Parabola

Given $y = ax^2 + bx + c$, $a, b, c \in R$, $a \neq 0$

1. The graph is a parabola.
2. If $a > 0$, the parabola has a minimum.
3. If $a < 0$, the parabola has a maximum.

(24)

Example: Find the maximum or minimum of $y = x^2 + 6x + 10$

Solution:

Line (1) $y = x^2 + 6x + 10$
Line (2) $y = x^2 + 6x + 9 + 10 + (-9)$ complete the square
Line (3) $y = (x + 3)^2 + 1$

(Frame 25, contd.)

The square of any number is greater than or equal to zero.
The lowest value $(x + 3)^2$ can take is zero thus the lowest value y can take is 1.
The minimum is 1; it occurs when $(x + 3)^2$ becomes zero or when $x = -3$.

(25)

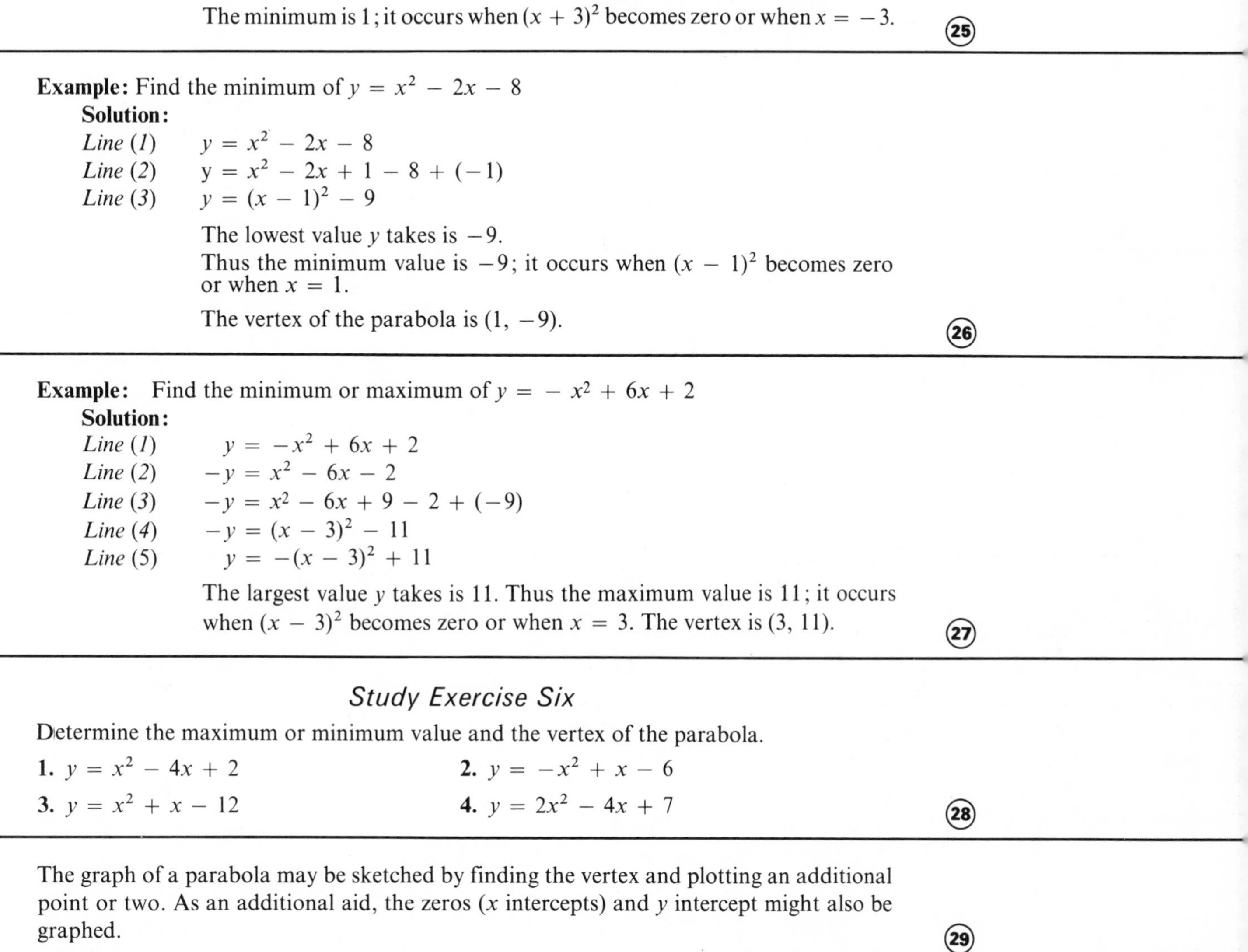

Example: Find the minimum of $y = x^2 - 2x - 8$

Solution:

Line (1) $y = x^2 - 2x - 8$
Line (2) $y = x^2 - 2x + 1 - 8 + (-1)$
Line (3) $y = (x - 1)^2 - 9$

The lowest value y takes is -9.
Thus the minimum value is -9; it occurs when $(x - 1)^2$ becomes zero or when $x = 1$.

The vertex of the parabola is $(1, -9)$.

(26)

Example: Find the minimum or maximum of $y = -x^2 + 6x + 2$

Solution:

Line (1) $y = -x^2 + 6x + 2$
Line (2) $-y = x^2 - 6x - 2$
Line (3) $-y = x^2 - 6x + 9 - 2 + (-9)$
Line (4) $-y = (x - 3)^2 - 11$
Line (5) $y = -(x - 3)^2 + 11$

The largest value y takes is 11. Thus the maximum value is 11; it occurs when $(x - 3)^2$ becomes zero or when $x = 3$. The vertex is $(3, 11)$.

(27)

Study Exercise Six

Determine the maximum or minimum value and the vertex of the parabola.

1. $y = x^2 - 4x + 2$ **2.** $y = -x^2 + x - 6$
3. $y = x^2 + x - 12$ **4.** $y = 2x^2 - 4x + 7$

(28)

The graph of a parabola may be sketched by finding the vertex and plotting an additional point or two. As an additional aid, the zeros (x intercepts) and y intercept might also be graphed.

(29)

The domain and range of the quadratic function can be given by knowing the vertex point.

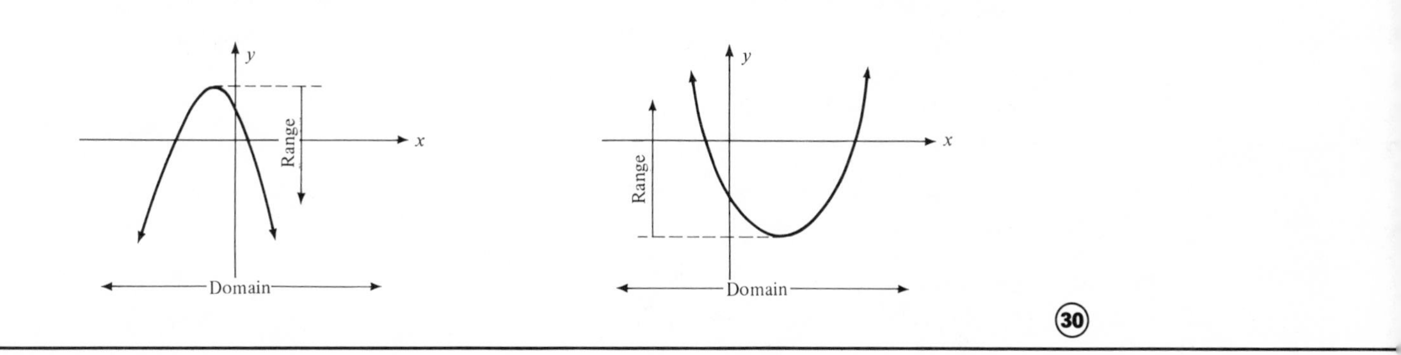

(30)

Example 5: Find the domain and range of $f(x) = x^2 + x - 6$ and sketch.

Solution: We will find intercepts first:

$0 = x^2 + x - 6$

$0 = (x + 3)(x - 2)$

The x intercepts are $(-3, 0)$ and $(2, 0)$.

The y intercept is $(0, -6)$.

Next we will find the vertex.

$y = x^2 + x + 1/4 - 6 + (-1/4)$

$y = (x + 1/2)^2 - 6\frac{1}{4}$

The vertex is $(-1/2, -6^1/_4)$

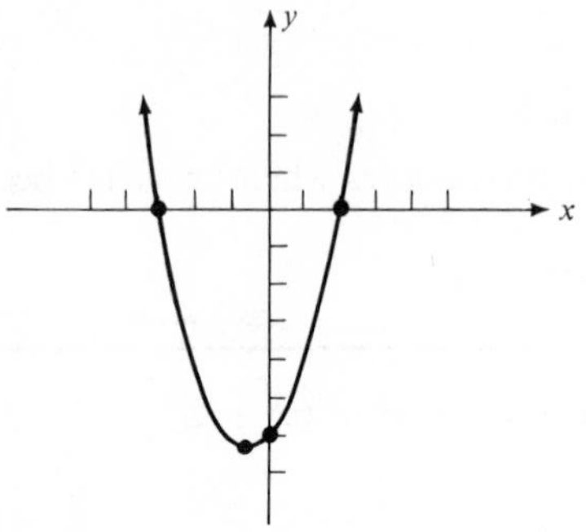

The domain is $\{x \mid x \in R\}$.

The range is $\{y \mid y \geqslant -6^1/_4\}$.

(31)

Study Exercise Seven

Find the domain and range and sketch the graph:

1. $f(x) = -x^2 + 8x - 12$

2. $y = x^2 + 3x$

(32)

REVIEW EXERCISES

A. True or False

1. The graph of $y = 3x^2 - 1$ opens up.
2. $y = 2 - x - x^2$ has a maximum value.
3. $f(x) = x^2$ has a y intercept of (0, 0).
4. A spill parabola has a highest point.
5. Every parabola must cross the x axis.

B. Miscellaneous

6. Find the x and y intercepts of $f(x) = x^2 - 3x - 10$.
7. Find the zeros of $y = x^2 - 4$.
8. Find the vertex of $y = -x^2 + 4x - 1$.
9. Find the minimum value of $f(x) = x^2 + x$.

C. Draw the graphs of the following quadratic functions and give the domain and range of each.

10. $f(x) = 2x^2 - 4$ **11.** $y = x^2 + 4x - 4$ **12.** $y = x^2$

33

SOLUTIONS TO REVIEW EXERCISES

A. **1.** true **2.** true **3.** true **4.** true **5.** false

B. **6.** $y = x^2 - 3x - 10$
The y intercept is $(0, -10)$.
$0 = (x - 5)(x + 2)$
The x intercepts are (5, 0) and $(-2, 0)$.

7. $0 = (x + 2)(x - 2)$
The zeros are 2 and -2.

8. $y = -x^2 + 4x - 1$
$-y = x^2 - 4x + 1$
$-y = x^2 - 4x + 4 + 1 - 4$
$-y = (x - 2)^2 - 3$
$y = -(x - 2)^2 + 3$
The vertex is (2, 3).

9. $y = x^2 + x + 1/4 + (-1/4)$
$y = (x + 1/2)^2 - 1/4$
The minimum value is $-1/4$.

C. **10.** **11.**

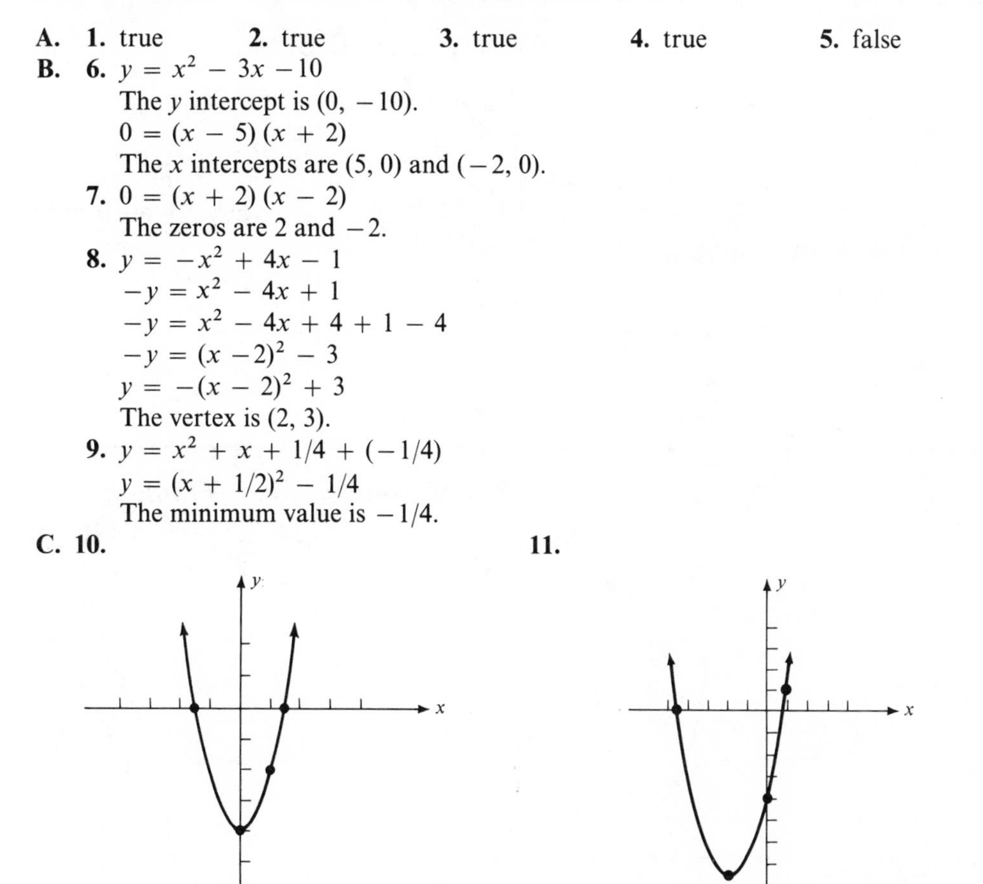

10. domain is $\{x | x \in R\}$
range is $\{y | y \geq -4\}$

11. domain is $\{x | x \in R\}$
range is $\{y | y \geq -8\}$

(Frame 34, contd.)

12.

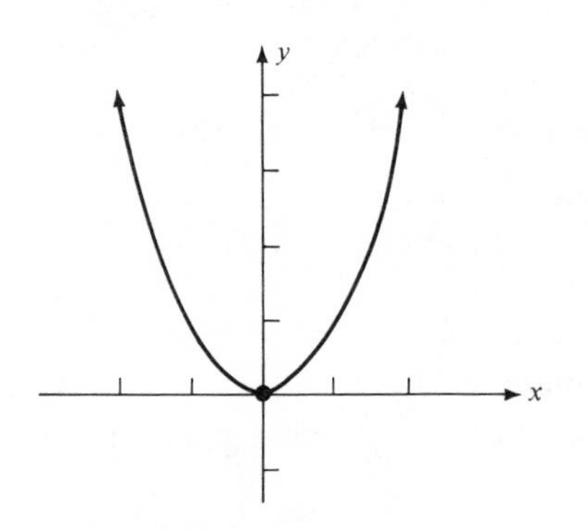

domain is $\{x|x \in R\}$
range is $\{y|y \geq 0\}$

(34)

SOLUTIONS TO STUDY EXERCISES

Study Exercise One (Frame 5)

1. $0 = (x - 3)(x + 1)$
$x - 3 = 0$ or $x + 1 = 0$
The zeros are 3 and -1.

2. 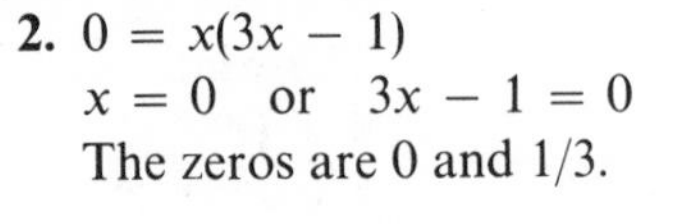
$0 = x(3x - 1)$
$x = 0$ or $3x - 1 = 0$
The zeros are 0 and 1/3.

3. $0 = x^2 + x + 2$

$$x = \frac{-1 \pm \sqrt{1^2 - 4(1)(2)}}{2}$$

no real roots
The function does not have any zeros.

(5A)

Study Exercise Two (Frame 9)

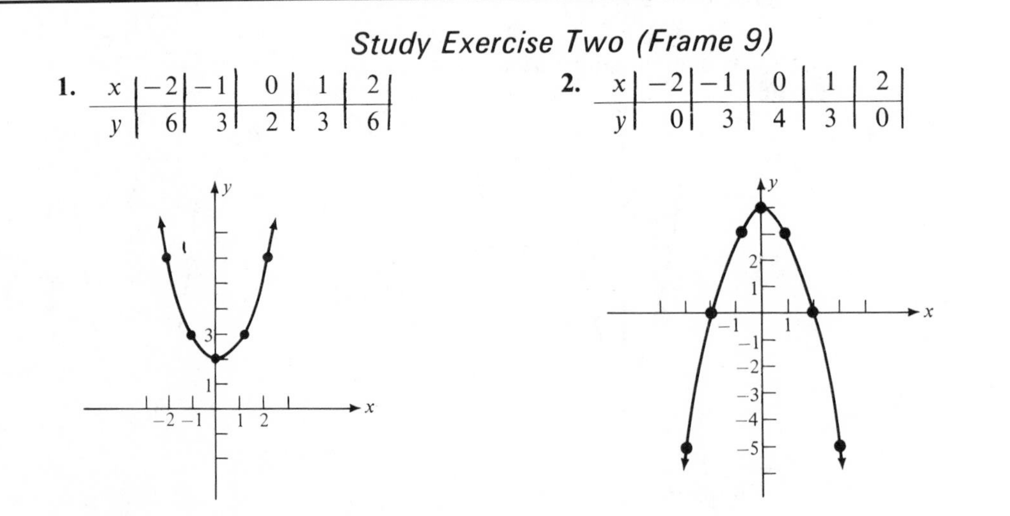

1.

x	−2	−1	0	1	2
y	6	3	2	3	6

2.

x	−2	−1	0	1	2
y	0	3	4	3	0

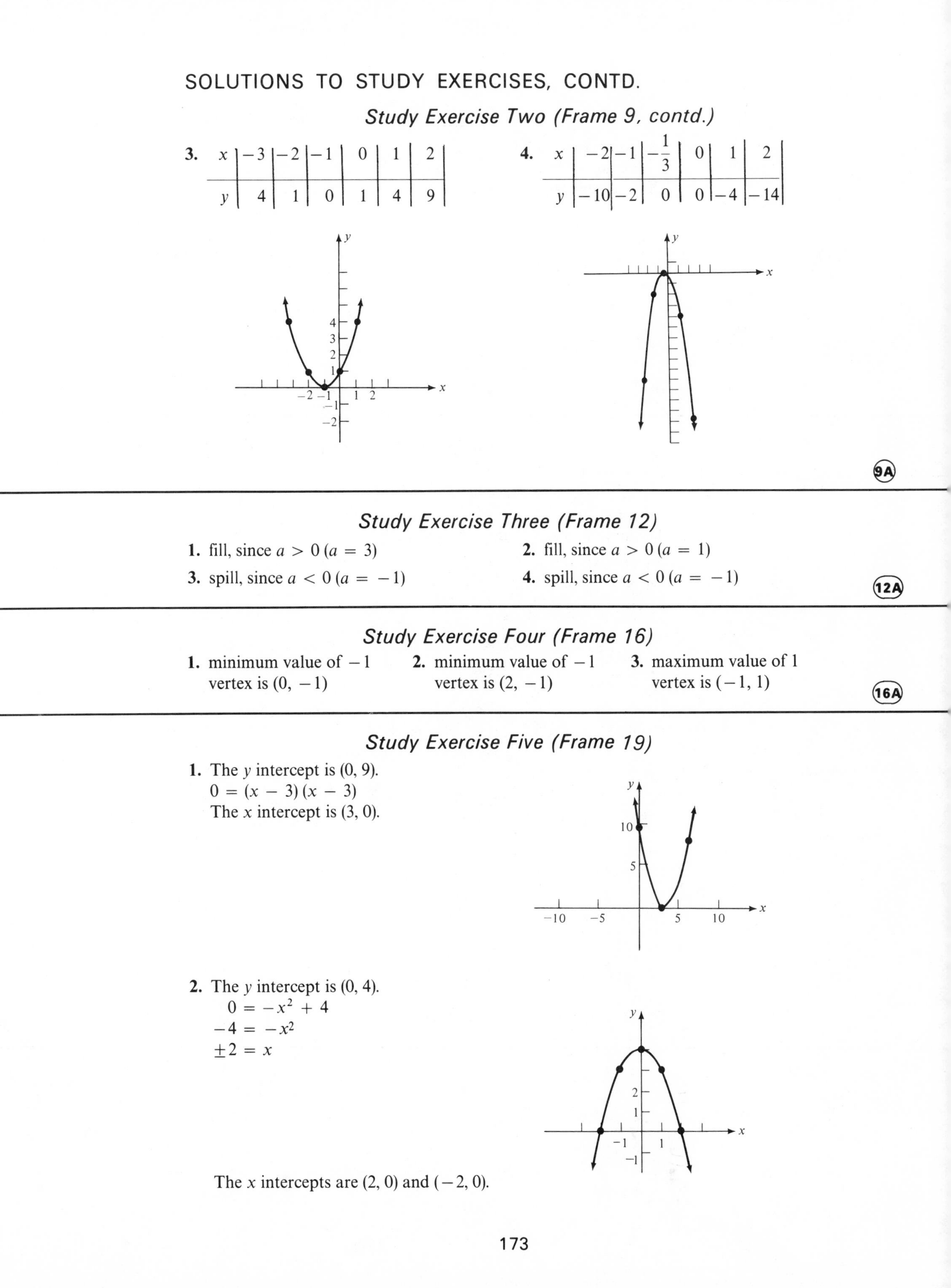

SOLUTIONS TO STUDY EXERCISES, CONTD.

Study Exercise Two (Frame 9, contd.)

3.

x	-3	-2	-1	0	1	2
y	4	1	0	1	4	9

4.

x	-2	-1	$-\frac{1}{3}$	0	1	2
y	-10	-2	0	0	-4	-14

9A

Study Exercise Three (Frame 12)

1. fill, since $a > 0$ $(a = 3)$

2. fill, since $a > 0$ $(a = 1)$

3. spill, since $a < 0$ $(a = -1)$

4. spill, since $a < 0$ $(a = -1)$

12A

Study Exercise Four (Frame 16)

1. minimum value of -1
vertex is $(0, -1)$

2. minimum value of -1
vertex is $(2, -1)$

3. maximum value of 1
vertex is $(-1, 1)$

16A

Study Exercise Five (Frame 19)

1. The y intercept is $(0, 9)$.
$0 = (x - 3)(x - 3)$
The x intercept is $(3, 0)$.

2. The y intercept is $(0, 4)$.

$$0 = -x^2 + 4$$
$$-4 = -x^2$$
$$\pm 2 = x$$

The x intercepts are $(2, 0)$ and $(-2, 0)$.

Study Exercise Five (Frame 19, contd.)

3. The y intercept is (0, 0).
$0 = (x)(x - 3)$
The intercepts are (0, 0) and (3, 0).

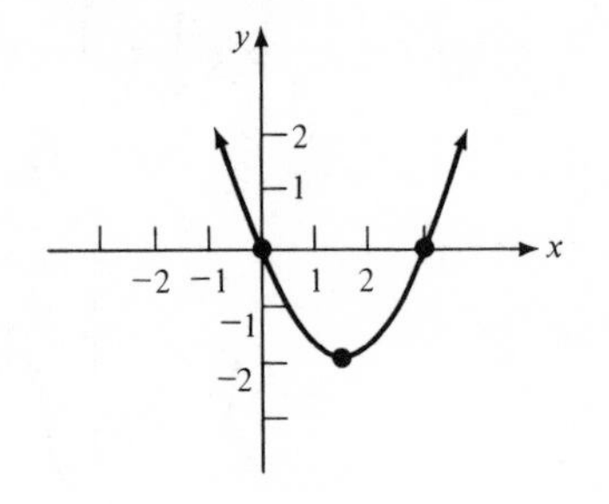

19A

Study Exercise Six (Frame 28)

1. $y = x^2 - 4x + 2$
$y = x^2 - 4x + 4 + 2 + (-4)$
$y = (x - 2)^2 - 2$
The minimum is -2; the vertex is $(2, -2)$.

2. $y = -x^2 + x - 6$
$-y = x^2 - x + 6$
$-y = x^2 - x + 1/4 + 6 - 1/4$
$-y = (x - 1/2)^2 + \frac{23}{4}$
$y = -(x - 1/2)^2 - \frac{23}{4}$
The maximum is $\frac{-23}{4}$; the vertex is $\left(\frac{1}{2}, \frac{-23}{4}\right)$.

3. $y = x^2 + x - 12$
$y = x^2 + x + 1/4 - 12 - 1/4$
$y = (x + 1/2)^2 - \frac{49}{4}$
The minimum is $\frac{-49}{4}$; the vertex is $\left(-1/2, \frac{-49}{4}\right)$.

4. $y = 2x^2 - 4x + 7$
$y/2 = x^2 - 2x + 7/2$
$y/2 = x^2 - 2x + 1 + 7/2 - 1$
$y/2 = (x - 1)^2 + 5/2$
$y = 2(x - 1)^2 + 5$
The minimum is 5; the vertex is (1, 5).

28A

SOLUTIONS TO STUDY EXERCISES, CONTD.

Study Exercise Seven (Frame 32)

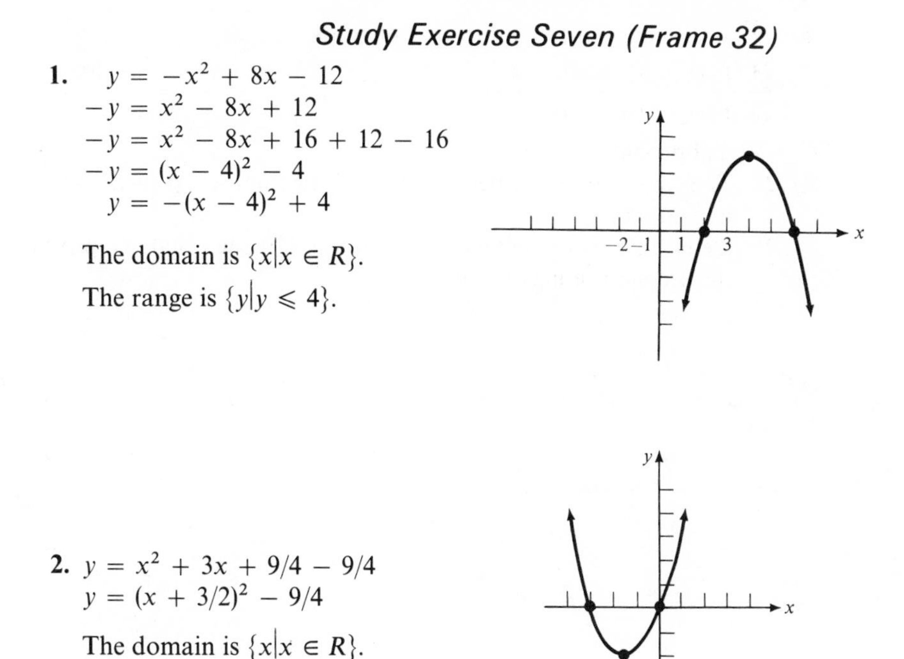

1. $y = -x^2 + 8x - 12$
$-y = x^2 - 8x + 12$
$-y = x^2 - 8x + 16 + 12 - 16$
$-y = (x - 4)^2 - 4$
$y = -(x - 4)^2 + 4$

The domain is $\{x|x \in R\}$.

The range is $\{y|y \leqslant 4\}$.

2. $y = x^2 + 3x + 9/4 - 9/4$
$y = (x + 3/2)^2 - 9/4$

The domain is $\{x|x \in R\}$.

The range is $\{y|y \geqslant -9/4\}$.

32A

UNIT 13—SUPPLEMENTARY PROBLEMS

A. Determine if the following parabolas are fill or spill type.

1. $y = 2x^2 + x - 3$ **2.** $y = 4x - x^2 + 6$ **3.** $y = x^2$

B. Find the x and y intercepts.

4. $f(x) = x^2 - 3x + 2$ **5.** $y = x^2 - 6x - 7$

6. $y = -x^2 - \frac{3}{2}x$ **7.** $y = -x^2 + 5x - 4$

C. Find the zeros.

8. $f(x) = x^2 - 36$ **9.** $f(x) = x^2 - 6x - 7$

10. $f(x) = x^2 + 5$ **11.** $f(x) = 4x^2 - 4$

D. Find the maximum or minimum value and state the coordinates of the vertex.

12. $y = x^2 + 6x - 1$ **13.** $y = 6 + x - x^2$

14. $y = x^2 - 2x$ **15.** $y = 2x^2 + 4x - 1$

16. $y = -x^2 + 6x - 1$

E. Graph the following functions.

17. $\{(x, y)|y = -x^2 + 5x\}$ **18.** $\{(x, y)|f(x) = 1 - x^2\}$

19. $\{(x, y)|f(x) = x^2 - 6x + 5\}$ **20.** $\left\{(x, y)|y = -\frac{1}{2}x^2\right\}$

F. Find the range of the following functions.

21. $f(x) = x^2 + 4x - 1$

22. $f(x) = -x^2 + 6x + 2$

23. $f(x) = 4x^2 + 4x$

G. Verbal problems.

24. The sum of two numbers is 30. Find the numbers if their product is to be as large as possible.

25. Determine two numbers whose sum is 120 such that the product of the first and the second is a maximum.

unit

14

Conic Sections

Objectives

1. Recognize the type of conic from its equation.
2. Graph a conic section from its equation by using its intercepts.

Terms
conic section ellipse hyperbola

①

Consider the statement $x^2 + y^2 = 16$.

Line (1) $y^2 = 16 - x^2$
Line (2) y would equal the two square roots of $(16 - x^2)$
Line (3) $y = +\sqrt{16 - x^2}$ or $y = -\sqrt{16 - x^2}$

Line (3) above is sometimes combined and written $y = \pm\sqrt{16 - x^2}$

②

In the previous frame, we solved $x^2 + y^2 = 16$ for y; the result was that $y = \pm\sqrt{16 - x^2}$.
Does this rule for y represent a function? The answer is no.

Does $y = +\sqrt{16 - x^2}$ represent a function? Does $y = -\sqrt{16 - x^2}$ represent a function?

③

Study Exercise One

In each of the following solve for y:

1. $x^2 + y^2 = 36$ **2.** $2x^2 + y^2 = 36$ **3.** $x^2 - y^2 = 25$

4. $xy = 4$ **5.** $x + y^2 = 9$

④

Consider the case $y = +\sqrt{16 - x^2}$.

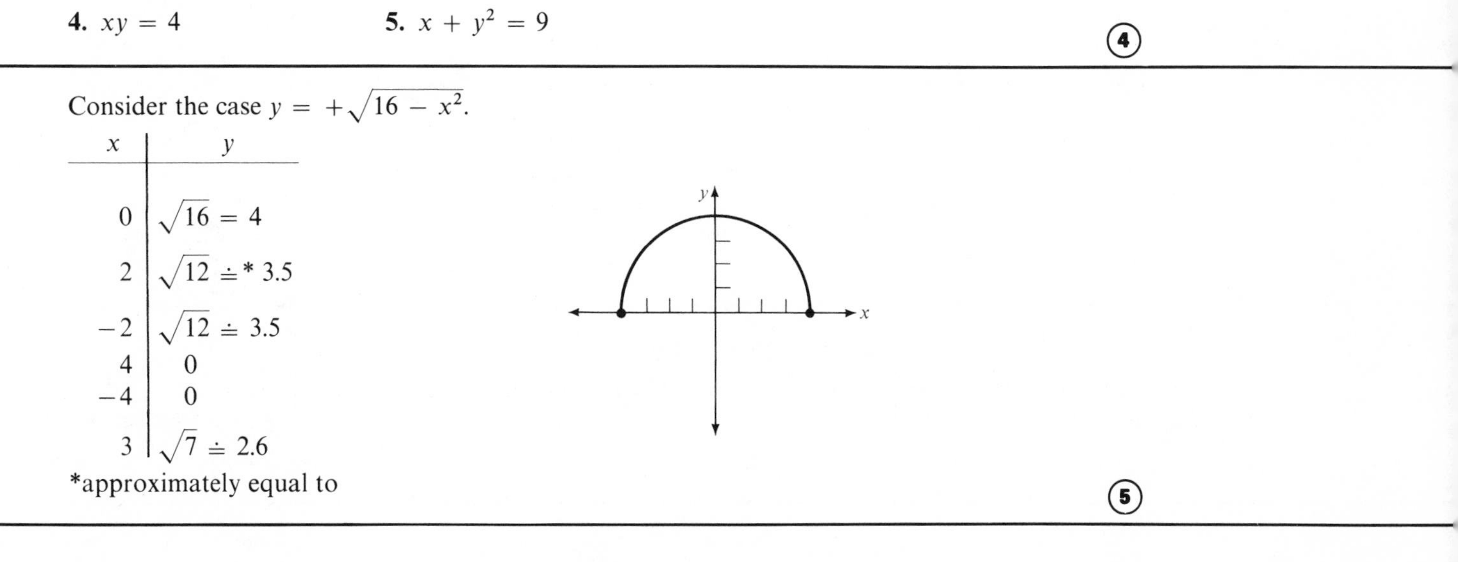

x	y
0	$\sqrt{16} = 4$
2	$\sqrt{12} \doteq$* 3.5
−2	$\sqrt{12} \doteq 3.5$
4	0
−4	0
3	$\sqrt{7} \doteq 2.6$

*approximately equal to

⑤

Now consider $y = -\sqrt{16 - x^2}$.

x	y
0	$-\sqrt{16} = -4$
2	$-\sqrt{12} \doteq -3.5$
−2	$-\sqrt{12} \doteq -3.5$
4	0
−4	0
3	$-\sqrt{7} \doteq -2.6$

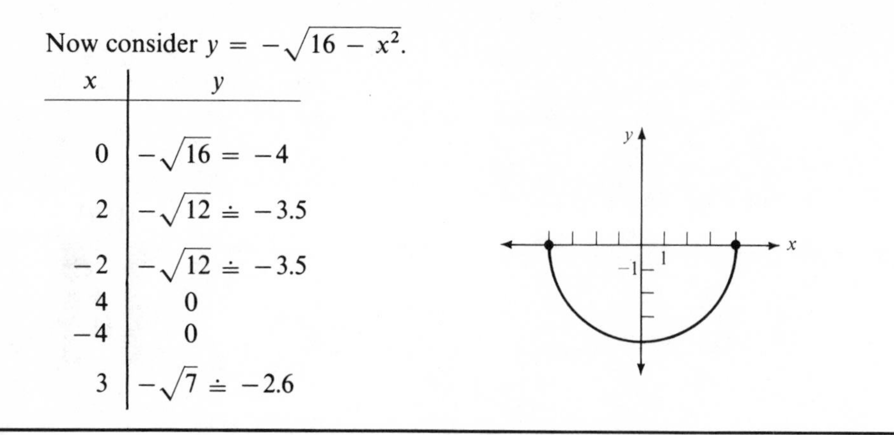

⑥

$$\{(x, y)|y = +\sqrt{16 - x^2}\} \cup \{(x, y)|y = -\sqrt{16 - x^2}$$
$$= \{(x, y)|y = \pm\sqrt{16 - x^2}\}$$
$$= \{(x, y)|x^2 + y^2 = 16\}$$

The graph is a circle with radius 4 and center at (0, 0).

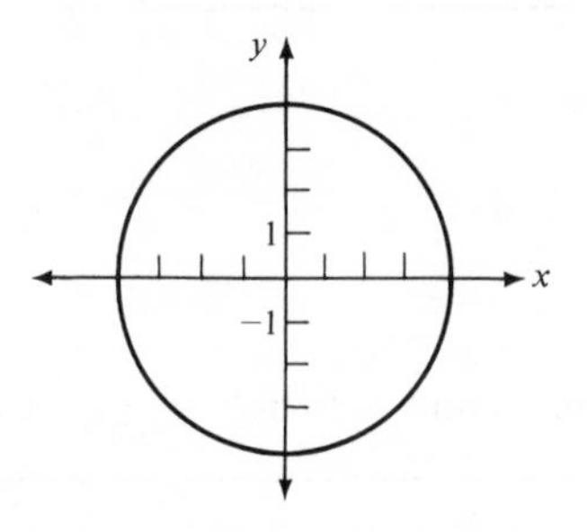

⑦

The function considered earlier, $f(x) = \sqrt{16 - x^2}$ has domain $\{x|-4 \leqslant x \leqslant 4\}$ and range $\{y|0 \leqslant y \leqslant 4\}$.

The function $g(x) = -\sqrt{16 - x^2}$ has domain $\{x|-4 \leqslant x \leqslant 4\}$ and range $\{y|-4 \leqslant y \leqslant 0\}$

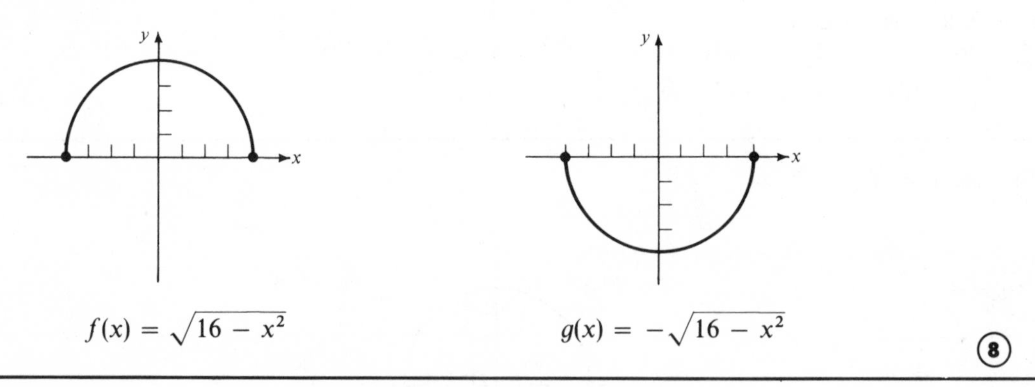

$f(x) = \sqrt{16 - x^2}$ $\qquad$ $g(x) = -\sqrt{16 - x^2}$

⑧

Study Exercise Two

Write each equation as two separate equations each of which defines a function and graph both equations on the same axes.

1. $x^2 + y^2 = 9$ $\qquad$ **2.** $2x^2 + 2y^2 = 50$

⑨

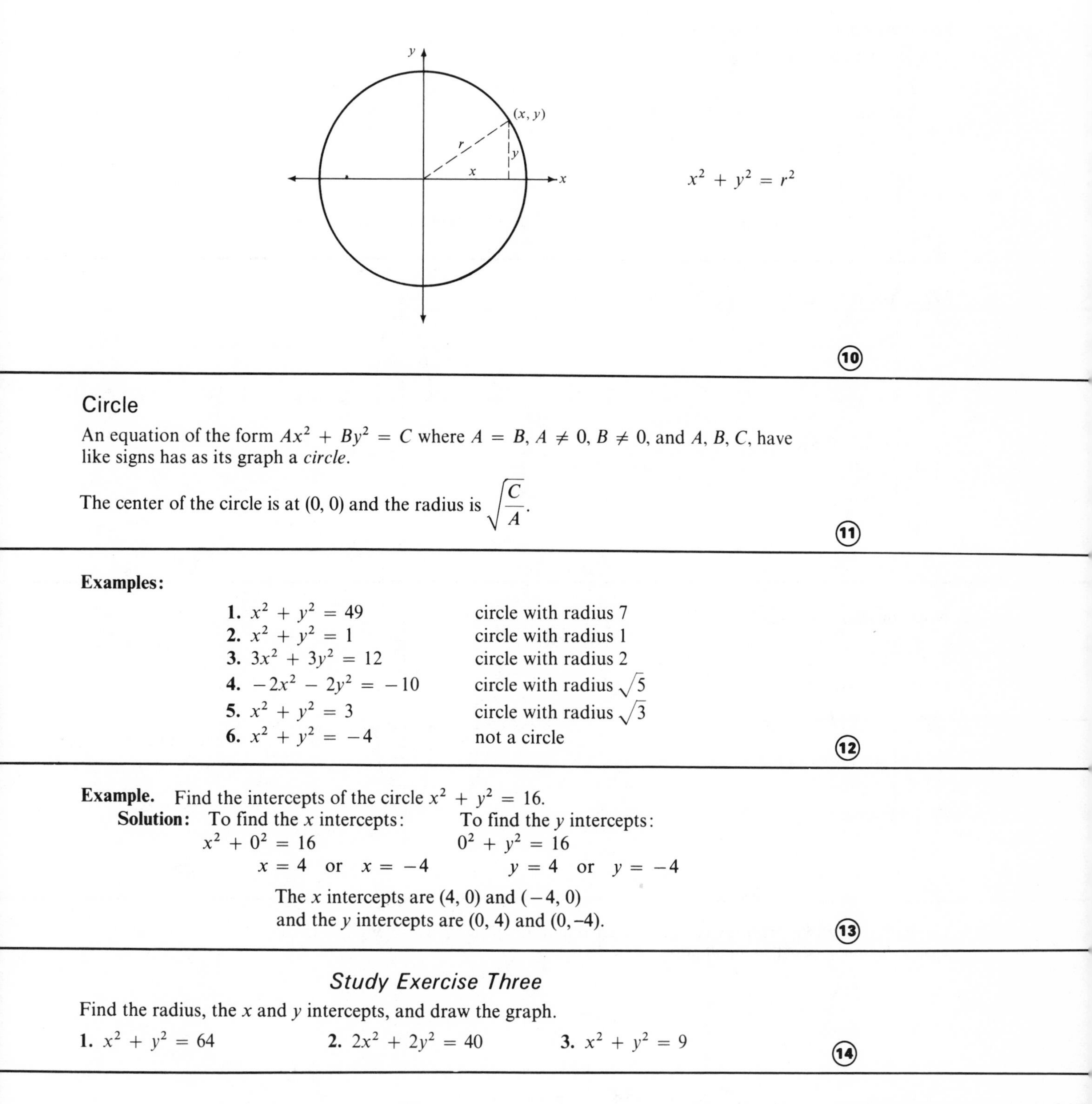

$$x^2 + y^2 = r^2$$

(10)

Circle

An equation of the form $Ax^2 + By^2 = C$ where $A = B$, $A \neq 0$, $B \neq 0$, and A, B, C, have like signs has as its graph a *circle*.

The center of the circle is at (0, 0) and the radius is $\sqrt{\dfrac{C}{A}}$.

(11)

Examples:

1. $x^2 + y^2 = 49$ — circle with radius 7
2. $x^2 + y^2 = 1$ — circle with radius 1
3. $3x^2 + 3y^2 = 12$ — circle with radius 2
4. $-2x^2 - 2y^2 = -10$ — circle with radius $\sqrt{5}$
5. $x^2 + y^2 = 3$ — circle with radius $\sqrt{3}$
6. $x^2 + y^2 = -4$ — not a circle

(12)

Example. Find the intercepts of the circle $x^2 + y^2 = 16$.

Solution: To find the x intercepts:

$$x^2 + 0^2 = 16$$
$$x = 4 \quad \text{or} \quad x = -4$$

To find the y intercepts:

$$0^2 + y^2 = 16$$
$$y = 4 \quad \text{or} \quad y = -4$$

The x intercepts are (4, 0) and $(-4, 0)$ and the y intercepts are (0, 4) and (0, −4).

(13)

Study Exercise Three

Find the radius, the x and y intercepts, and draw the graph.

1. $x^2 + y^2 = 64$ **2.** $2x^2 + 2y^2 = 40$ **3.** $x^2 + y^2 = 9$

(14)

Let us now consider $\{(x, y) | 9x^2 + y^2 = 36\}$.
This is not a circle since $9 \neq 1$.

Line (*1*) $9x^2 + y^2 = 36$
Line (*2*) $y^2 = 36 - 9x^2$
Line (*3*) $y = \pm\sqrt{36 - 9x^2}$

$y = \pm\sqrt{36 - 9x^2}$ is not a function, but

$y = +\sqrt{36 - 9x^2}$, $y = -\sqrt{36 - 9x^2}$ do represent functions.

(15)

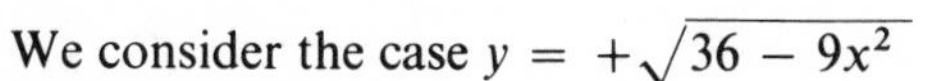

We consider the case $y = +\sqrt{36 - 9x^2}$

x	y
0	6
2/3	$\sqrt{32} \doteq 5.6$
1	$\sqrt{27} \doteq 5.2$
2	0
-1	$\sqrt{27} \doteq 5.2$
-2	0

(16)

Now consider $y = -\sqrt{36 - 9x^2}$

x	y
0	-6
2/3	$-\sqrt{32} \doteq -5.6$
1	$-\sqrt{27} \doteq -5.2$
2	0
-1	$-\sqrt{27} \doteq -5.2$
-2	0

(17)

Combining the two previous results, we get that $\{(x, y) | 9x^2 + y^2 = 36\}$ has the graph shown below. It is called an *ellipse*.

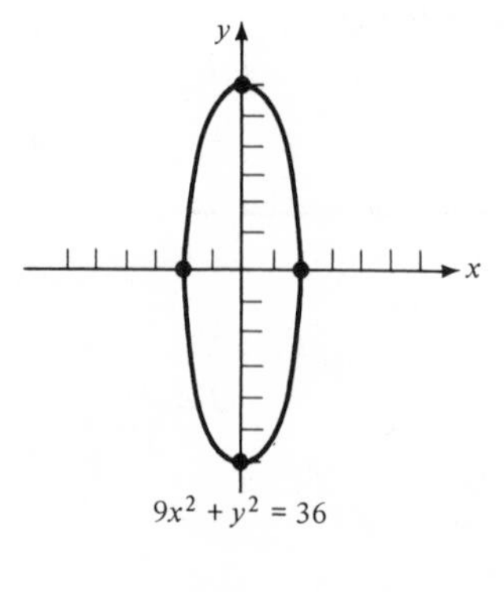

(18)

Ellipse

An equation of the form $Ax^2 + By^2 = C$ has as its graph an ellipse provided $A \neq B$ and A, B, C have the same sign. The center of the ellipse is at (0, 0)

Some examples of equations of ellipses include $4x^2 + y^2 = 1$, $9x^2 + 16y^2 = 144$, and $-16x^2 - 4y^2 = -16$.

(19)

To find the x intercepts of $Ax^2 + By^2 = C$, let $y = 0$. Thus $Ax^2 = C$. This implies that $x^2 = \frac{C}{A}$ and $x = \pm\sqrt{\frac{C}{A}}$. The x intercepts are $\left(\sqrt{\frac{C}{A}}, 0\right)$ and $\left(-\sqrt{\frac{C}{A}}, 0\right)$.

To find the y intercepts of $Ax^2 + By^2 = C$, let $x = 0$. Thus $By^2 = C$. This implies that $y^2 = \frac{C}{B}$ and $y = \pm\sqrt{\frac{C}{B}}$. The y intercepts are $\left(0, \sqrt{\frac{C}{B}}\right)$ and $\left(0, -\sqrt{\frac{C}{B}}\right)$.

(20)

Example Graph $\{(x, y)|9x^2 + 25y^2 = 225\}$

Solution The graph is an ellipse. To find the x intercepts: let $y = 0$. Then $9x^2 = 225$, $x^2 = 25$ and $x = \pm 5$. The x intercepts are (5,0) and (−5, 0). To find the y intercepts, let $x = 0$. Then $25y^2 = 225$, $y^2 = 9$ and $y = \pm 3$. The y intercepts are (0, 3) and (0, −3).

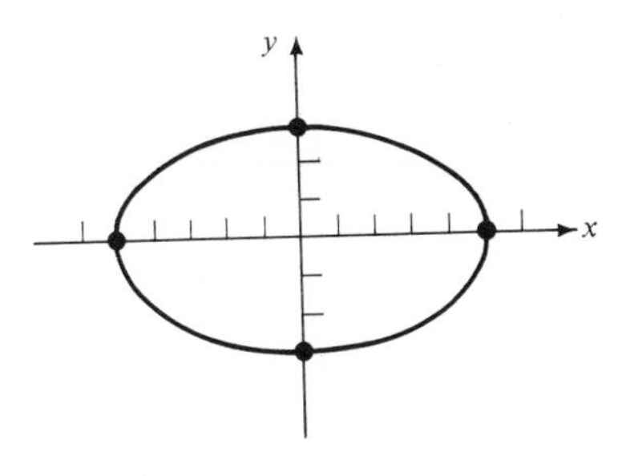

(21)

Study Exercise Four

Graph each of the following:

1. $\{(x, y)|25x^2 + 9y^2 = 225\}$

2. $\{(x, y)|x^2 + y^2 = 16\}$

(22)

We now consider $x^2 - 4y^2 = 16$.
This is not an ellipse since A and B do not have the same sign.

Line (1) $x^2 - 4y^2 = 16$

Line (2) $x^2 - 16 = 4y^2$

Line (3) $4y^2 = x^2 - 16$

Line (4) $y^2 = \frac{1}{4}x^2 - 4$

(Frame 23, contd.)

Line (5) $y = \pm\sqrt{\frac{1}{4}x^2 - 4}$

$y = \pm\sqrt{\frac{1}{4}x^2 - 4}$ is not a function, but

$y = +\sqrt{\frac{1}{4}x^2 - 4}$, $y = -\sqrt{\frac{1}{4}x^2 - 4}$ do represent functions.

(23)

We consider the case $y = +\sqrt{\frac{1}{4}x^2 - 4}$

x	y
0	does not exist
4	0
5	3/2
6	$\sqrt{5} \doteq 2.2$
-4	0
-5	3/2
-6	$\sqrt{5} \doteq 2.2$

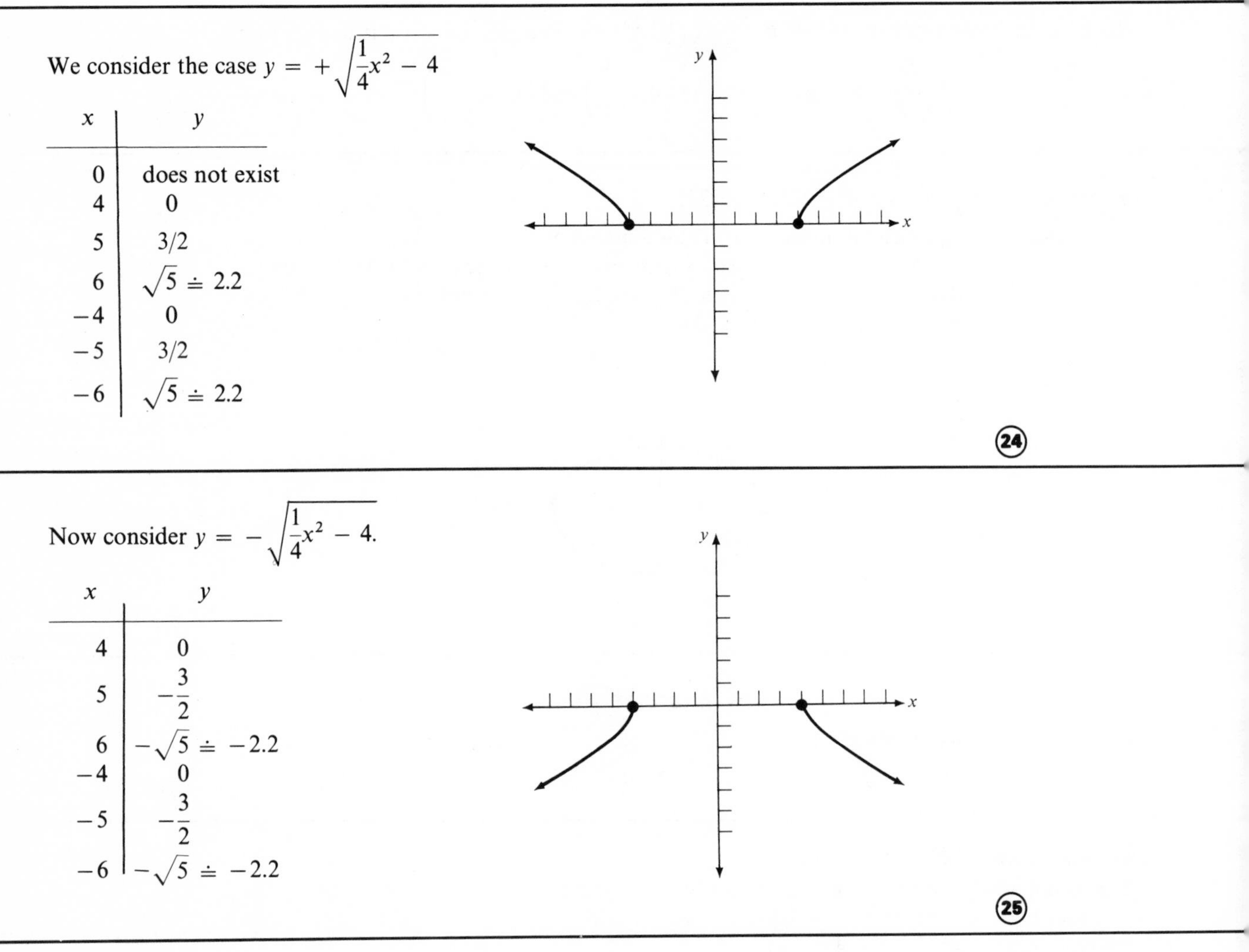

(24)

Now consider $y = -\sqrt{\frac{1}{4}x^2 - 4}$.

x	y
4	0
5	$-\frac{3}{2}$
6	$-\sqrt{5} \doteq -2.2$
-4	0
-5	$-\frac{3}{2}$
-6	$-\sqrt{5} \doteq -2.2$

(25)

Combining the two previous results, we see that $\{(x, y)|x^2 - 4y^2 = 16\}$ has the graph shown below. It is called a *hyperbola*.

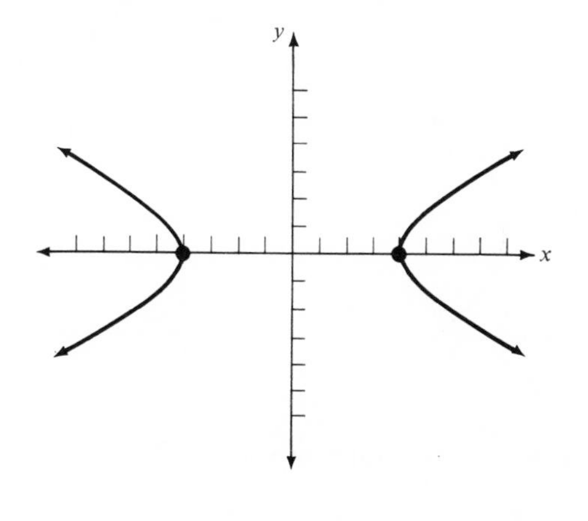

(26)

To graph $\{(x, y)|4y^2 - x^2 = 16\}$, we first solve for y.

Line (1) $4y^2 - x^2 = 16$

Line (2) $4y^2 = x^2 + 16$

Line (3) $y^2 = \frac{1}{4}x^2 + 4$

Line (4) $y = \pm\sqrt{\frac{1}{4}x^2 + 4}$

x	y
0	± 2
2	$\pm\sqrt{5} \doteq \pm 2.2$
-2	$\pm\sqrt{5} \doteq \pm 2.2$
4	$\pm\sqrt{8} \doteq \pm 2.8$
-4	$\pm\sqrt{8} \doteq \pm 2.8$

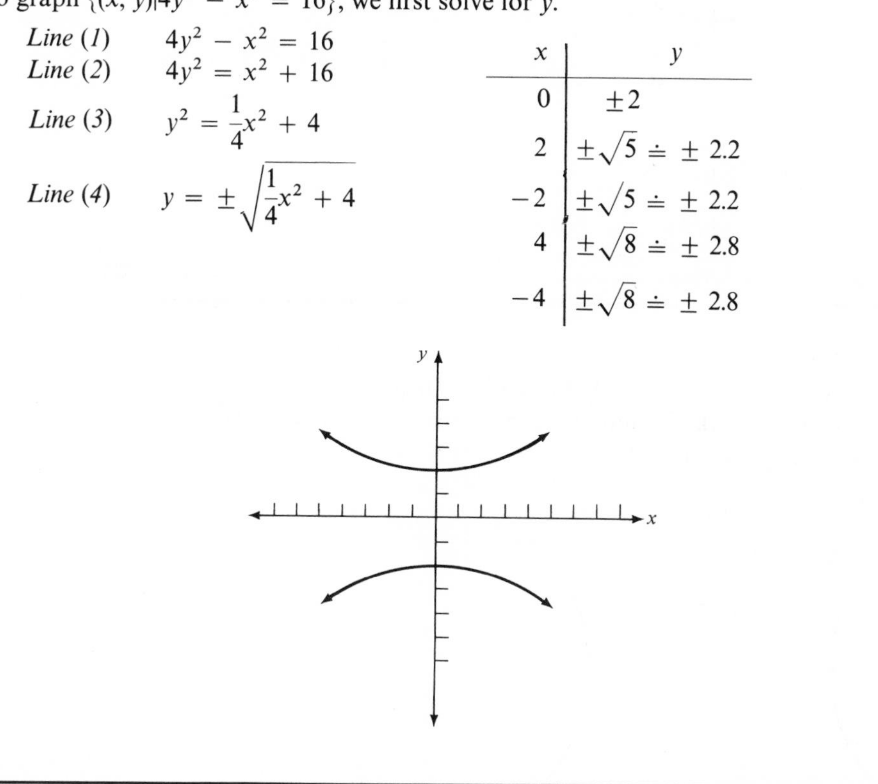

(27)

Hyperbola

An equation of the form $Ax^2 - By^2 = C$ if $A, B, C > 0$ graphs into a *hyperbola.*

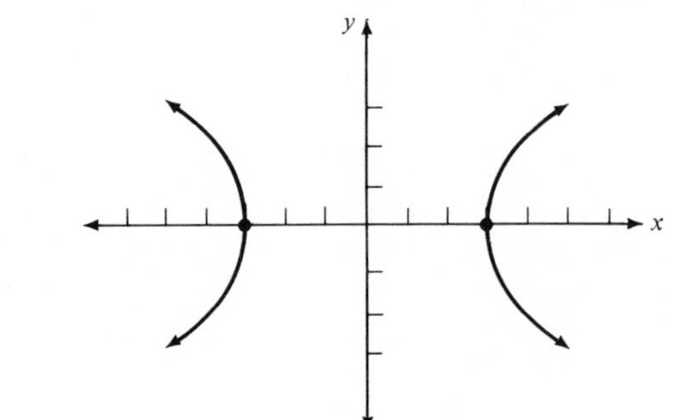

An equation of the form $By^2 - Ax^2 = C$ if $A, B, C > 0$ graphs into a *hyperbola.*

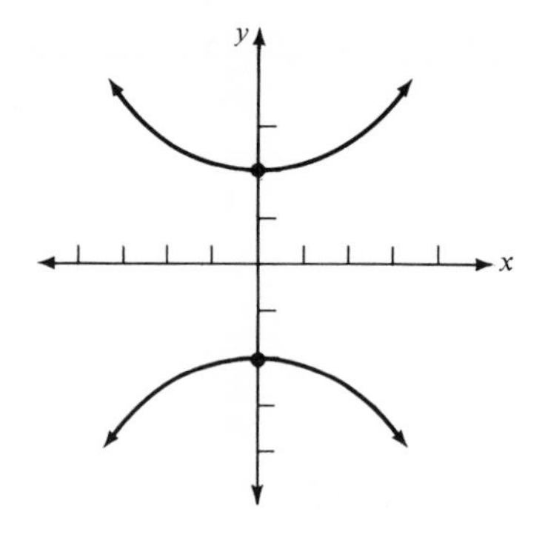

(28)

Example: Graph $\{(x, y) | x^2 - y^2 = 16\}$

Solution: To find the x intercepts, we let $y = 0$. Then $x^2 = 16$ and $x = \pm 4$. The x intercepts are $(4, 0)$ $(-4, 0)$. To find any y intercepts, we let $x = 0$. Then $-y^2 = 16$, thus there are no y intercepts.

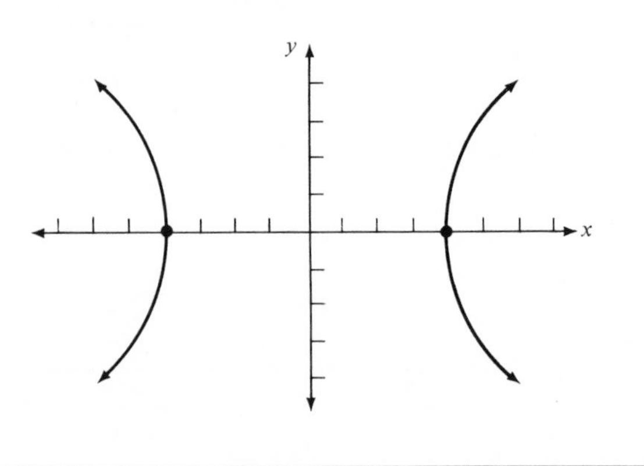

(29)

Graph $\{(x, y) | xy = 4\}$

x	y
1	4
-1	-4
2	2
-2	-2
4	1
-4	-1
1/2	8

The graph is also a hyperbola.

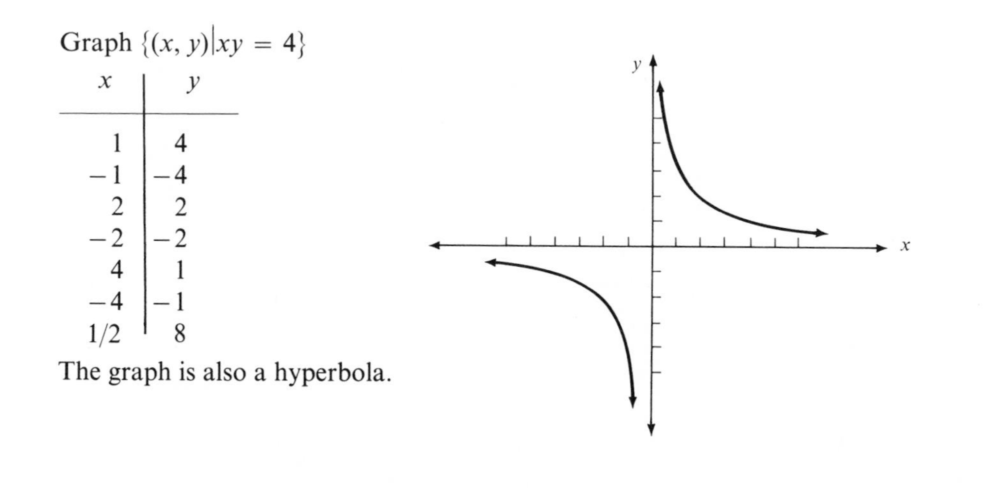

(30)

An equation of the form $xy = c$, $c \in R$, $c \neq 0$ graphs into a hyperbola. If $c > 0$, the hyperbola is in the first and third quadrants, and if $c < 0$, the hyperbola is in the second and fourth quadrants.

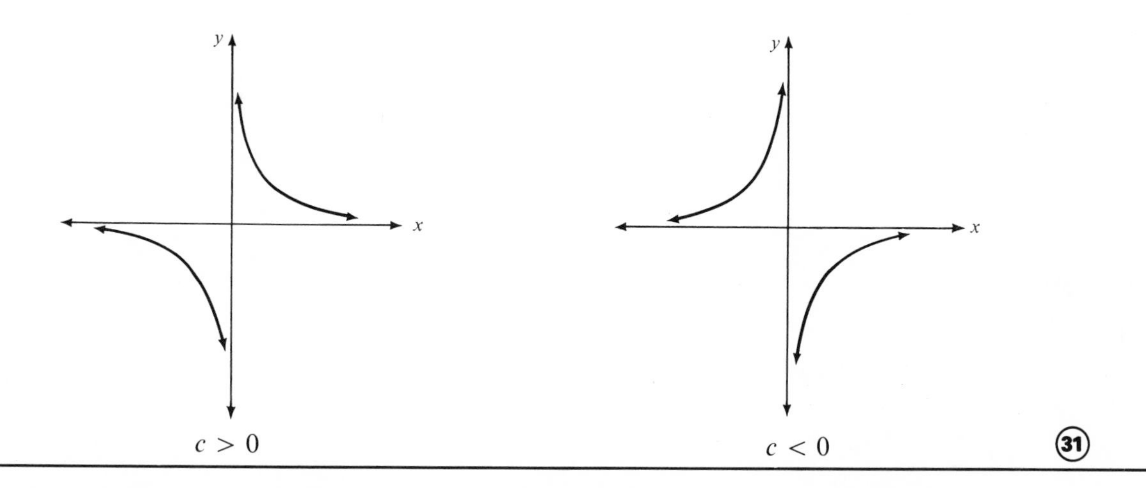

$c > 0$ $\qquad\qquad$ $c < 0$

(31)

As a special case, we consider $xy = 0$. An ordered pair satisfies $xy = 0$, if either $x = 0$ or $y = 0$. Some ordered pairs include (1, 0) (0, 1) (2, 0) (0, 2) (3, 0) (0, 3). If $x = 0$, y is any real number and if $y = 0$, x is any real number. The graph consists of both the x and y axes.

(32)

Study Exercise Five

Graph each of the following:

1. $\{(x, y) | xy = 6\}$

2. $\{(x, y) | xy = -4\}$

3. $\{(x, y) | 4x^2 - 9y^2 = 36\}$

4. $\{(x, y) | 9x^2 - 4y^2 = 36\}$

5. $\{(x, y) | 4y^2 - 9x^2 = 36\}$

(33)

Summary

1.	$y = ax^2 + bx + c, a \neq 0$	*Parabola*
2.	$Ax^2 + By^2 = C, A = B, A \neq 0, B \neq 0$ A, B, C with like signs	*Circle*
3.	$Ax^2 + By^2 = C, A \neq B, A, B, C > 0$	*Ellipse*
4.	$Ax^2 - By^2 = C, A, B, C > 0$ $By^2 - Ax^2 = C, A, B, C > 0$	*Hyperbola*
5.	$xy = c \qquad c \neq 0$	*Hyperbola*
6.	$xy = c \qquad c = 0$	*2 Straight Lines*

(34)

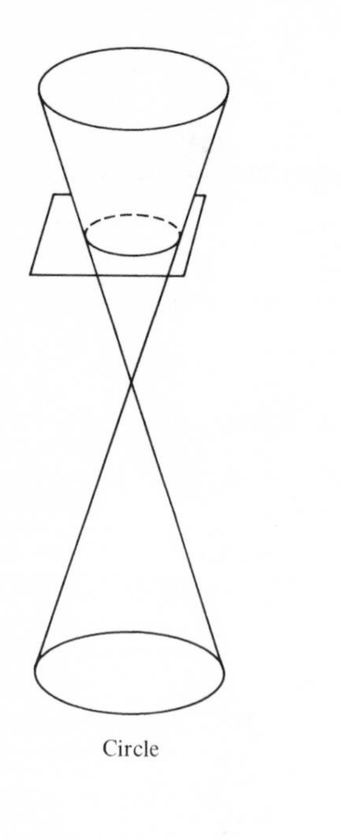

Circle

Ellipse

Parabola

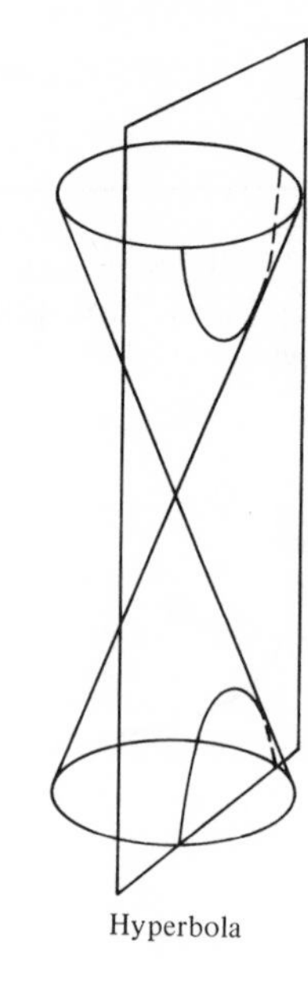

Hyperbola

Conic Sections

(35)

REVIEW EXERCISES

A. Without graphing, name the conic section represented by each of the following.

1. $x^2 - 4y^2 = 4$ **2.** $4x^2 + y^2 = 4$ **3.** $x^2 + 4y^2 = 4$

4. $4x^2 + 4y^2 = 4$ **5.** $x^2 + 4y = 4$ **6.** $xy = -4$

7. $x^2 + y^2 = 4$ **8.** $x^2 - y^2 = 4$ **9.** $xy = 4$

10. $2x^2 + y^2 = 4$

B. Draw the graph of each of the following.

11. $\{(x, y)|x^2 = 9 - y^2\}$ **12.** $\{(x, y)|4x^2 + 25y^2 = 100\}$

13. $\{(x, y)|x^2 - y^2 = 16\}$

C. Miscellaneous:

14. Solve for y in $4x^2 - y^2 = 16$

15. Write your answer to problem fourteen as two separate equations each of which defines a function.

16. State the domain of each of the functions in problem fifteen.

(36)

SOLUTIONS TO REVIEW EXERCISES

A. **1.** hyperbola **2.** ellipse **3.** ellipse **4.** circle

5. parabola **6.** hyperbola **7.** circle **8.** hyperbola

9. hyperbola **10.** ellipse

B. 11.

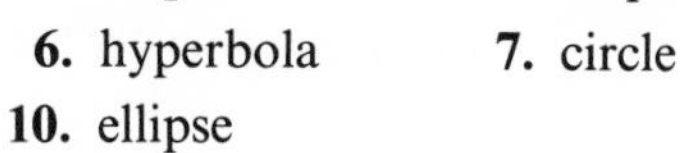

12.

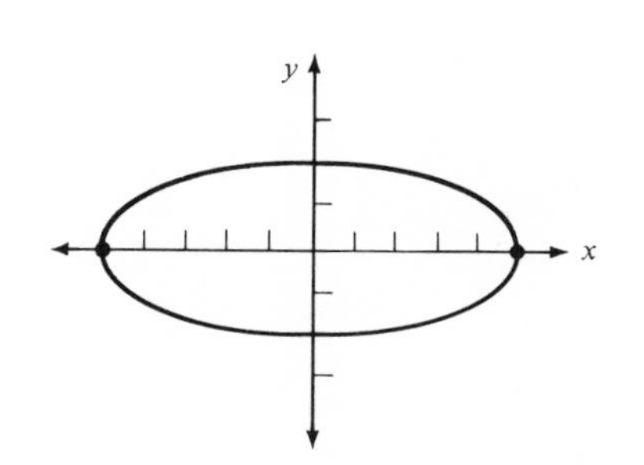

SOLUTIONS TO REVIEW EXERCISES, CONTD.

(Frame 37, contd.)

13.

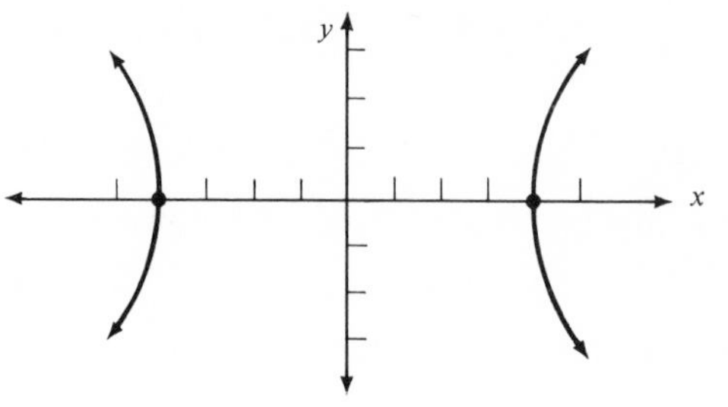

C. 14. $y = \pm\sqrt{4x^2 - 16}$ or $\pm 2\sqrt{x^2 - 4}$

15. $y = +\sqrt{4x^2 - 16}$ or $2\sqrt{x^2 - 4}$

$y = -\sqrt{4x^2 - 16}$ or $-2\sqrt{x^2 - 4}$

16. domain of each is $\{x | x \geqslant 2 \text{ or } x \leqslant -2\}$

(37)

SOLUTIONS TO REVIEW EXERCISES

Study Exercise One (Frame 4)

1. $y^2 = 36 - x^2$

$y = \pm\sqrt{36 - x^2}$

2. $y^2 = 36 - 2x^2$

$y = \pm\sqrt{36 - 2x^2}$

3. $-y^2 = -x^2 + 25$

$y^2 = x^2 - 25$

$y = \pm\sqrt{x^2 - 25}$

4. $y = \dfrac{4}{x}\ x \neq 0$

5. $y^2 = 9 - x$

$y = \pm\sqrt{9 - x}$

(4A)

Study Exercise Two (Frame 9)

1. $x^2 + y^2 = 9$

$y^2 = 9 - x^2$

$y = \sqrt{9 - x^2}$ $\quad y = -\sqrt{9 - x^2}$

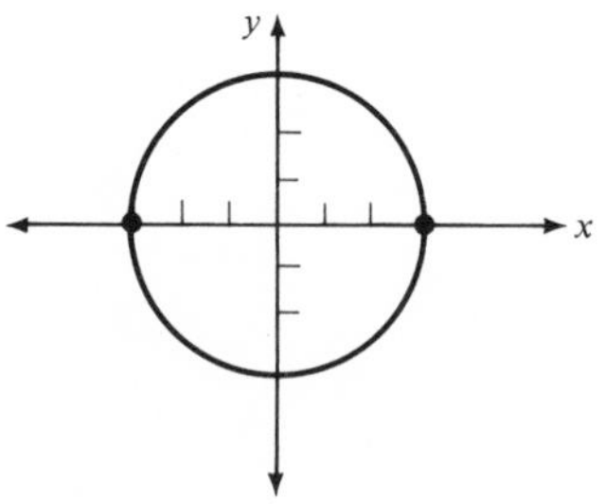

SOLUTIONS TO STUDY EXERCISES, CONTD.

Study Exercise Two (Frame 9, contd.)

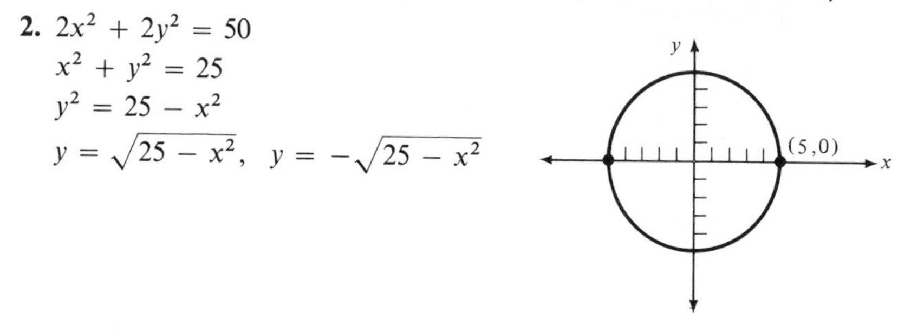

2. $2x^2 + 2y^2 = 50$

$x^2 + y^2 = 25$

$y^2 = 25 - x^2$

$y = \sqrt{25 - x^2}, \quad y = -\sqrt{25 - x^2}$

9A

Study Exercise Three (Frame 14)

1. radius is 8;

x intercepts are (8, 0) (−8, 0)

y intercepts are (0, 8) (0, −8).

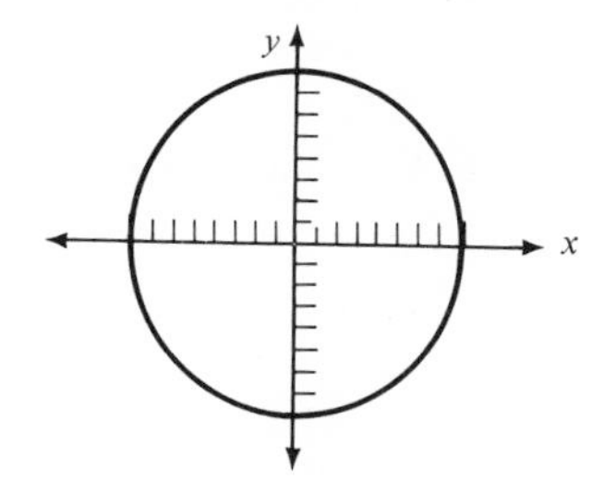

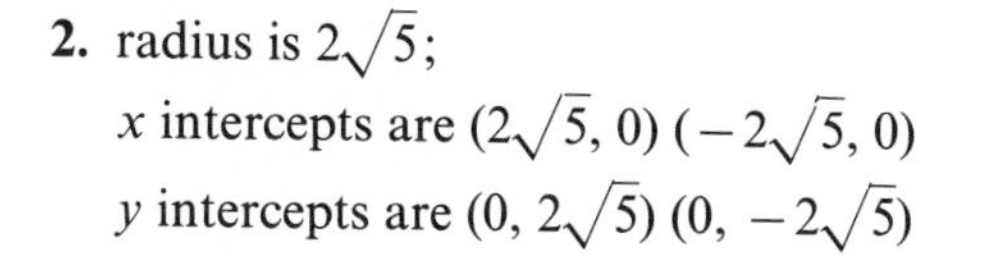

2. radius is $2\sqrt{5}$;

x intercepts are $(2\sqrt{5}, 0)$ $(-2\sqrt{5}, 0)$

y intercepts are $(0, 2\sqrt{5})$ $(0, -2\sqrt{5})$

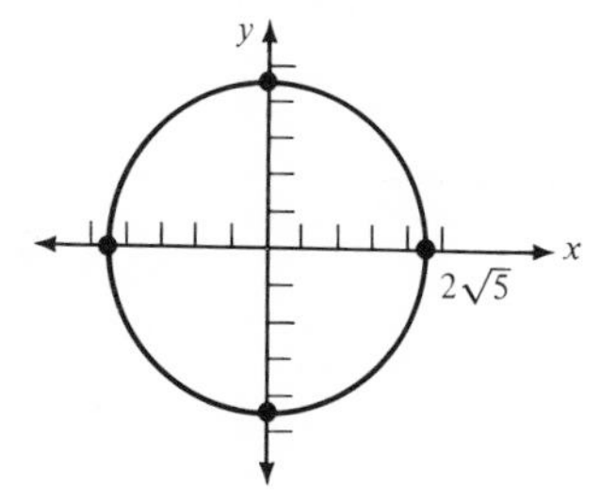

3. radius is 3;

x intercepts are (3, 0) (−3, 0)

y intercepts are (0, 3) (0, −3)

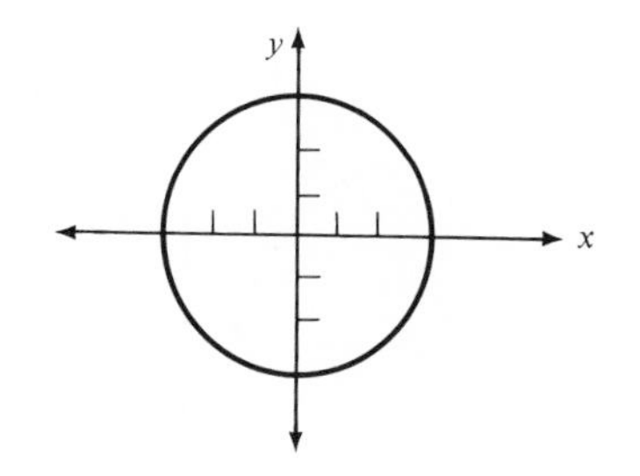

14A

Study Exercise Four (Frame 22)

1. x intercepts are $(3, 0)$, $(-3, 0)$
   y intercepts are $(0, 5)$, $(0, -5)$.

2. x intercepts are $(4, 0)$, $(-4, 0)$
   y intercepts are $(0, 4)$, $(0, -4)$.

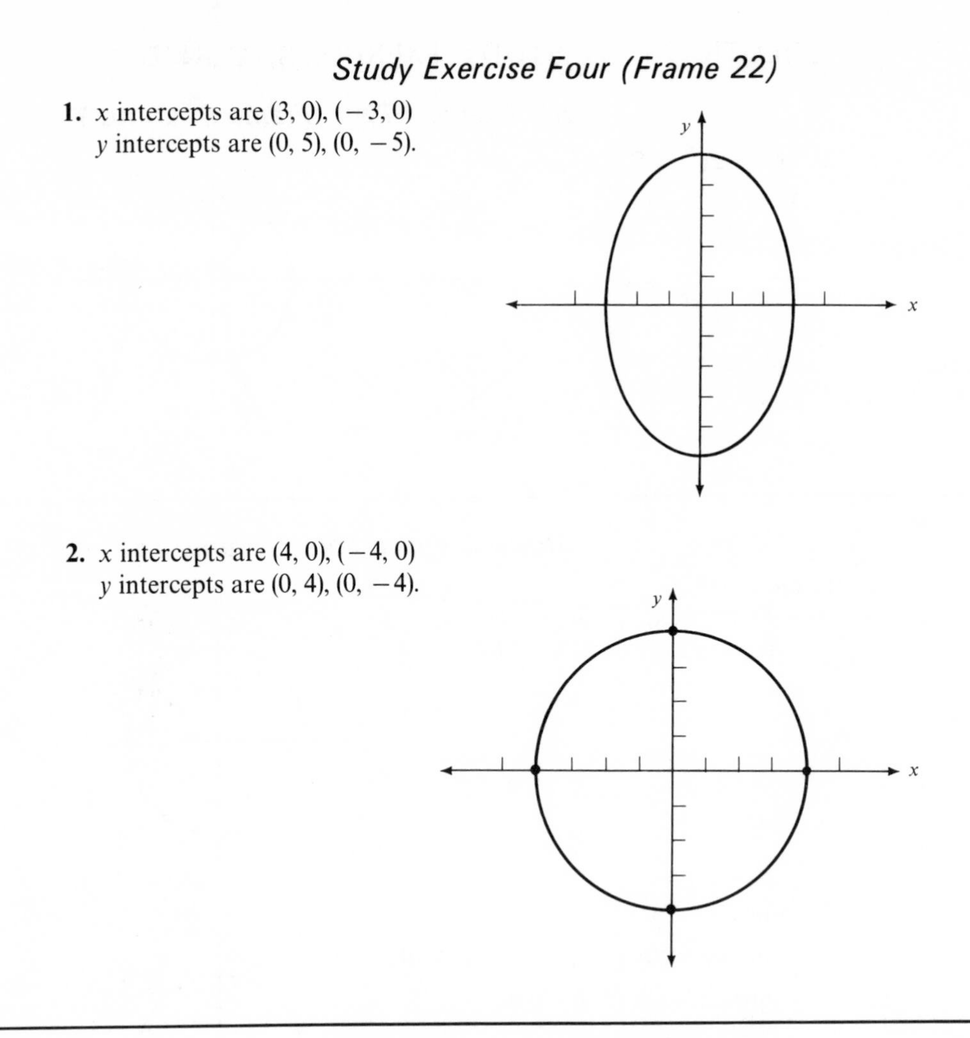

22A

Study Exercise Five (Frame 33)

1.

2.

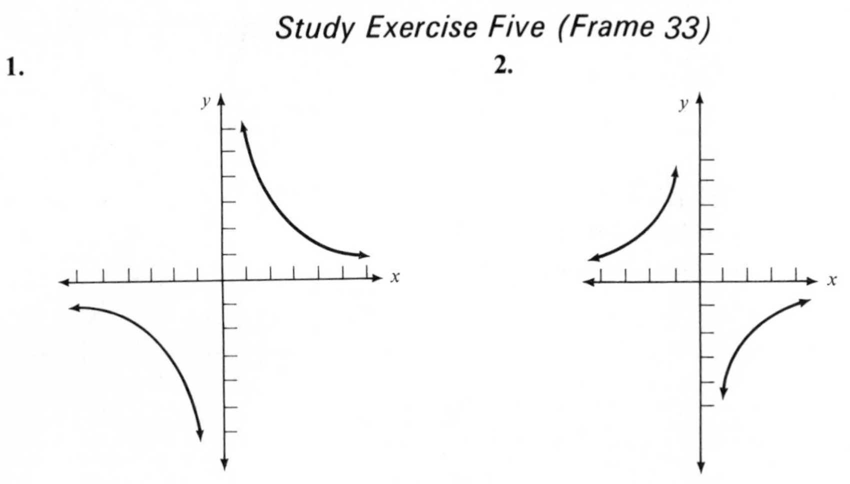

Study Exercise Five (Frame 33, contd.)

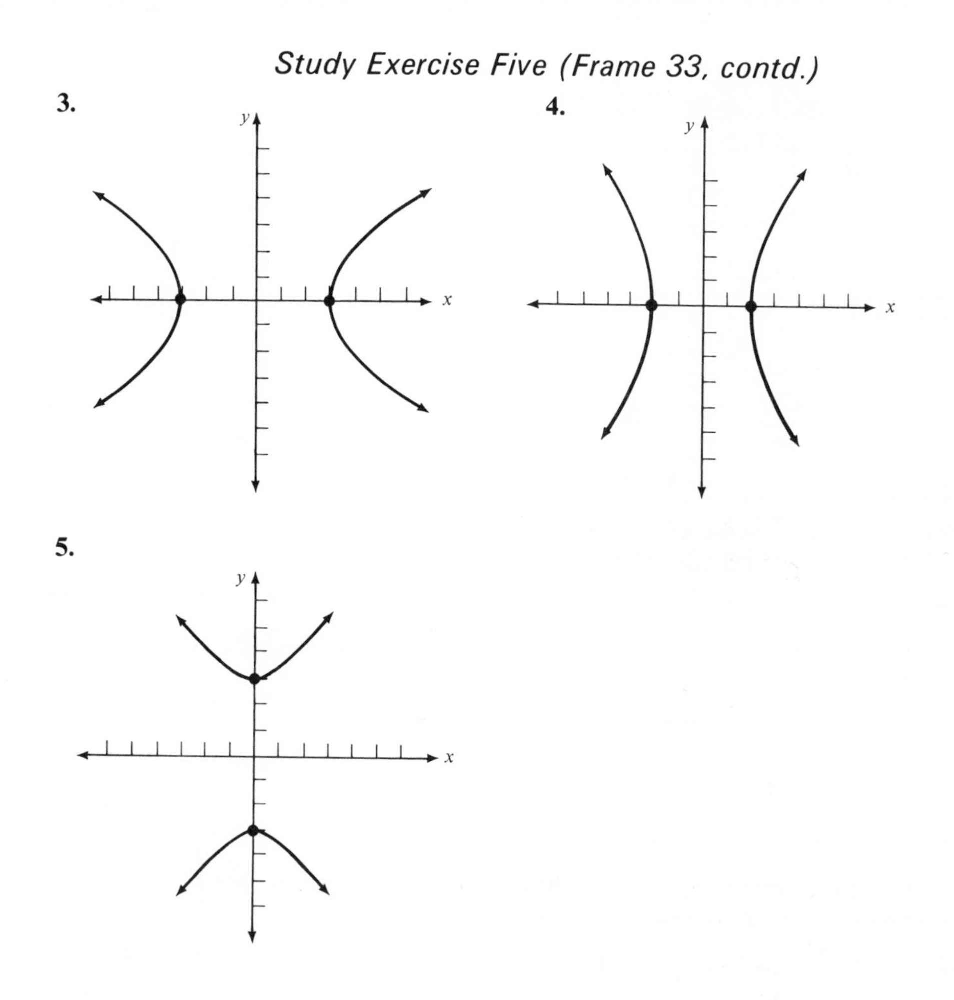

33A

UNIT 14—SUPPLEMENTARY PROBLEMS

A. Identify the graph of each of the following:

1. $x^2 + y^2 = 8$ **2.** $x^2 + 2y^2 = 8$ **3.** $x^2 = 8y$

4. $4x^2 + 16y^2 = 64$ **5.** $x - 36y = 36$ **6.** $y^2 - x^2 = 16$

7. $x^2 - 36y^2 = 36$ **8.** $xy = 10$ **9.** $4x^2 + 4y^2 = 1$

10. $x^2 + y^2 = 0$ (careful)

B. Graph each of the following:

11. $\{(x, y)|x^2 - 9y^2 = 81\}$ **12.** $\{(x, y)|y = 4 - 2x^2\}$

13. $\{x, y)|y = -6/x\}$ **14.** $\{(x, y)|4y^2 - 3x^2 = 12\}$

15. $\{(x, y)|x^2 + y^2 = 16\}$

16. $\{(x, y)|x^2 + y^2 = 16\} \cup \{(x, y)|x^2 - y^2 = 4\}$

17. $\{(x, y)|x^2 + y^2 = 25\} \cup \{(x, y)|4x^2 + y^2 = 36\}$

C. Miscellaneous:

18. Rewrite $4x^2 + 16y^2 = 64$ as two separate functions of x.

19. Give the domain and range of each of the functions in problem 18.

20. Rewrite $x^2 - y^2 = 4$ as two separate functions.

21. Give the domain and range of each function in problem 20.

22. Graph $\{(x, y)|xy = 6\}$ and give the domain and range.

unit

15

Variation

Objectives

1. Understand direct, inverse, and joint variation.
2. Be able to write mathematical relationships involving these three types of variation.

(1)

Recall that the graph of $y = mx + b$, $m, b \in R$ is a straight line with slope m and y intercept $(0, b)$.

(2)

If $b = 0$ in $y = mx + b$, then the graph of $y = mx$ is a line with y intercept $(0, 0)$. Therefore, $y = mx$ passes through the origin. Also, $y = mx$ is a family of lines through the origin.

(3)

$y = mx$ has a special importance because the relationship between two physical quantities is often expressed by a rule of this simple form.

For example, $d = 40t$ is a rule relating distance and time for a car traveling at 40 miles per hour.

(4)

If x and y are variables and $y = mx$, $m = 0$, the relationship between x and y is described by saying y varies directly as x or y is proportional to x.

In physical situations, the letter k is usually used instead of m. Hence, y varies directly as x is then written $y = k \cdot x$.

(5)

If two variables x and y are so related that the ratio of y to x is a fixed constant k, then y is said to vary directly as x and this is written:

$$\frac{y}{x} = k \quad \text{or} \quad y = kx.$$

(6)

If two variables x and y are so related that the ratio of y to x^2 is a fixed constant k, then y is said to vary directly as x^2 and is indicated by:

$$\frac{y}{x^2} = k \quad \text{or} \quad y = kx^2.$$

(7)

To write that T varies directly as M, write $T = k \cdot M$.

To write that B varies directly as the square of T, write $B = k \cdot T^2$.

To write that A is proportional to the square of B, write $A = k \cdot B^2$.

(8)

Example: y varies directly as x^2 and $y = 9$ when $x = 3$. Find y when $x = 5$.

Solution:

Line (1) $y = kx^2$ statement of direct variation.

Line (2) $9 = k \cdot 3^2$ substitute $y = 9$ and $x = 3$

Line (3) $9 = 9k$

Line (4) $1 = k$ constant is 1

Line (5) $y = 1 \cdot x^2$ statement of direct variation with $k = 1$

Line (6) $y = 1 \cdot 5^2$

Line (7) $y = 25$

(9)

Example: y varies directly as x and $y = 10$ when $x = 4$. Find x when $y = 8$.

Solution:

Line (1) $y = kx$

Line (2) $10 = k \cdot 4$

Line (3) $10/4 = k$ or $k = 5/2$

Line (4) $y = 5/2x$

Line (5) $8 = 5/2x$

Line (6) $16/5 = x$

(10)

Study Exercise One

1. If n varies directly as d and $n = 10$ when $d = 2$, find n when d is 8.
2. The distance in feet an object will fall from rest varies directly as the square of the time in seconds. A body will fall 16 feet the first second. How far will it fall in 4 seconds?
3. The time required for a pendulum of a clock to make a vibration varies as the square root of its length. A pendulum 100 cm. long makes a vibration in 1 second. Find the time of vibration of a pendulum 81 cm. long.

(11)

If two variables x and y are so related that the product of x and y is always a constant k, then we say that y varies *inversely* as x or that y is *inversely proportional* to x.

$$\text{We write } y = \frac{k}{x}.$$

(12)

Recall that the graph of $xy = k$, $k \in R$, $k \neq 0$ is a hyperbola.
If $k > 0$, the two branches are in the first and third quadrants, and if $k < 0$, the two branches are in the second and fourth quadrants.

(13)

Almost all physical quantities are restricted by their nature to positive values. Thus the positive branch of the hyperbola in the first quadrant is the characteristic graph of an inverse variation relationship between two quantities.

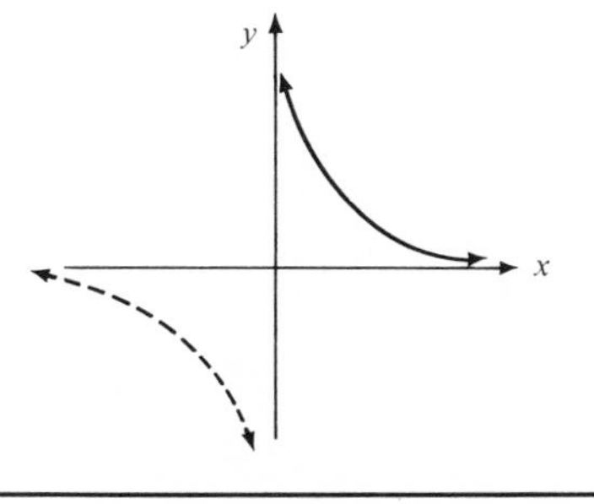

(14)

If z varies inversely as t, write $z = \frac{k}{t}$

If z varies inversely as the square of M, write $z = \frac{K}{M^2}$

If z varies inversely as the principle square root of t, write $z = \frac{k}{\sqrt{t}}$

(15)

Combinations of direct and inverse variation are also possible. If y varies directly as x and inversely as z, write $y = \frac{k \cdot x}{z}$

(16)

Example: The volume of a gas varies inversely as the pressure. Gas in a cylinder occupies 60 cubic inches when the pressure upon it is 8 pounds. What will be the volume if the pressure is increased to 42 pounds?

Solution:

Line (1) $v = \frac{k}{p}$

Line (2) $60 = k/8$

Line (3) $k = 480$

Line (4) $v = \frac{480}{p}$

Line (5) $v = \frac{480}{42}$

Line (6) $v = 11\frac{3}{7}$ *cubic inches*

(17)

Study Exercise Two

1. When air is pumped into an automobile tire, the pressure required varies inversely as the volume. If the pressure is 40 pounds when the volume is 140 cubic inches, find the pressure when the volume is 100 cubic inches.
2. If F varies directly as the square of v and inversely as r. Also, if $F = 160$ when $v = 8$ and $r = 5$, find the formula for F in terms of v and r.
3. The force required to open a gate varies inversely as the distance of the force from the line of hinges. The least force that can be used to open a certain gate 8 feet wide is 10 pounds. What force applied at a distance of 15 inches from the hinges would be needed to open the gate?

(18)

If the variable z varies directly as x when y is held constant, and varies directly as y when x is held constant, then z is said to vary *jointly* as x and y and is written $z = k\,x\,y$.

(19)

Study Exercise Three

In the following it may be assumed that there is a variation relationship between the quantities for which corresponding pairs of values are given. Write the variation statement which describes the relationship and write the formula expressing the upper variable in terms of the lower.

1.

y	2	4	6
x	30	15	10

2.

c	9	18	45
n	2	4	10

3. If p varies jointly as q and r, and $p = 5$ when $q = 2$ and $r = 7$, find r when $q = 14$ and $p = 6$.
4. Write as a variation statement: The volume of a rectangular box of fixed depth varies jointly as its length and width.

(20)

REVIEW EXERCISES

1. The weight of a body is inversely proportional to the square of its distance from the center of the earth. If a man weighs 147 pounds on the earth's surface, what will he weigh 200 miles above the earth? (Assume the radius of the earth is 4,000 miles)
2. If C varies directly as D and $C = 12\pi$ when $D = 12$, find C when $D = 3$.
3. If y varies jointly as r, s, and t and $y = 30$ when $r = 3, s = 4$, and $t = 5$, what is r when $y = 40, s = 5$, and $t = 8$?
4. If y varies directly as x, what is the resulting change in y when x is doubled?
5. If y varies inversely as x, what is the resulting change in y when x is doubled?

(21)

SOLUTIONS TO REVIEW EXERCISES

1.

$$w = \frac{k}{d^2}$$

$$147 = \frac{k}{(4{,}000)^2}$$

$$k = (4{,}000)^2\,(147)$$

$$w = \frac{(4{,}000)^2\,(147)}{d^2}$$

$$w = \frac{(4{,}000)\,(4{,}000)\,(147)}{(4200)\,(4200)}$$

$$w = \frac{400}{3} \text{ or } 133^1/_3 \text{ pounds}$$

2.

$$c = k \cdot D$$
$$12\pi = k \cdot 12$$
$$\pi = k$$
$$C = \pi d$$
$$C = 3\pi$$

3.

$$y = krst$$
$$30 = k(3)\,(4)\,(5)$$
$$1/2 = k$$
$$y = 1/2rst$$
$$40 = 1/2r(5)\,(8)$$
$$40 = 20r$$
$$2 = r$$

4.

$$y = kx$$
$$y/x = k$$

If the denominator is doubled, the numerator must also be doubled to maintain the same constant.

5. $y = k/x$

$xy = k$

If x is doubled, y must be halved.

(22)

SOLUTIONS TO STUDY EXERCISES

Study Exercise One (Frame 11)

1.

$$n = k \cdot d \qquad n = 5 \cdot d$$
$$10 = k \cdot 2 \qquad n = 5 \cdot 8$$
$$5 = k \qquad n = 40$$

2.

$$D = k \cdot t^2 \qquad D = 16t^2$$
$$16 = k \cdot 1^2 \qquad D = 16 \cdot 4^2$$
$$16 = k \qquad D = 256 \text{ feet}$$

3.

$$T = k \cdot \sqrt{l} \qquad T = 1/10\sqrt{l}$$
$$1 = k\sqrt{100} \qquad T = 1/10\sqrt{81}$$
$$1 = 10k \qquad T = 9/10 \text{ seconds}$$
$$1/10 = k$$

(11A)

Study Exercise Two (Frame 18)

1.

$$p = \frac{k}{v} \qquad p = \frac{5{,}600}{v}$$

$$40 = \frac{k}{140} \qquad p = \frac{5{,}600}{100}$$

$$5600 = k \qquad p = 56 \text{ pounds}$$

2.

$$F = \frac{k \cdot v^2}{r} \qquad F = \frac{\frac{25}{2} \cdot v^2}{r}$$

$$160 = \frac{k \cdot 8^2}{5} \qquad F = \frac{25v^2}{2r}$$

$$800 = 64k$$

$$\frac{25}{2} = k$$

3.

$$F = \frac{k}{d} \qquad F = \frac{80}{d} \quad \text{(15 inches = 5/4 feet)}$$

$$10 = \frac{k}{8} \qquad F = \frac{80}{5/4}$$

$$80 = k \qquad F = \frac{80 \cdot 4}{\frac{5}{4} \cdot 4}$$

$$F = \frac{320}{5}$$

$$F = 64 \text{ pounds}$$

18A

Study Exercise Three (Frame 20)

1. $xy = k$ or $xy = 60$

$y = \frac{60}{x}$, inverse variation

2. $\frac{c}{n} = k$ or $\frac{c}{n} = 4^1/_2$

$c = 4^1/_2 n$, direct variation

3.

$$p = k \cdot q \cdot r \qquad p = \frac{5}{14} \cdot q \cdot r$$

$$5 = k \cdot 2 \cdot 7 \qquad 6 = \frac{5}{14} \cdot 14 \cdot r$$

$$5 = 14k \qquad 6 = 5r$$

$$\frac{5}{14} = k \qquad 6/5 = r$$

4. $v = k \cdot L \cdot W$

20A

UNIT 15—SUPPLEMENTARY PROBLEMS

1. If z varies jointly as x^2 and y, and $z = 24$ when $x = 2$ and $y = 3$, find z when $x = 3$ and $y = 5$.
2. If y varies directly as x and $y = 4$ when $x = 2$, write the rule relating y to x. Draw the graph.
3. The surface area of a sphere varies directly as the square of the radius. If the surface area is 36π square inches when the radius is 3 inches, what is the surface area when the radius is 10 inches?
4. At constant temperature, the resistance of a wire varies directly as its length and inversely as the square of its diameter. If a piece of wire .1 inch in diameter and 50 feet long has a resistance of .1 ohm, what is the resistance of a piece of wire of the same material 2,000 feet long and .2 inch in diameter?
5. The current I in amperes in an electric circuit varies inversely as the resistance R in ohms when the electromotive force is constant. In a certain circuit I is 15 amperes where R is 2 ohms. Find I when R is .2 ohms.
6. The force p of the wind on a plane surface varies directly as the square of the speed s of the wind. If the force is 2 pounds per square foot when the speed of the wind is 20 miles per hour, find the force when the wind's speed is 50 miles per hour.
7. When a ball is thrown vertically upward, the height reached varies directly as the square of the speed with which the ball is thrown. A ball thrown with a speed of 40 feet per second reaches a height of 25 feet. Find the height reached when the ball is thrown with a speed of 60 feet per second.
8. The weight of a piece of copper wire varies jointly as the length and as the square of the diameter. If 50 feet of copper wire with a certain diameter weighs $1^1/_4$ pounds, find the weight of 175 feet of copper wire having diameter 80% of that of the first wire.
9. If S varies directly as the cube of x, what change in S results from doubling x?
10. If S varies directly as the square of x and inversely as y, what change in S results from doubling x and tripling y?
11. If y varies directly as x and inversely as t, what is the effect on y of doubling x and t?
12. If y varies inversely as x and x varies inversely as t, state the relationship between y and t.

unit

16

Inverse Functions

Objectives

Know what is meant by an inverse relation. Be able to find the rule for the inverse relation if the rule of the relation is given. Know under what conditions a function has an inverse which is a function. Know the relationship between the graph of a relation and the graph of its inverse.

Terms

Inverse relation | Inverse function

(1)

Consider the relation $\{(1, 4), (2, 4), (3, -1), (5, 0)\}$.
If the first and second components are interchanged, we get the relation $\{(4, 1), (4, 2), (-1, 3), (0, 5)\}$. This relation is called the *inverse relation* of the original.

(2)

Relation. $\{(1, 4), (2, 4), (3, -1), (0, 5)\}$.
Inverse relation. $\{(4, 1), (4, 2), (-1, 3), (5, 0)\}$.

Notice that the original relation is also a function but that the inverse relation is not a function.

(3)

Let us consider the function f given by $\{(1, 2), (2, 3), (4, 1), (5, 5)\}$.
The domain of f is $\{1, 2, 4, 5\}$.
The range of f is $\{2, 3, 1, 5\}$.

Furthermore, $f(1) = 2$
$f(2) = 3$
$f(4) = 1$
$f(5) = 5$

(4)

---

f
(1, 2)
(2, 3)
(4, 1)
(5, 5)

If the components of the ordered pairs were interchanged, would the resulting relation be a function?

(5)

---

f	*Interchange Components*
(1, 2)	(2, 1)
(2, 3)	(3, 2)
(4, 1)	(1, 4)
(5, 5)	(5, 5)

The resulting relation is also a function. We call this new function the *inverse function of f*.

⑥

The notation for the inverse function of f is f^{-1}.
Notice that in the inverse function the roles of the domain and range are interchanged from the original function.

⑦

---

f	f^{-1}
(1, 2)	(2, 1)
(2, 3)	(3, 2)
(4, 1)	(1, 4)
(5, 5)	(5, 5)

Domain of f is $\{1, 2, 4, 5\}$
Range of f is $\{2, 3, 1, 5\}$

Domain of f^{-1} is $\{2, 3, 1, 5\}$
Range of f^{-1} is $\{1, 2, 4, 5\}$

⑧

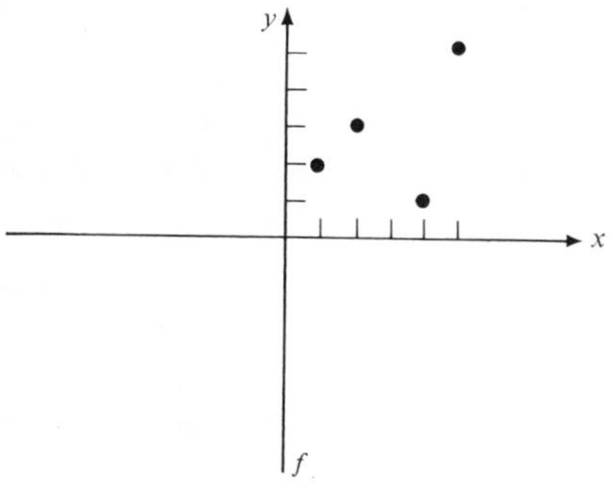

---

y
x
f

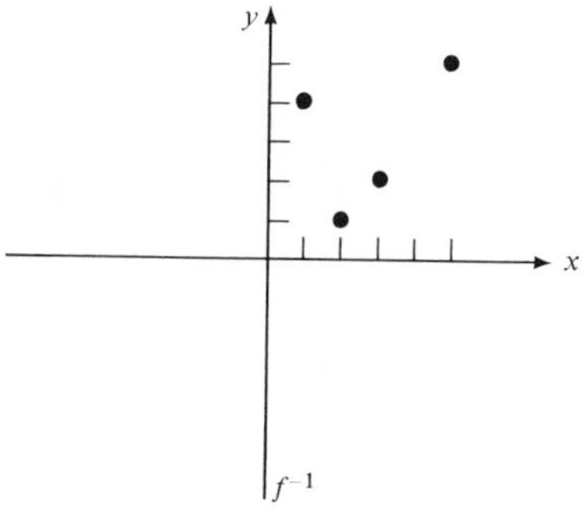

y
x
f^{-1}

f	f^{-1}
(1, 2)	(2, 1)
(2, 3)	(3, 2)
(4, 1)	(1, 4)
(5, 5)	(5, 5)

⑨

Consider the function g given by $\{(1, 2), (3, 2), (4, 1)\}$.
Does g have an inverse which is also a function?

g	*inverse relation of g*
(1, 2)	(2, 1)
(3, 2)	(2, 3)
(4, 1)	(1, 4)

The inverse does not represent a function.

⑩

Given a function f. The inverse function, f^{-1}, is the set of ordered pairs obtained by interchanging the first and second components of each ordered pair of f; the inverse function exists if and only if f has no two ordered pairs with the same second component.

(11)

Example:

f	f^{-1}
(0, 1)	(1, 0)
(1, 2)	(2, 1)
(2, 3)	(3, 2)
(−1, 0)	(0, −1)
(−2, −1)	(−1, −2)
(3, 4)	(4, 3)

$f(2) = 3$ $f^{-1}(2) = 1$

$f(3) = 4$ $f^{-1}(4) = 3$

(12)

Study Exercise One

A. Which of the following have inverses which are functions?

1. $\{(1, 1), (2, 2), (3, 3), (4, 4), (5, 5)\}$ **2.** $\{(1, 0), (2, 0), (3, 0)\}$

3. $\{(1, -1), (2, -2), (3, -3), (4, -4)\}$ **4.** $\{(-1, 5), (5, -1), (6, 2), (2, 6)\}$

5. $\{(1, 1)\}$

B. Miscellaneous:

6. If g is the function represented by $\{(1, 5), (6, 7), (2, 4), (8, 1), (0, 0)\}$, find;

(a) $g(8)$ **(b)** $g^{-1}(4)$ **(c)** $g^{-1}(0)$ **(d)** $g[g^{-1}(4)]$

7. For some function f, f^{-1} exists, the domain of f is $\{x|0 \leqslant x \leqslant 1\}$ and the range of f is $\{y|1 \leqslant y \leqslant 4\}$.
What is the domain of f^{-1}?

(13)

$f(x) = x + 1$

x	0	1	2	3	4	−1	−2	...
$f(x)$	1	2	3	4	5	0	−1	...

then $f^{-1}(x)$ has ordered pairs.

x	1	2	3	4	5	0	−1	...
$f^{-1}(x)$	0	1	2	3	4	−1	−2	...

Can you find the rule for these ordered pairs?

(14)

---

f: $\{(x, y)|y = x + 1\}$

f^{-1} is obtained by interchanging ordered pairs.

f^{-1}: $\{(x, y)|x = y + 1\} = \{(x, y)|y = x - 1\}$

(15)

To find the inverse of $y = x + 1$;

(1) interchange x with y and y with x.

(2) solve the resulting equation for y.

Line (1) $y = x + 1$

Line (2) $x = y + 1$ (interchange x and y)

Line (3) $y = x - 1$

The inverse of $y = x + 1$ is $y = x - 1$.

(16)

Example: Let f be given by $y = 3x + 1$ with domain $\{x|0 \leqslant x \leqslant 2\}$. Find f^{-1} and its domain and range.

Solution:

$$y = 3x + 1$$
$$x = 3y + 1 \quad \text{(interchanging } x \text{ and } y\text{)}$$
$$x - 1 = 3y$$
$$3y = x - 1$$
$$y = \frac{x - 1}{3}$$
$$f^{-1}(x) = \frac{x - 1}{3}$$

Domain of $f^{-1}(x)$ is $\{x|1 \leqslant x \leqslant 7\}$.

Range of $f^{-1}(x)$ is $\{y|0 \leqslant y \leqslant 2\}$.

(17)

Example: Let $f(x) = \sqrt{x}$. Find $f^{-1}(x)$ and its domain and range.

Solution: $f(x)$ has domain $\{x|x \geqslant 0\}$ and range $\{y|y \geqslant 0\}$.

$$y = \sqrt{x}$$
$$x = \sqrt{y} \quad \text{(interchanging } x \text{ and } y\text{)}$$
$$x^2 = y$$
$$y = x^2$$

$f^{-1}(x) = x^2$ with domain $\{x|x \geqslant 0\}$ and range $\{y|y \geqslant 0\}$.

(18)

Study Exercise Two

The inverse of each of the following functions is a function.
Find the inverse function and its domain and range.

1. $f(x) = x + 1, x \geqslant 0$

2. $f(x) = 3x - 2$

3. $f(x) = \sqrt{x - 1}$

4. $f(x) = x^2, x \geqslant 0$

(19)

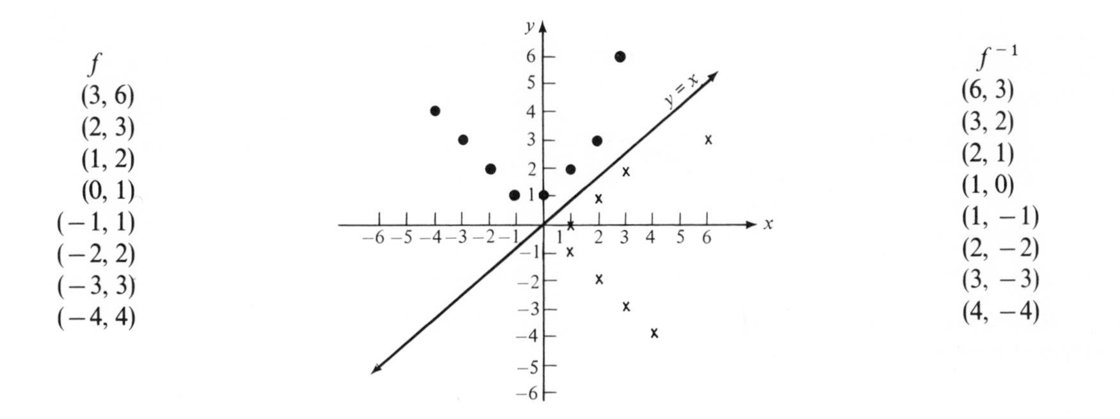

The inverse points seem to be reflections of the original points across the line $y = x$.

20

Point (a, b) is the reflection of point (b, a) and vice versa across the line $y = x$. The line $y = x$ is the perpendicular bisector of the line segment joining (a, b) and (b, a).

21

The procedure for obtaining the graph of an inverse statement from the graph of the original statement is to first draw the line $y = x$. Then, from a selection of points on the given graph draw line segments perpendicular to $y = x$ which extend an equal distance on the opposite side of this line.

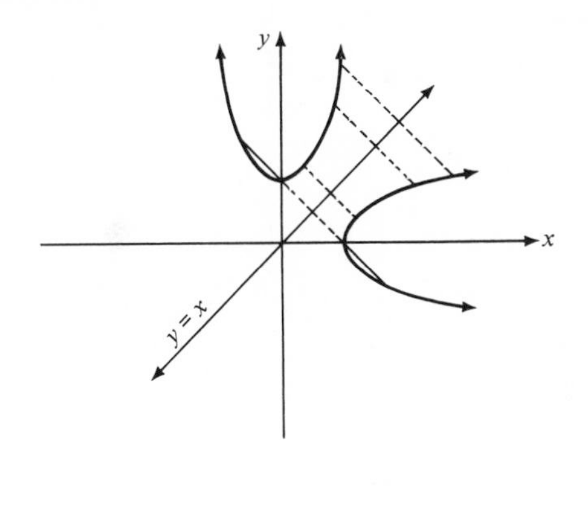

22

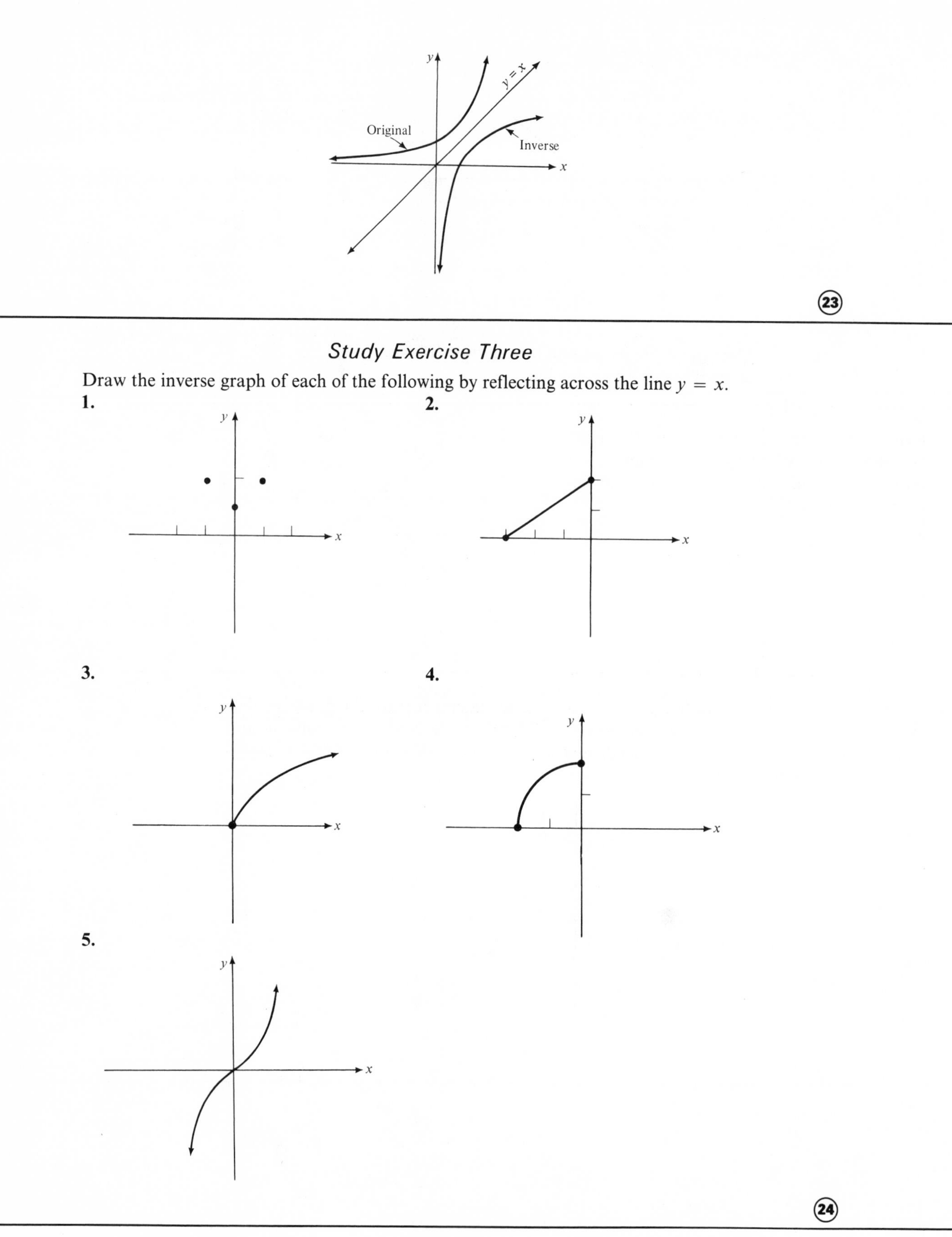

Study Exercise Three

Draw the inverse graph of each of the following by reflecting across the line $y = x$.

1.

2.

3.

4.

5.

24

REVIEW EXERCISES

A. Determine which of the following functions have inverses which are also functions.

1. $\{(0, 0), (1, 1), (-1, 2), (2, -1)\}$ **2.** $\{(3, -2), (4, -2), (5, 2)\}$

3. $\{(x, y)|y = 2x\}$ **4.** $\{x, y)|y = |x|\}$

B. The inverse of each of the following functions is also a function. Find the inverse function and also its domain and range.

5. $\{(x, y)|y = 5x - 2\}$ **6.** $f(x) = \frac{1}{x}, x \neq 0$

7. $y = \frac{12 - 3x}{4}$ **8.** $y = \sqrt{x - 2}$

C. Find the inverse graphically by reflecting across $y = x$ and state if the inverse is also a function.

9.

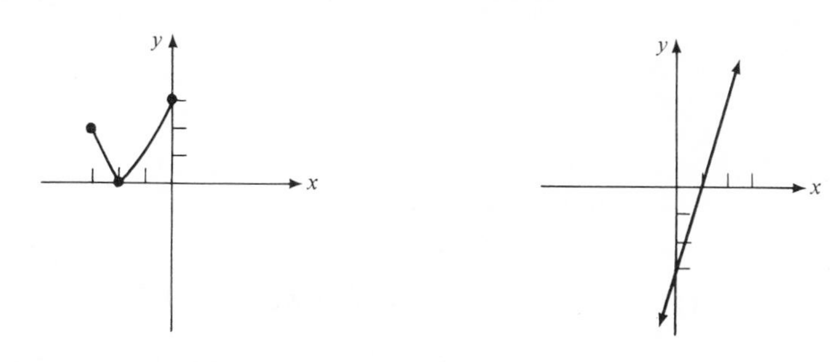

25

SOLUTIONS TO REVIEW EXERCISES

A. 1. The inverse is a function.

2. The inverse is not a function (the second component is the same for two ordered pairs).

3. The inverse is a function.

4. The inverse is not a function (for example $(-2, 2)$ and $(2, 2)$).

B. 5. $y = 5x - 2$

$x = 5y - 2$ (interchange)

$y = \frac{x + 2}{5}$

Domain of f^{-1}: reals
Range of f^{-1}: reals

6. $y = \frac{1}{x}$

$x = \frac{1}{y}$ (interchange)

$xy = 1$

$y = \frac{1}{x}$

$f^{-1}(x) = \frac{1}{x}$

Domain of f^{-1}: $\{x|x \neq 0\}$
Range of f^{-1}: $\{y|y \neq 0\}$

7. $y = \frac{12 - 3x}{4}$

$x = \frac{12 - 3y}{4}$ (interchange)

$4x = 12 - 3y$

$y = \frac{-4x + 12}{3}$

$f^{-1}(x) = \frac{-4x + 12}{3}$

Domain of f^{-1}: reals
Range of f^{-1}: reals

(Frame 26, contd.)

8. $y = \sqrt{x-2}$

$x = \sqrt{y-2}$ (interchange)

$x^2 = y - 2$

$y = x^2 + 2$

$f^{-1}(x) = x^2 + 2$

Domain of f: $\{x|x \geqslant 2\}$

Range of f: $\{y|y \geqslant 0\}$

Domain of f^{-1}: $\{x|x \geqslant 0\}$

Range of f^{-1}: $\{y|y \geqslant 2\}$

9.

inverse, not a function

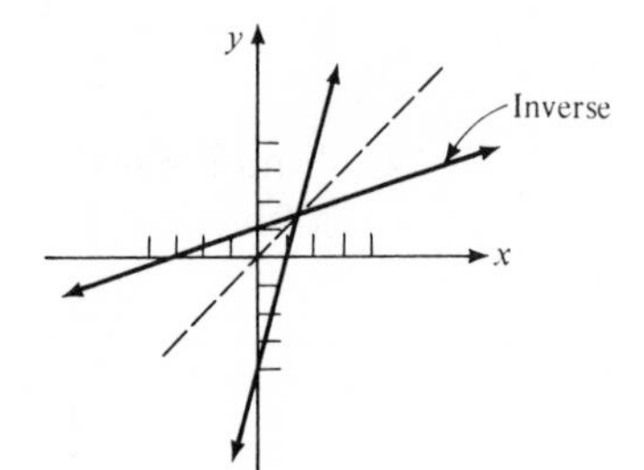

inverse is a function

(26)

SOLUTIONS TO STUDY EXERCISES

Study Exercise One (Frame 13)

A. 1. Inverse is a function.

2. Inverse is not a function.

3. Inverse is a function.

4. Inverse is a function.

5. Inverse is a function.

B. 6. **(a)** 1 **(b)** 2 **(c)** 0 **(d)** 4

7. All real numbers between 1 and 4 inclusive.

(13A)

Study Exercise Two (Frame 19)

1. $y = x + 1$

$x = y + 1$ (interchange)

$y = x - 1$

$f^{-1}(x) = x - 1$

Domain is $\{x|x \geqslant 1\}$;

range is $\{y|y \geqslant 0\}$.

2. $y = 3x - 2$

$x = 3y - 2$ (interchange)

$y = \dfrac{x+2}{3}$

$f^{-1}(x) = \dfrac{x+2}{3}$

Domain is the set of reals;

range is the set of reals.

3. $y = \sqrt{x-1}$

$x = \sqrt{y-1}$ (interchange)

$x^2 = y - 1$

$y = x^2 + 1$

$f^{-1}(x) = x^2 + 1$

Domain is $\{x|x \geq 0\}$;

range is $\{y|y \geqslant 1\}$.

4. $y = x^2, x \geqslant 0$

$x = y^2$ (interchange)

$y = \sqrt{x}$

$f^{-1}(x) = \sqrt{x}$

Domain is $\{x|x \geqslant 0\}$.

range is $\{y|y \geqslant 0\}$.

(19A)

SOLUTIONS TO STUDY EXERCISES, CONTD.

Study Exercise Three (Frame 24)

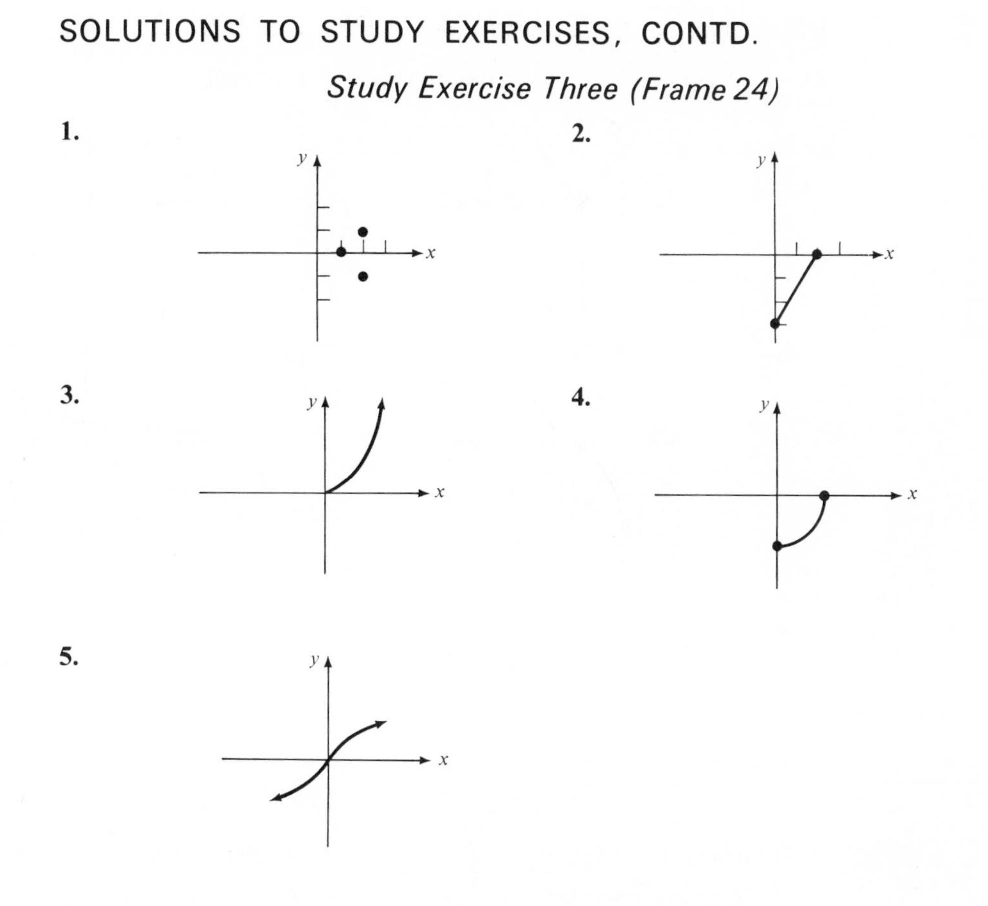

24A

UNIT 16—SUPPLEMENTARY PROBLEMS

A. True or false:

1. Every function is a relation.

2. Every function has an inverse which is a function.

3. Every relation is a function.

4. The inverse of the set of points forming a horizontal line is not a function.

B. 5. Find the inverse relation of $\{(2, 3), (4, 2), (5, 6), (-1, 0)\}$.

6. What is the domain and range of the inverse relation found in problem five?

7. Is the inverse of problem five a function?

8. If f is $\{(4, 2)\ (-1, 3)\ (3, -2)\ (7, 1)\}$, find;

(a) $f(-1)$ **(b)** $f^{-1}(-1)$ **(c)** $f^{-1}(2)$

(d) $f^{-1}(4)$ **(e)** $f[f^{-1}(1)]$ **(f)** $f[f(3)]$

9. If $f(x) = 3x + 1$, find $f^{-1}(x)$ and find;

(a) $f^{-1}(3)$ **(b)** $f[f^{-1}(x)]$ **(c)** $f^{-1}[f(x)]$

C. The inverses of the following functions are also functions. Find the inverse function and its domain and range.

10. $f(x) = 2x$ **11.** $f(x) = (x - 1)^2, \quad x \geqslant 1$

12. $f(x) = 4x - 1$ **13.** $f(x) = x^3$

14. $f(x) = \sqrt{x + 1}$

D. Reflect about $y = x$ and find the inverse graphically.

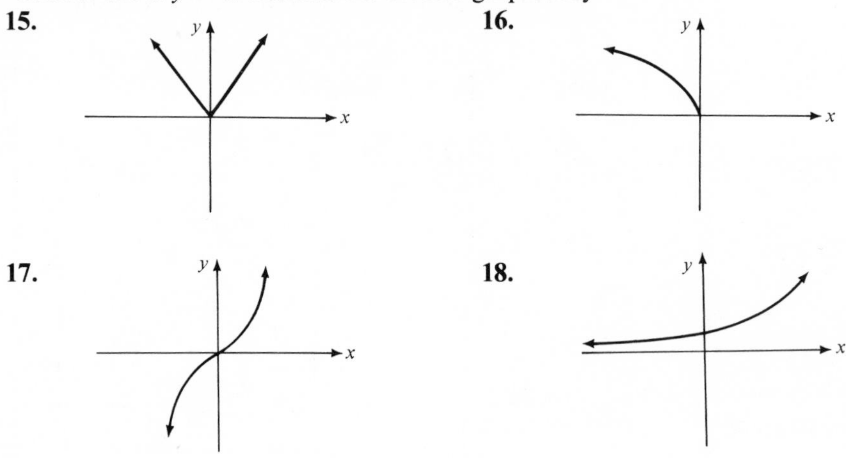

unit

17

Exponential Functions

Objectives

1. Know what an exponential function is.
2. Be able to graph the exponential function.

Terms

Strictly increasing　　　　Strictly decreasing

①

Let b be any real number. We are familiar with b^x for certain numbers x. As an example, when $b = 2$, 2 means 2 is used as a factor x times provided x is a natural number. 2^0 stands for 1. 2^x if $x < 0$ and x is any integer means $\frac{1}{2^{-x}}$.

2^x if $x > 0$ and $x \in I$ takes the same meaning as the natural number exponent.

②

Thus 2^x has meaning when $x \in N$ and when $x \in I$;

2^x also has meaning when $x \in F$. Let $x = \frac{m}{n}$ where $m, n \in I, n \neq 0$.

$$\text{Then } 2^{m/n} = \sqrt[n]{2^m} \quad \text{or} \quad (\sqrt[n]{2})^m$$

③

But what about 2^x when x is an irrational number?

What sort of meaning can be given to $2^{\sqrt{2}}$ or $2^{\sqrt{3}}$?

④

$\sqrt{2}$ can be thought of as being found in a succession of better and better approximations.

$\sqrt{2}$ is about 1.4, 1.41, 1.414, 1.4142

⑤

$2^{\sqrt{2}}$ will be thought of as approximated by $2^{1.4}$, $2^{1.41}$, $2^{1.414}$, $2^{1.4142}$, etc.

Each of these exponents is now rational.

⑥

Irrational Exponents

We will assume without proof that a^x has meaning when $x \in R$ and that the following laws apply:

1. $a^x \cdot a^y = a^{x+y}, \quad a > 0$　　**2.** $\frac{a^x}{a^y} = a^{x-y}, \quad a > 0$　　**3.** $(a^x)^y = a^{xy}, \quad a > 0$

⑦

Graph of $y = 2^x, x \in R$

x	y
2	4
1	2
0	1
-1	1/2
-2	1/4
-3	1/8
1/4	1.19
1/3	1.26
2/5	1.32
1/2	1.41
3/5	1.52

(8)

Some Properties of $y = 2^x, x \in R$.

1. $y = 2^x, x \in R$ is a function.
2. The domain is $\{x | x \in R\}$.
3. The range is $\{y | y > 0 \text{ and } y \in R\}$.
4. If $x_1 < x_2$, then we notice from the graph that $f(x_1) < f(x_2)$ and we say that the function is *strictly increasing*.

(9)

Study Exercise One

1. Sketch the graph of $y = 3^x$.
2. Sketch $y = (1/2)^x$.
3. Sketch $y = 1^x$.
4. Sketch $y = 10^x$.
5. From the graphs of problems 1 through 4 determine whether or not each represents a function. Also, find the domain and range of each.

(10)

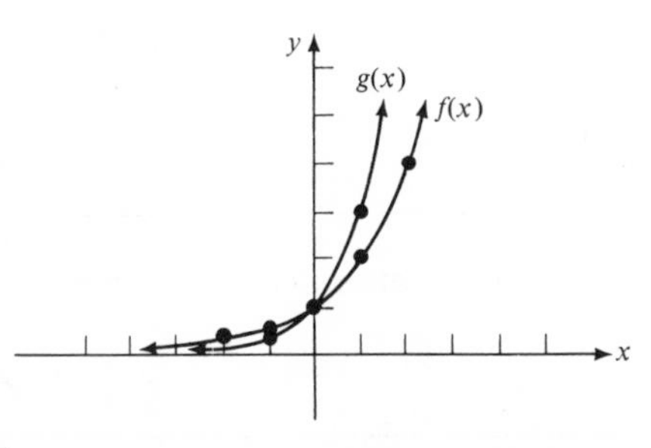

$f(x) = 2^x$ and $g(x) = 3^x$.
Both functions are strictly increasing.
Both functions have value 1 when $x = 0$.
Both functions are positive for all values of x.

(11)

$$y = a^x, a > 1$$

Properties:

1. The function is positive for all values of x.
2. The function passes through (0, 1).
3. The function is strictly increasing.
4. The graph of $y = a^x, a > 1$ resembles

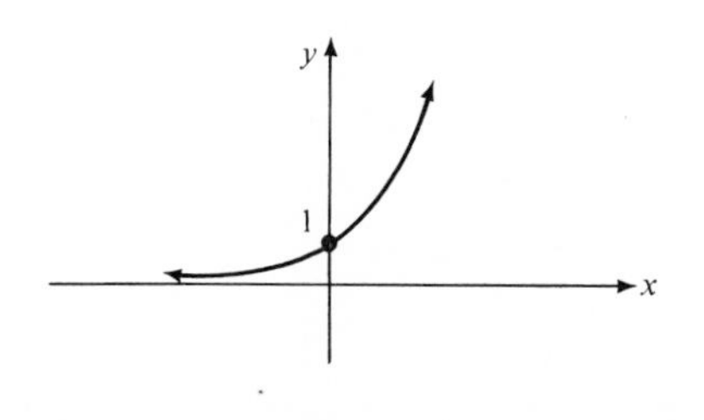

⑫

$$y = a^x, a = 1$$

Properties:

1. The function is constant; its y value is always 1.
2. The graph of $y = a^x,\ a = 1$ is

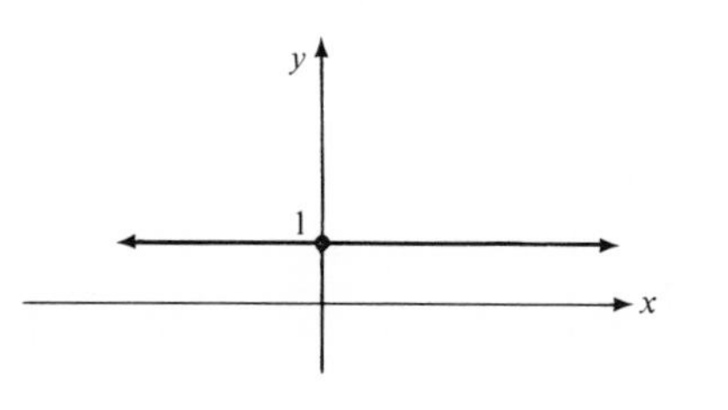

⑬

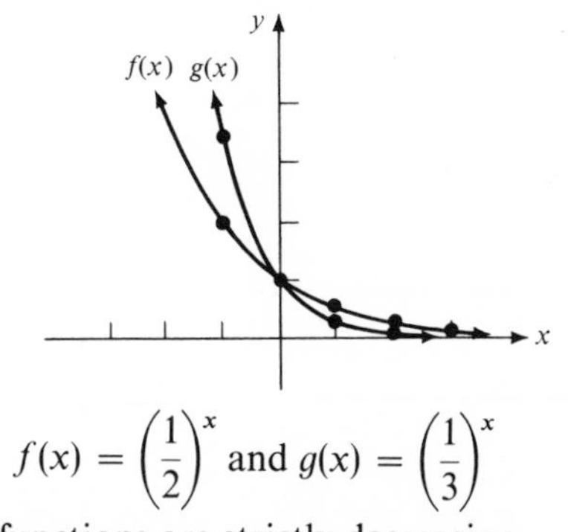

$$f(x) = \left(\frac{1}{2}\right)^x \text{ and } g(x) = \left(\frac{1}{3}\right)^x$$

Both functions are strictly decreasing.
Both functions have value 1 when $x = 0$.
Both functions are positive for all values of x.

⑭

$$y = a^x, 0 < a < 1$$

Properties:

1. The function is positive for all values of x.
2. The function passes through (0, 1).
3. The function is strictly decreasing.
4. The graph of $y = a^x$ where $0 < a < 1$ is

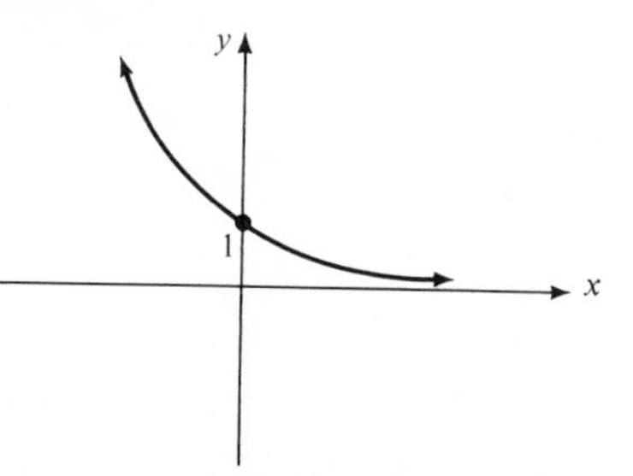

(15)

Study Exercise Two

Complete the table of y values.

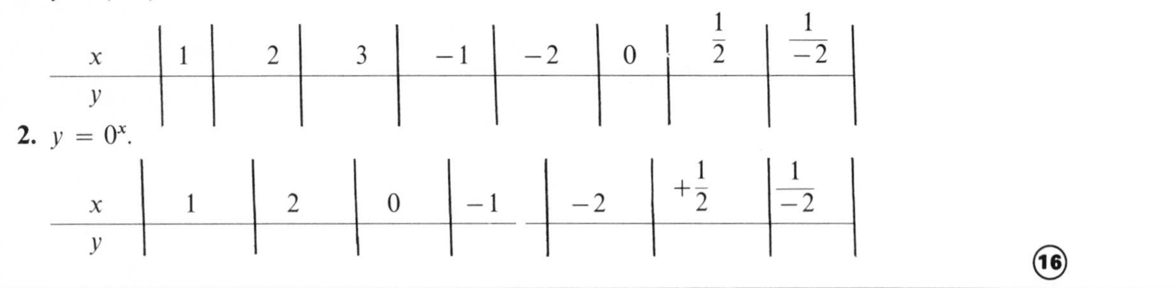

1. $y = (-2)^x$.

x	1	2	3	-1	-2	0	$\frac{1}{2}$	$\frac{1}{-2}$
y								

2. $y = 0^x$.

x	1	2	0	-1	-2	$+\frac{1}{2}$	$\frac{1}{-2}$
y							

(16)

What about $y = a^x$ if $a < 0$?

Since y will not exist when a is negative and x is of the form $1/n$, $n \in I$ and n is even, we will not define $y = a^x$ if $a < 0$.

What about $y = a^x$ if $a = 0$?

Since y will not exist when x is negative or zero and y will be zero for any positive x, we will not consider this case either.

(17)

Summary

$y = a^x$ has as graph:

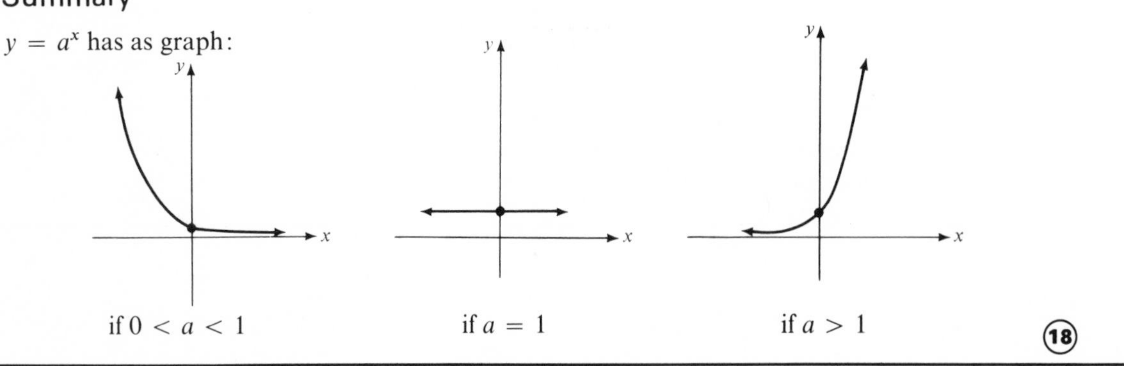

if $0 < a < 1$ if $a = 1$ if $a > 1$

(18)

Study Exercise Three

Find the inverses of the three graphs below by reflecting about the line $y = x$. That is, draw the line $y = x$ and find all the points symmetric about that line.
Which of the three inverses are also functions?

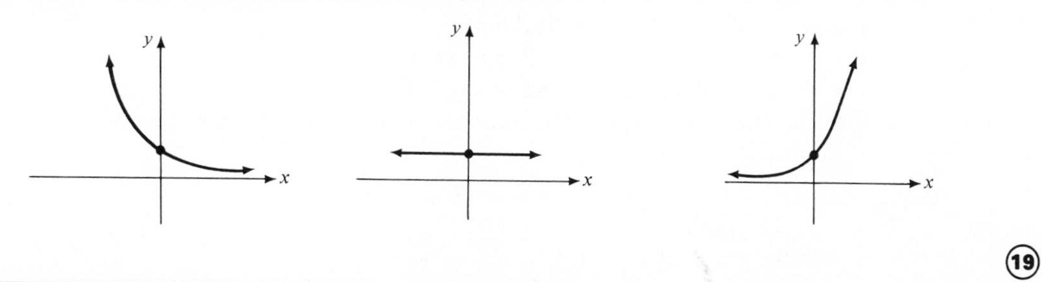

(19)

Conclusion

Thus we find that for the *exponential function*, $y = a^x, a > 0$, in order for its inverse to also be a function a must be positive but not one.

$f(x) = a^x, a > 0$ and $a \neq 1$
Domain is $\{x|x \in R\}$.
Range is $\{y|y > 0 \text{ and } y \in R\}$.

The inverse of $f(x)$ is also a function with domain positive reals and range all reals.

(20)

REVIEW EXERCISES

A. True or False:

1. The graph of $y = 4^x$ is strictly increasing.
2. The range of the exponential function is the set of reals.
3. The graph of $y = 1^x$ is a vertical line.
4. The inverse of $y = 2^x$ is a strictly increasing function.
5. The exponential function passes through (0, 1)
6. If the inverse of the exponential function is to be a function, a must be positive and not one.

B.

7. Sketch the graph of $y = 4^x$.
8. Sketch the graph of $y = (1/3)^x$.
9. Sketch the inverse of $y = 10^x$ by sketching $y = 10^x$ and then reflecting through the line $y = x$.

(21)

SOLUTIONS TO REVIEW EXERCISES

A. **1.** True

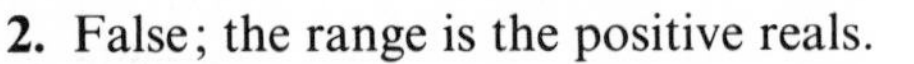

2. False; the range is the positive reals.

3. False; the graph is a horizontal line.

4. True

5. True

6. True

B. **7.**

8.

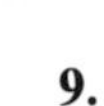

9.

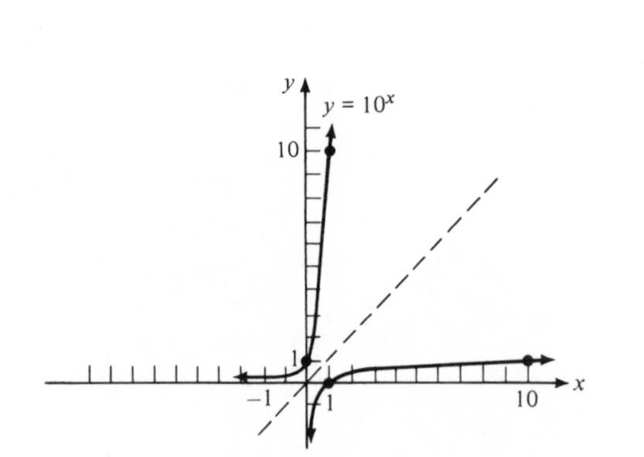

(22)

SOLUTIONS TO STUDY EXERCISES

Study Exercise One (Frame 10)

1.

x	0	1	2	−1	−2
y	1	3	9	$\frac{1}{3}$	$\frac{1}{9}$

2.

x	0	1	2	−1	−2
y	1	$\frac{1}{2}$	$\frac{1}{4}$	2	4

3.

x	0	1	2	−1	−2	−3
y	1	1	1	1	1	1

4.

x	0	1	2	−1	−2
y	1	10	100	$\frac{1}{10}$	$\frac{1}{100}$

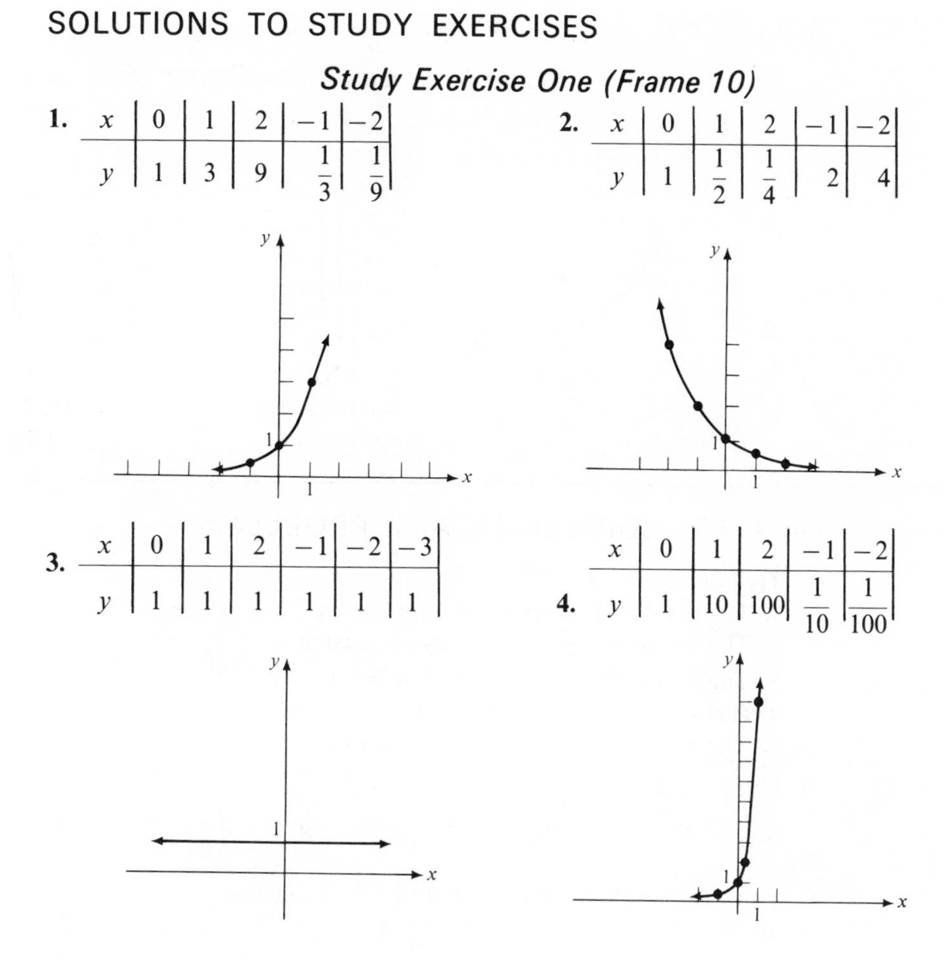

5. All are functions.

	$y = 3^x$	$y = \left(\frac{1}{2}\right)^x$	$y = 1^x$	$y = 10^x$
Domain	Reals	Reals	Reals	Reals
Range	Positive reals	Positive reals	1	Positive reals

(10A)

Study Exercise Two (Frame 16)

1.

y	−2	4	−8	−1/2	1/4	1	Does not exist	Does not exist

2.

y	0	0	Does not exist	Does not exist	Does not exist	0	Does not exist

(16A)

SOLUTIONS TO STUDY EXERCISES, CONTD.

Study Exercise Three (Frame 19)

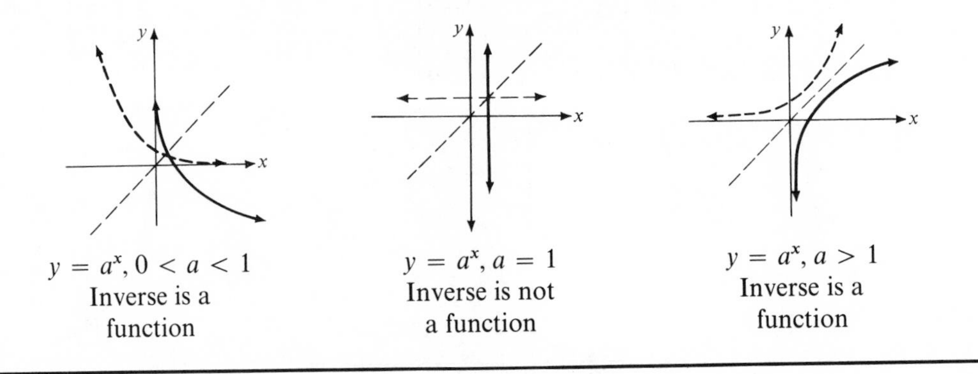

$y = a^x, 0 < a < 1$ Inverse is a function	$y = a^x, a = 1$ Inverse is not a function	$y = a^x, a > 1$ Inverse is a function

19A

UNIT 17—SUPPLEMENTARY PROBLEMS

A. True or false:

1. The point (0, 1) is on the graph of $y = a^x$, $a > 0$, $a \neq 1$.
2. The function $y = 2^x$ is strictly decreasing.
3. The range of $f(x) = a^x$, $a > 0$, $a \neq 1$ is $\{y|y \geqslant 0\}$.
4. $f(x) = 10^x$ is a strictly increasing function.
5. The inverse of $y = 10^x$ is a strictly increasing function.

B. Draw the graphs:

6. $y = 4^x$ **7.** $y = 2^{-x}$ **8.** $y = \left(\frac{1}{2}\right)^x$ **9.** $y = -2^x$

C. Draw the graph of the inverse of the given function.

10. $y = 2^x$ **11.** $y = 3^x$ **12.** $y = 2^{-x}$

unit

18

Properties of Logarithms

Objectives

1. Be able to change from exponential to logarithmic form and vice versa.
2. Know and be able to use the three laws of logarithms

Terms logarithm

①

In the previous unit we studied the exponential function $y = a^x, a > 0, a \neq 1$. The graph of this function is *strictly increasing* if $a > 1$ and *strictly decreasing* if $0 < a < 1$.

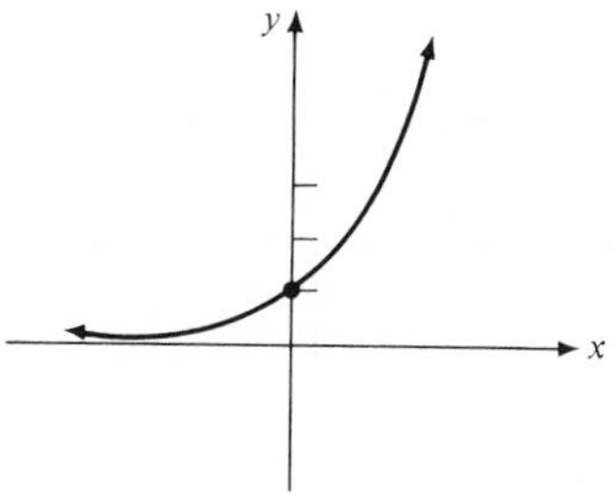

$y = a^x, \quad a > 1$

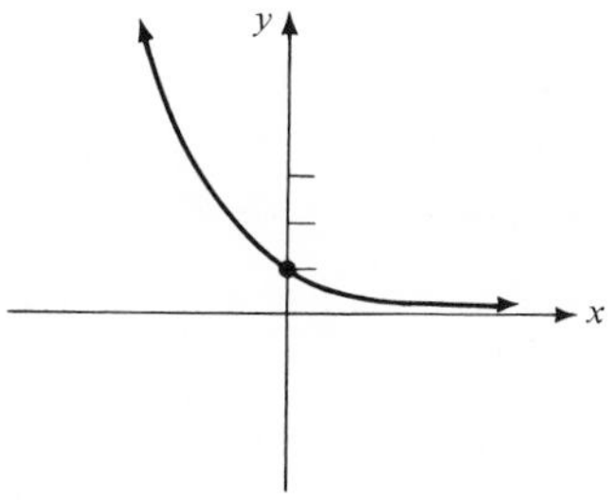

$y = a^x, \quad 0 < a < 1$

②

Let us consider the case $y = a^x, a > 1$:

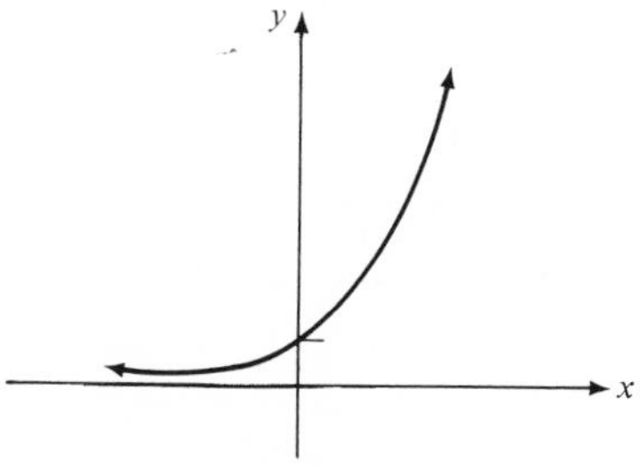

The domain is $\{x|x \in R\}$.
The range is $\{y|y > 0 \text{ and } y \in R\}$.

③

Does the function $y = a^x, a > 1$ have an inverse that is also a function? To answer the question we must reflect $y = a^x, a > 1$ across the line $y = x$.

④

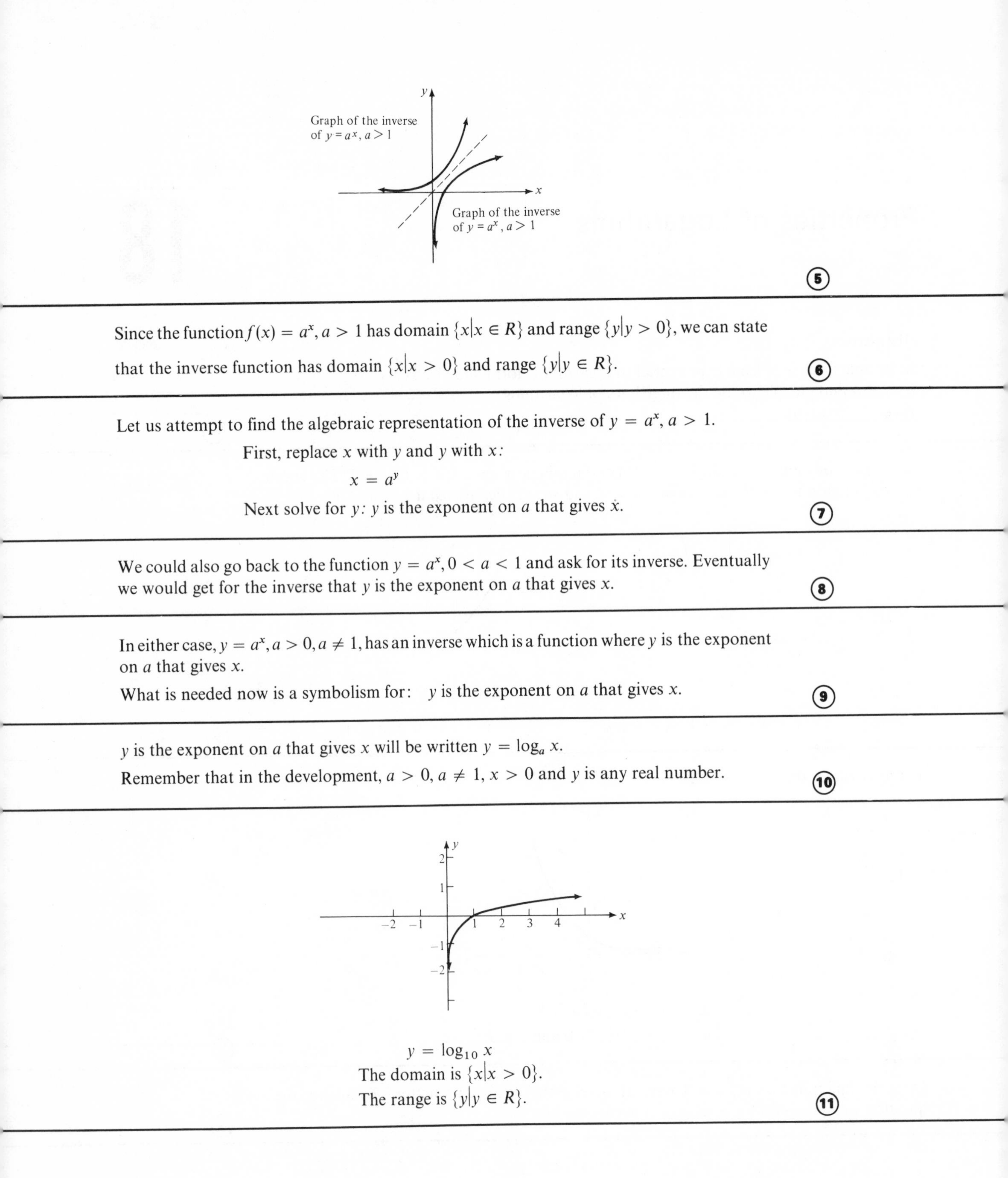

(5)

Since the function $f(x) = a^x, a > 1$ has domain $\{x|x \in R\}$ and range $\{y|y > 0\}$, we can state that the inverse function has domain $\{x|x > 0\}$ and range $\{y|y \in R\}$.

(6)

Let us attempt to find the algebraic representation of the inverse of $y = a^x, a > 1$.

First, replace x with y and y with x:

$$x = a^y$$

Next solve for y: y is the exponent on a that gives x.

(7)

We could also go back to the function $y = a^x, 0 < a < 1$ and ask for its inverse. Eventually we would get for the inverse that y is the exponent on a that gives x.

(8)

In either case, $y = a^x, a > 0, a \neq 1$, has an inverse which is a function where y is the exponent on a that gives x.

What is needed now is a symbolism for: y is the exponent on a that gives x.

(9)

y is the exponent on a that gives x will be written $y = \log_a x$.

Remember that in the development, $a > 0$, $a \neq 1$, $x > 0$ and y is any real number.

(10)

$$y = \log_{10} x$$

The domain is $\{x|x > 0\}$.

The range is $\{y|y \in R\}$.

(11)

$$y = \log_b x \text{ if and only if } x = b^y.$$

$y = \log_b x$ and $x = b^y$ basically express the same relationship between the same three quantities; one emphasizes the role of exponent (or *logarithm*), the other the role of the power.

(12)

Definition of Logarithm

If $n = b^x$, when b is any positive real except one, and x any real number, then x is the *logarithm* of n to base b.

The logarithm of a number n to a base b is the exponent x to which the base is raised to give the number.

$$x = \log_b n, b > 0, b \neq 1, n > 0$$

(13)

Since $n = b^x$ if and only if $x = \log_b n$, a logarithm is an exponent.

$n = b^x$ is called the *exponential form* and $x = \log_b n$ is called the log form.

(14)

Example: Express each of the log forms in exponential form.

1. $\log_{17} 1 = 0$
Solution: $17^0 = 1$

2. $\log_6 36 = 2$
Solution: $6^2 = 36$

3. $\log_8 32 = 5/3$
Solution: $8^{5/3} = 32$

(15)

Example: Express each of the exponential forms in log form.

1. $3^2 = 9$
Solution: $2 = \log_3 9$

2. $7^{-2} = \frac{1}{49}$
Solution: $-2 = \log_7 \frac{1}{49}$

3. $10^0 = 1$
Solution: $0 = \log_{10} 1$

In each case check your answer:

$0 = \log_{10} 1$ $\qquad$ $10^0 = 1$

(16)

Exponential Form	*Log Form*
$b^x = n$	$x = \log_b n$
1. $2^3 = 8$	**1.** $3 = \log_2 8$
2. $(1/2)^2 = 1/4$	**2.** $2 = \log_{1/2} 1/4$
3. $4^2 = 16$	**3.** $2 = \log_4 16$

(17)

Study Exercise One

A. Express in log form:

1. $4^3 = 64$ **2.** $81^{1/4} = 3$ **3.** $5^{-2} = 1/25$ **4.** $10^0 = 1$

B. Express in exponential form:

5. $\log_{10} 1/10 = -1$ **6.** $\log_{64} 16 = 2/3$ **7.** $\log_{17} 1 = 0$ **8.** $\log_4 8 = 3/2$

(18)

If we wish to find $\log_2 8$, we may proceed as follows:

$$\begin{aligned} \text{let } \log_2 8 &= x \\ \text{then } \quad 2^x &= 8 \\ x &= 3 \end{aligned}$$

(19)

Study Exercise Two

Find the exact value:

1. $\log_3 9$ **2.** $\log_2 32$ **3.** $\log_{10} 1000$ **4.** $\log_4 16$ **5.** $\log_{10} .01$

(20)

Laws of Logarithms

Law I. The logarithm of a product is equal to the sum of the logarithms of its factors, all logarithms taken to the same base.

$$\log_b (m \cdot n) = \log_b m + \log_b n$$

(21)

$$\log_b (m \cdot n) = \log_b m + \log_b n$$

Proof: Let $x = \log_b m$

$y = \log_b n$

Write in exponential form $m = b^x$

$n = b^y$

Multiply m by n $\quad m \cdot n = b^x \cdot b^y$

$= b^{x+y}$

Write in log form $\quad \log_b (m \cdot n) = x + y$

Replace x and y by their values $\quad \log_b (m \cdot n) = \log_b m + \log_b n$

(22)

Example 1: Using Law I expand $\log_b 5x$ into sums.

Solution: $\log_b 5 \cdot x = \log_b 5 + \log_b x$

Example 2: Using Law I expand $\log_{10} 100x$ into sums.

Solution: $\log_{10} 100 \cdot x = \log_{10} 100 + \log_{10} x$

$= 2^* + \log_{10} x$

*since $\log_{10} 100 = 2$

(23)

Laws of Logarithms

Law II. The logarithm of a quotient is equal to the logarithm of the dividend minus the logarithm of the divisor, all logarithms taken to the same base.

$$\log_b \frac{m}{n} = \log_b m - \log_b n$$

(24)

$$\log_b \frac{m}{n} = \log_b m - \log_b n$$

Proof. Let $x = \log_b m$

$y = \log_b n$

Write in exponential form $\quad m = b^x$

$n = b^y$

Divide m by n $\quad m/n = \dfrac{b^x}{b^y}$

$= b^{x-y}$

Write in log form $\quad \log_b m/n = x - y$

Replace x and y by their values $\quad \log_b m/n = \log_b m - \log_b n$

(25)

Example 1: Using Law II (and/or Law I) expand $\log_6 \left(\frac{xy}{z}\right)$ into sums.

Solution: $\log_6 \left(\frac{xy}{z}\right) = \log_6 (xy) - \log_6 z$
$= \log_6 x + \log_6 y - \log_6 z$

Example 2: Using Law II (and/or Law I) expand $\log_4 \frac{x}{x+3}$ into sums.

Solution: $\log_4 \frac{x}{x+3} = \log_4 x - \log_4 (x+3)$

(26)

Laws of Logarithms

Law III. The logarithm of a power of a number is equal to the exponent times the logarithm of the number, all logarithms taken to the same base.

$$\log_b n^k = k \log_b n$$

Proof: *Let* $x = \log_b n$

Write in exponential form $n = b^x$

Raise to k power $n^k = (b^x)^k$
$= b^{kx}$

Write in log form $\log_b n^k = kx$

Replace x by its value $\log_b n^k = k \log_b n$

(27)

Example 1: Using the laws of logs expand into sums $\log_3 \sqrt[5]{4} \cdot y$.

Solution: $\log_3 \sqrt[5]{4} \cdot y = \log_3 4^{1/5} \cdot y$
$= \log_3 4^{1/5} + \log_3 y$
$= \frac{1}{5} \log_3 4 + \log_3 y$

Example 2: Expand $\log_{10} \frac{2^5}{3}$ into sums.

Solution: $\log_{10} \frac{2^5}{3} = \log_{10} 2^5 - \log_{10} 3$
$= 5 \log_{10} 2 - \log_{10} 3$

(28)

Example: Expand $\log_4 2\pi \sqrt{\frac{l}{g}}$ into sums.

Solution: $\log_4 2\pi \sqrt{\frac{l}{g}} = \log_4 (2\pi) + \log_4 \left(\frac{l}{g}\right)^{1/2}$
$= \log_4 2 + \log_4 \pi + \log_4 \left(\frac{l}{g}\right)^{1/2}$
$= \log_4 2 + \log_4 \pi + 1/2 \log_4 \frac{l}{g}$
$= \log_4 2 + \log_4 \pi + 1/2 \log_4 l - 1/2 \log_4 g$

(29)

More examples. Express each as a single logarithm.

1. $\log_a y - \log_a 2$

Solution: $\log_a y - \log_a 2 = \log_a \frac{y}{2}$

(Frame 30, contd.)

2. $\log_{10} \pi + 2 \log_{10} r + \log_{10} h - \log_{10} 3$

Solution: $\log_{10} \pi + 2 \log_{10} r + \log_{10} h - \log_{10} 3$

$$= \log_{10} \pi r^2 h - \log_{10} 3$$
$$= \log_{10} \frac{\pi r^2 h}{3}$$

(30)

Study Exercise Three

A. Using the laws of logarithms expand into sums.

1. $\log_a (2 \cdot 3^5 \cdot 7)$ **2.** $\log_b (\sqrt{7} \cdot x)$ **3.** $\log_{10} \pi r^2$

B. Using the laws of logarithms write as a single logarithm.

4. $\log_{10} 2 + \log_{10} \pi + \log_{10} r$ **5.** $\frac{1}{3} \log_a r - 2 \log_a t$ **6.** $-\log_{10} k - \log_{10} M$

(31)

If we are given that $\log_{10} 2 = .3010$, we may compute the $\log_{10} 20$ without tables.

$$\begin{aligned} \log_{10} 20 &= \log_{10} (10) \cdot 2 \\ &= \log_{10} 10 + \log_{10} 2 \\ &= 1 + \log_{10} 2 \\ &= 1 + .3010 \\ &= 1.3010 \end{aligned}$$

(32)

Example: If $\log_{10} 2 = .3010$, find $\log_{10} .2$ without tables.

Solution:
$$\begin{aligned} \log_{10} .2 &= \log_{10} (.1)(2) \\ &= \log_{10} .1 + \log_{10} 2 \\ &= \log_{10} \frac{1}{10} + \log_{10} 2 \\ &= -1 + \log_{10} 2 \\ &= -1 + .3010 \\ &= -.6990 \end{aligned}$$

(33)

Example: Find $\log_{10} 40$ without tables if $\log_{10} 2 = .3010$

Solution:
$$\begin{aligned} \log_{10} 40 &= \log_{10} (4 \cdot 10) \\ &= \log_{10} (2^2)(10) \\ &= \log_{10} 2^2 + \log_{10} 10 \\ &= 2 \log_{10} 2 + 1 \\ &= 2(.3010) + 1 \\ &= .6020 + 1 \\ &= 1.6020 \end{aligned}$$

(34)

Study Exercise Four

If $\log_{10} 3 = .4771$, find the following logarithms without tables.

1. $\log_{10} 30$ **2.** $\log_{10} .3$ **3.** $\log_{10} 27$ **4.** $\log_{10} \sqrt{300}$

(35)

Equations involving logarithms may be solved by using the definition of logarithm.

Example: Solve $\log_{10} x = -2$

Solution: $10^{-2} = x$

$$\frac{1}{100} = x$$

Example: Solve $\log_{10} 10^x = 7$

Solution: $10^7 = 10^x$

$$x = 7$$

(36)

Study Exercise Five

Solve for the variable.

1. $\log_{10} x = -4$ **2.** $\log_{10} .001 = x$ **3.** $\log_3 81 = y$

4. $\log_b 25 = 2$ **5.** $\log_{10} .0001 = t$

(37)

Logarithms of positive numbers to base ten may be estimated from a carefully drawn graph of $y = \log_{10} x$.

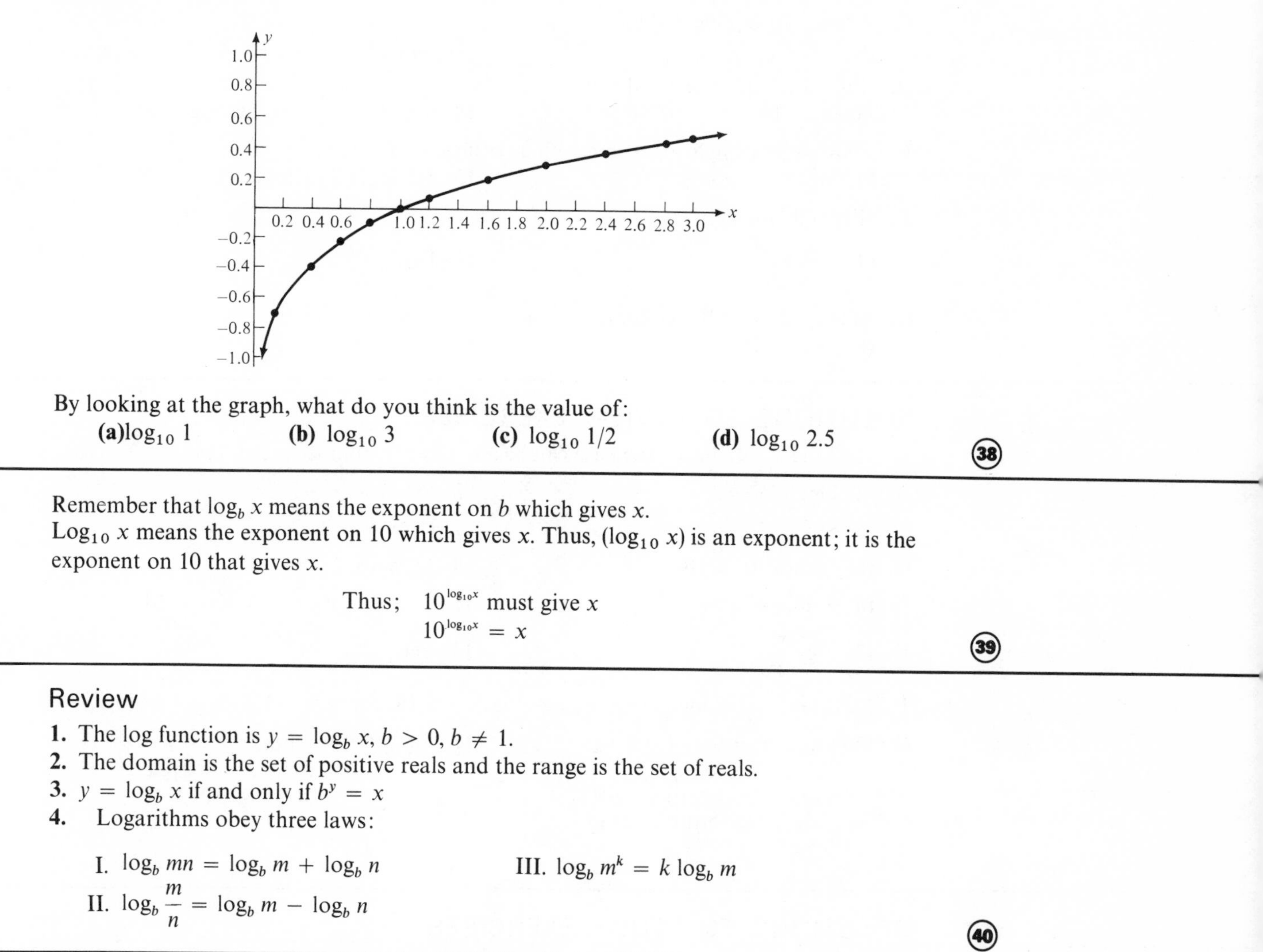

By looking at the graph, what do you think is the value of:

(a) $\log_{10} 1$ **(b)** $\log_{10} 3$ **(c)** $\log_{10} 1/2$ **(d)** $\log_{10} 2.5$

(38)

Remember that $\log_b x$ means the exponent on b which gives x.
$\log_{10} x$ means the exponent on 10 which gives x. Thus, $(\log_{10} x)$ is an exponent; it is the exponent on 10 that gives x.

Thus; $10^{\log_{10} x}$ must give x

$$10^{\log_{10} x} = x$$

(39)

Review

1. The log function is $y = \log_b x$, $b > 0$, $b \neq 1$.
2. The domain is the set of positive reals and the range is the set of reals.
3. $y = \log_b x$ if and only if $b^y = x$
4. Logarithms obey three laws:

I. $\log_b mn = \log_b m + \log_b n$

II. $\log_b \frac{m}{n} = \log_b m - \log_b n$

III. $\log_b m^k = k \log_b m$

(40)

REVIEW EXERCISES

A. True or False:

1. Logarithms of negative numbers are not defined.
2. The base of a logarithm can be any positive number.
3. The logarithm of a positive number may be negative.

B. Change to log form.

4. $5^3 = 125$ **5.** $2^{-2} = 1/4$

C. Change to exponential form.

6. $\log_8 2 = 1/3$ **7.** $\log_{1/4} 4 = -1$

D. Find the exact numerical value.

8. $\log_{10} 10$ **9.** $\log_{25} 5$ **10.** $\log_6 1$ **11.** $\log_{10} .01$

E. Express as a single logarithm with a coefficient of one.

12. $\log_2 xy - 2\log_2 z$ **13.** $1/2(\log_{10} x + \log_{10} y)$

F. Expand into sums.

14. $\log_5 \dfrac{x^2y^2}{3a}$ **15.** $\log_{10} \dfrac{x\sqrt[3]{y}}{a}$

G. If $\log_{10} 2 = .3010$ and $\log_{10} 3 = .4771$ find without tables:

16. $\log_{10} 18$ **17.** $\log_{10} 1.5$

(41)

SOLUTIONS TO REVIEW EXERCISES

A. **1.** True **2.** False; the base can be any positive number except one. **3.** True

B. **4.** $\log_5 125 = 3$ **5.** $\log_2 1/4 = -2$

C. **6.** $8^{1/3} = 2$ **7.** $(1/4)^{-1} = 4$

D. **8.** 1 since $10^1 = 10$ **9.** 1/2 since $25^{1/2} = 5$

10. 0 since $6^0 = 1$ **11.** -2 since $10^{-2} = 1/100 = .01$

E. **12.** $\log_2 \dfrac{xy}{z^2}$ **13.** $\log_{10} \sqrt{xy}$

F. **14.** $2\log_5 x + 2\log_5 y - \log_5 3 - \log_5 a$ **15.** $\log_{10} x + 1/3\log_{10} y - \log_{10} a$

G. **16.**
$$\begin{aligned}\log_{10} 18 &= \log_{10}(2 \cdot 3^2)\\ &= \log_{10} 2 + 2\log_{10} 3\\ &= (.3010) + 2(.4771)\\ &= .3010 + .9542\\ &= 1.2552\end{aligned}$$

17.
$$\begin{aligned}\log_{10} 1.5 &= \log_{10} 3/2\\ &= \log_{10} 3 - \log_{10} 2\\ &= .4771 - .3010\\ &= .1761\end{aligned}$$

(42)

SOLUTIONS TO STUDY EXERCISES

Study Exercise One (Frame 18)

A. **1.** $3 = \log_4 64$ **2.** $1/4 = \log_{81} 3$ **3.** $-2 = \log_5 1/25$ **4.** $0 = \log_{10} 1$

B. **5.** $10^{-1} = 1/10$ **6.** $64^{2/3} = 16$ **7.** $17^0 = 1$ **8.** $4^{3/2} = 8$

(18A)

Study Exercise Two (Frame 20)

1. 2 **2.** 5 **3.** 3 **4.** 2 **5.** -2

(20A)

Study Exercise Three (Frame 31)

A. 1. $\log_a (2 \cdot 3^5 \cdot 7) = \log_a 2 + \log_a 3^5 + \log_a 7$
$= \log_a 2 + 5 \log_a 3 + \log_a 7$

2. $\log_b \sqrt{7} \cdot x = \log_b \sqrt{7} + \log_b x$
$= \log_b 7^{1/2} + \log_b x$
$= 1/2 \log_b 7 + \log_b x$

3. $\log_{10} \pi r^2 = \log_{10} \pi + \log_{10} r^2$
$= \log_{10} \pi + 2 \log_{10} r$

B. 4. $\log_{10} 2 + \log_{10} \pi + \log_{10} r = \log_{10} 2\pi r$

5. $1/3 \log_a r - 2 \log_a t = \log_a r^{1/3} - \log_a t^2$
$= \log_a \sqrt[3]{r} - \log_a t^2$
$= \log_a \frac{\sqrt[3]{r}}{t^2}$

6. $-\log_{10} k - \log_{10} M = -(\log_{10} k + \log_{10} M)$
$= -1 \log_{10} kM$
$= \log_{10} (kM)^{-1}$
$= \log_{10} \frac{1}{kM}$

31A

Study Exercise Four (Frame 35)

1. $\log_{10} 30 = \log_{10} (3 \cdot 10)$
$= \log_{10} 3 + \log_{10} 10$
$= .4771 + 1$
$= 1.4771$

2. $\log_{10} .3 = \log_{10} (.1)(3)$
$= \log_{10} .1 + \log_{10} 3$
$= -1 + .4771$
$= -.5229$

3. $\log_{10} (27) = \log_{10} 3^3$
$= 3 \log_{10} 3$
$= 3(.4771)$
$= 1.4313$

4. $\log_{10} \sqrt{300} = \log_{10} 10\sqrt{3}$
$= \log_{10} 10 + 1/2 \log_{10} 3$
$= 1 + 1/2(.4771)$
$= 1 + .2386$
$= 1.2386$

35A

Study Exercise Five (Frame 37)

1. $\log_{10} x = -4$
$10^{-4} = x$
$\frac{1}{10000} = x$

2. $\log_{10} .001 = x$
$10^x = .001$
$x = -3$

3. $\log_3 81 = y$
$3^y = 81$
$y = 4$

4. $\log_b 25 = 2$
$b^2 = 25$
$b = 5$
(-5 is not an answer since the base of a logarithm cannot be negative).

5. $\log_{10} .0001 = t$
$10^t = .0001$
$t = -4$

37A

UNIT 18—SUPPLEMENTARY PROBLEMS

A. **1.** Give the definition of $y = log_b N$
2. For what values of N is $\log_b N$ not defined?
3. For what values of b is $\log_b N$ not defined?

B. Find the numerical value of each of the following logarithms:
4. $\log_6 36$ **5.** $\log_2 64$ **6.** $\log_8 32$ **7.** $\log_8 8$
8. $\log_3 9$ **9.** $\log_3 27$ **10.** $\log_3 1$ **11.** $\log_1 8$

C. True or false:
12. $\log_2 8 - \log_{1/2} 8 = 6$
13. $\log_{10} 10 + \log_{10} 100 + \log_{10} 1000 = 6$
14. $\log_2 4 + \log_4 2 = 1$
15. $\log_b (m + n) = \log_b m + \log_b n, m, n > 0, b > 0, b \neq 1$
16. $\log_b \frac{m}{n} = \frac{\log_b m}{\log_b n}, m, n > 0, b > 0, b \neq 1$
17. $\log_b (ax)^x = x \log_b a + x, a, x > 0, b > 0, b \neq 1$

D. Change to exponential form.
18. $\log_8 16 = 4/3$ **19.** $\log_5 \frac{1}{125} = -3$ **20.** $\log_8 1 = 0$

E. Change to log form:
21. $6^3 = 216$ **22.** $4^{-3} = 1/64$ **23.** $64^{-1/3} = 1/4$

F. Using the laws of logarithms, expand into sums:
24. $\log_{10} \left(\frac{RT}{V}\right)$ **25.** $\log_c \frac{4\pi r^2}{3}$ **26.** $\log_{10} 2\pi r$

G. Express as a single logarithm:
27. $1/2 \log_2 5 - \log_2 7 + 2 \log_2 14$ **28.** $\log_b 2x + 3(\log_b x - \log_b 4)$
29. $1/2 \log_b m - 2/3 \log_b n$

H. Find the value of the variable:
30. $\log_5 N = 3$ **31.** $\log_4 N = 3/2$
32. $\log_{25} 125 = x$ **33.** $\log_8 4 = N$
34. $\log_b 81 = 2$ **35.** $n = \log_2 8^3$
36. $\log_{10} x - 2 \log_{10} 3 = \log_{10} 2$

I. If $\log_{10} 2 = .3010$, $\log_{10} 3 = .4771$, find without tables:
37. $\log_{10} 6$ **38.** $\log_{10} 8$ **39.** $\log_{10} 81$
40. $\log_{10} 10$ **41.** $\log_{10} (2/3)$ **42.** $\log_{10} 15$

J. **43.** Prove that $\log_b 1/x = -\log_b x$.

unit

19

Computation With Logarithms Part 1

Objectives

1. Be able to use a 4 place table of common logarithms.
2. Be able to interpolate

Terms

characteristic   mantissa   antilogarithm

①

Laws of Logarithms

Law I. $\log_b (m \cdot n) = \log_b m + \log_b n$
The logarithm of a product equals the sum of the logarithms of its factors.

Law II. $\log_b \left(\frac{m}{n}\right) = \log_b m - \log_b n$
The logarithm of a quotient equals the logarithm of the dividend minus the logarithm of the divisor.

Law III. $\log_b (n^k) = k \log_b n$
The logarithm of a power of a number equals the exponent times the logarithm of the number.

②

Common Logarithms

Definition. Logarithms to the base 10 are called *common logarithms.*

Common logarithms are used for ordinary computations. When the base is not indicated, it is understood to be 10. Thus, $\log n = \log_{10} n$.

③

Bases for Logarithms

Base 10 is the common base for computational usage. Logarithms to base 10 are called *common or Briggsian logarithms*, named for Henry Briggs (1560–1631).

Base e, where e stands for an irrational number approximately equal to 2.718, is exclusively used in theoretical mathematics and in many scientific applications. Logarithms to base e are called *natural or Napierian logarithms*, named for John Napier (1550–1617).

④

Study Exercise One

Using the definition of logarithm and the three Laws of Logatithms, find the value of the following logs.

1. $\log 100$ **2.** $\log .0001$ **3.** $\log 10^{2.13}$
4. $\log 10^{\pi}$ **5.** $\log 4 + \log 25$ (Hint: apply Law I)

⑤

Scientific Notation

Scientific notation writes any number N as the product of a number M between 1 and 10 or equal to 1 by a power of 10.

$$N = M \cdot 10^c \text{ where } 1 \leqslant M < 10,\ c \in \text{Integers}$$

(6)

Number	*Scientific notation form of the number*
37.5	$(3.75)(10^1)$
.0043	$(4.3)(10^{-3})$
4,800	$(4.8)(10^3)$
1	$(1)(10^0)$
200	$(2)(10^2)$

(7)

Study Exercise Two

Write the following numbers in scientific notation form.

1. 135 **2.** 72,854 **3.** .00123 **4.** 1,000 **5.** 7.314 **6.** .074

(8)

Logarithms of Integral Powers of Ten

Recall that $\log_b N = x$ if and only if $b^x = N$. Thus,

$\log_{10} 10 = 1$ (since $10^1 = 10$)
$\log_{10} 100 = 2$ (since $10^2 = 100$)
$\log_{10} 1000 = 3$ (since $10^3 = 1000$)
$\log_{10} 1 = 0$ (since $10^0 = 1$)
$\log_{10} .1 = -1$ (since $10^{-1} = 1/10$)
$\log_{10} .01 = -2$ (since $10^{-2} = 1/100$)

(9)

Since $\log_{10} 1 = 0$ and $\log_{10} 10 = 1$, a number between 1 and 10 has a logarithm of more than zero but less than 1.

$\log_{10} 1 = 0$
$\log_{10} M$ is more than zero but less than one, where $1 < M < 10$
$\log_{10} 10 = 1$

(10)

Exponential Form	*Log Form*
$10^3 = 1{,}000$	$\log 1000 = 3$
	←$\log 354 = 2 + \text{decimal}$
$10^2 = 100$	$\log 100 = 2$
$10^1 = 10$	$\log 10 = 1$
	←$\log 3.54 = 0 + \text{decimal}$
$10^0 = 1$	$\log 1 = 0$
$10^{-1} = .1$	$\log .1 = -1$
	←$\log .034 = -2 + \text{decimal}$
$10^{-2} = .01$	$\log .01 = -2$
$10^{-3} = .001$	$\log .001 = -3$

(11)

Recall that if $N > 0$, $N = M \cdot 10^c$ in scientific notation where $1 \leqslant M < 10$:

Line (1) $\log N = \log (M \cdot 10^c)$
Line (2) $\log N = \log M + \log 10^c$
Line (3) $\log N = \log M + c \log 10$

(Frame 12, contd.)

Line (*4*) $\log N = \log M + c$

Line (*5*) $\log N = c + \log M$

Thus the logarithm of a positive number is made up of two parts: an integer plus the logarithm of a number between 1 and 10.

⑫

In the previous frame we concluded with:

$$\log N = c + \log M \text{ where } 1 \leqslant M < 10$$

c ↑ integer part; $\log M$ ↑ positive or zero decimal portion

c will be called the *characteristic* and the positive or zero decimal portion will be called the *mantissa*.

⑬

Conclusions

1. The logarithm of an integral power of 10 is an integer.

2. The logarithm of a number, not an integral power of 10, consists of two parts:

(1) an integer part (*characteristic*) and

(2) a positive decimal part (*mantissa*).

⑭

Example: Find the characteristic of log 145.

Solution:

Line (*1*) $\log (145) = \log (1.45)(10^2)$

Line (*2*) $= \log (1.45) + \log 10^2$

Line (*3*) $= \log (1.45) + 2 \log 10$

Line (*4*) $= \log 1.45 + 2$

Line (*5*) $= 2 + \log 1.45$

↑ characteristic (the 2)

The characteristic is 2.

⑮

Further Examples

Find the characteristic of each of the following logarithms.

1. log (23.5):

Solution:
$$\log (23.5) = \log (2.35)(10^1) = \log 2.35 + \log 10 = \log 2.35 + 1 = 1 + \log 2.35$$

The characteristic is 1.

2. log (.013):

Solution:
$$\log (.013) = \log (1.3)(10^{-2}) = \log (1.3) + \log 10^{-2} = \log (1.3) + (-2) \log 10 = -2 + \log 1.3$$

The characteristic is -2.

⑯

Study Exercise Three

Find the characteristic.

1. log 3,500 **2.** log .1123 **3.** log .0007 **4.** log 451.3

(17)

To find the characteristic of the logarithm of a number

1. write the number in scientific notation

2. the power of 10 is the characteristic.

Example: log (531.2) = log (5.312) (10^2)

characteristic is 2

(18)

Study Exercise Four

Give the characteristic at sight.

1. log (55.2) **2.** log (.34) **3.** log (827)

4. log (.0062) **5.** log (4,250) **6.** log (3.245)

(19)

Examples:

	logarithm	*characteristic*	*mantissa*
1.	0.4782	0	.4782
2.	2.8209	2	.8209
3.	1.0000	1	.0000
4.	$-3 + .4782$ or (-2.5218)	-3	.4782
5.	$-2 + .3971$ or (-1.6029)	-2	.3971
6.	$-1 + .3971$ or $(-.6029)$	-1	.3971

(20)

Warning.

In the logarithm (-1.6029), the decimal part is not the mantissa since it is not positive.

To find the mantissa, add to (-1.6029) the smallest integer that would make (-1.6029) positive and then subtract off that integer.

$$\begin{aligned} -1.6029 &= 2 + (-1.6029) + (-2) \\ &= .3971 + (-2) \end{aligned}$$

The mantissa is .3971 and the characteristic is -2.

(21)

Example: Find the characteristic and mantissa if the logarithm of a number is $-.4638$.

Solution:

$$\begin{aligned} -.4638 &= (1) + (-.4638) + (-1) \\ &= .5362 + (-1) \\ &= -1 + .5362 \end{aligned}$$

The characteristic is -1 and the mantissa is .5362.

(22)

Study Exercise Five

Assume the following represent logarithms. Give the characteristic and mantissa.

1. 3.4122 **2.** 0.5162 **3.** $-.4123$

4. $-2 + .4682$ **5.** -1.2400 **6.** 4.0000

(23)

Notations for Negative Characteristics

When the characteristic is negative, it is more convenient to use an equivalent form. We will write a negative characteristic as the difference of two positive numbers of which the second is a multiple of ten.

Write -2 as $8 - 10$ *or*
$18 - 20$ *or*
$28 - 30$ *or*
$38 - 40$ etc.

Write -1 as $9 - 10$ *or*
$19 - 20$ *or*
$29 - 30$ etc.

(24)

Characteristic	*commonly written as*
-1	$9 - 10$
-2	$8 - 10$
-3	$7 - 10$
-4	$6 - 10$
-5	$5 - 10$

(25)

A logarithm like $-2 + .4771$ will first be written as $8 - 10 + .4771$. Then the 8 and .4771 will be combined into $8.4771 - 10$.

In this form, the mantissa is .4771 and the characteristic is $8 - 10$ or -2.

(26)

Examples:

	logarithm	*characteristic*	*mantissa*
1.	3.6027	3	.6027
2.	$9.1402 - 10$	-1	.1402
3.	$6.0345 - 10$	-4	.0345
4.	6.0345	6	.0345
5.	0.6027	0	.6027

(27)

The logarithm of a number is negative if and only if its characteristic is negative.

(28)

Tables of Logarithms

1. Tables of logarithms will show only mantissas; they will not show characteristics.
2. The mantissas of the logarithms of most numbers are non repeating, non ending decimals (irrational numbers).
3. The mantissas may be approximated to any number of decimal places.
4. A four place table implies that the mantissas are approximated to four decimal places.

(29)

Four Place Log Table

N	0	1	2	3	4	5	6	7	8	9
10	0000	0043	0086	0128	0170	0212	0253	0294	0334	0374
11	0414	0453	0492	0531	0569	0607	0645	0682	0719	0755
12	0792	0828	0864	0899	0934	0969	1004	1038	1072	1106
13	1139	1173	1206	1239	1271	1303	1335	1367	1399	1430

The decimal point which precedes the mantissa must be supplied. It is not printed in the table.

(30)

Remember also that the table does not give the characteristics.

Thus, log 132 = 2.1206

found in the table

determined after writing 132 in scientific notation

(31)

Study Exercise Six

Using a four place common log table, find:

1. log 134 **2.** log 6.23 **3.** log .0381 **4.** log 1200

5. log .1790 **6.** log 500 **7.** log .911

(32)

Let us now try the reverse process. Suppose log N = 1.6542.

What is the number that has 1.6542 as a logarithm?

The number 1 is characteristic and is not found in the table.

The mantissa .6542 is located in the table of mantissas. The number corresponding is 451.

Since the characteristic is 1, the number is $(4.51)(10^1)$ or 45.1.

(33)

Antilogarithm

The number corresponding to a given logarithm is called the antilogarithm of the given logarithm.

Antilog (1.3118) means find the number N such that log N = 1.3118.

(34)

Example 1: Find antilog 0.7642

Solution: the characteristic is zero
the matissa is .7642
the number corresponding to .7642 is 581
Since the characteristic is zero the number is $(5.81)(10^0)$ or 5.81

Example 2: Find antilog (8.4955 − 10)

Solution: the characteristic is −2
the number corresponding to .4955 is 313
Since the characteristic is −2, the number is $(3.13)(10^{-2})$ or .0313

(35)

Study Exercise Seven

Use a four place log table to determine N.

1. $\log N = 2.5211$ **2.** $\log N = 8.7348 - 10$ **3.** $\log N = 3.7959$

4. antilog $(.6693) = N$ **5.** $\log N = 9.3909 - 10$

(36)

Our table allows us to find log 3.26. However, is there a way to find log 3.265? We could locate a table that had logs of numbers with four digits or we could approximate or guess what the value should be.

(37)

Linear Interpolation

Interpolation is a procedure where an estimate of the value of a logarithm is made between two known values.

Linear interpolation assumes the function being interpolated is a straight line between known values.

(38)

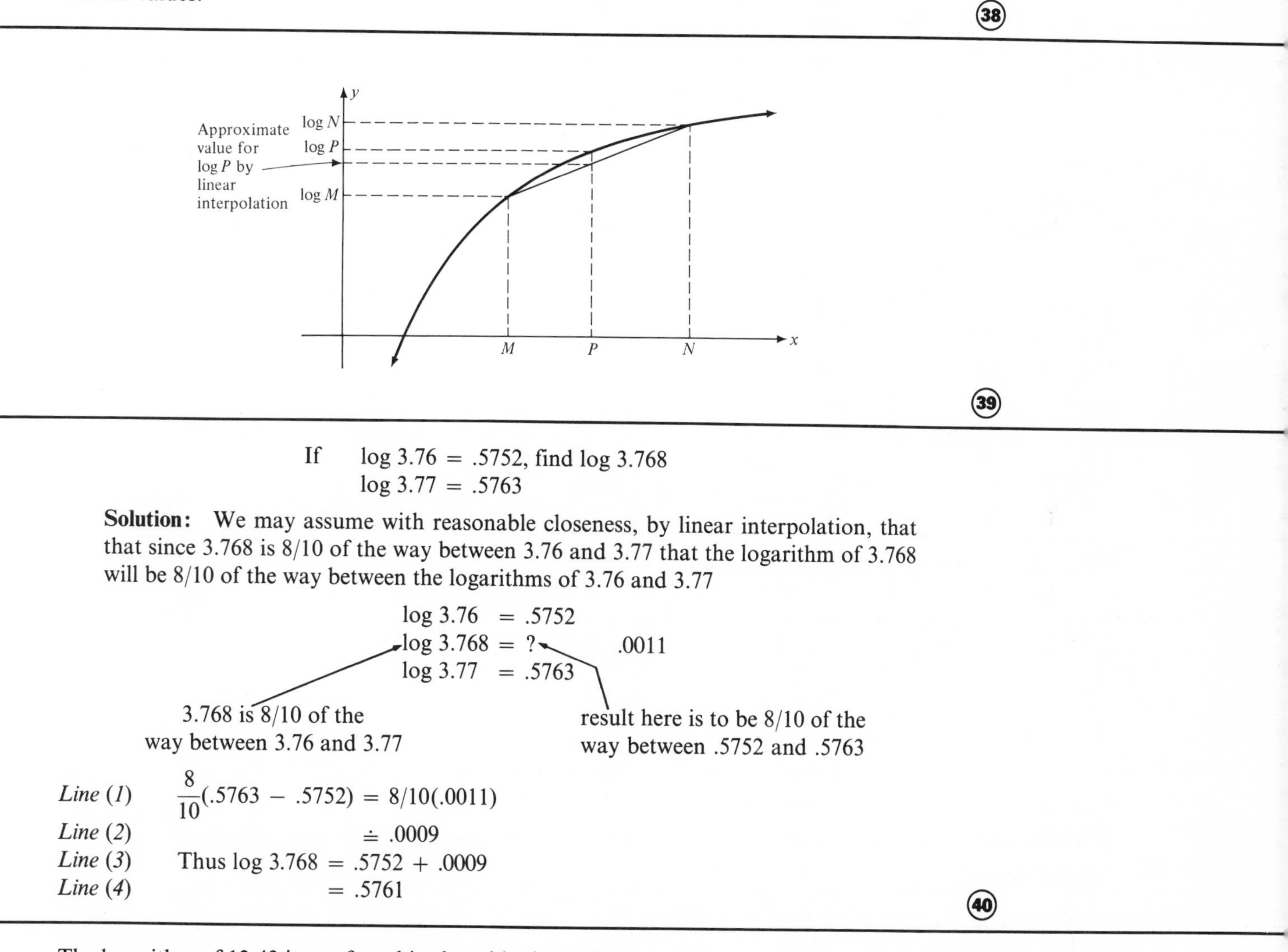

(39)

If $\log 3.76 = .5752$, find $\log 3.768$
$\log 3.77 = .5763$

Solution: We may assume with reasonable closeness, by linear interpolation, that that since 3.768 is 8/10 of the way between 3.76 and 3.77 that the logarithm of 3.768 will be 8/10 of the way between the logarithms of 3.76 and 3.77

$$\log 3.76 = .5752$$
$$\log 3.768 = ?$$
$$\log 3.77 = .5763$$

.0011

3.768 is 8/10 of the way between 3.76 and 3.77

result here is to be 8/10 of the way between .5752 and .5763

Line (1) $\frac{8}{10}(.5763 - .5752) = 8/10(.0011)$

Line (2) $\doteq .0009$

Line (3) Thus $\log 3.768 = .5752 + .0009$

Line (4) $= .5761$

(40)

The logarithm of 12.43 is not found in the table since 12.43 is a 4 digit number.

Between which two entries in the table is log 12.43 found?

Log 12.43 would be found between log 12.40 and log 12.50.

(41)

Example 1: Between which two entries in the log table would log .9876 be found?
Solution: between log .9870 and log .9880.

Example 2: Between which two entries in the log table would log 1432 be found?
Solution: between log 1430 and log 1440.

(42)

Study Exercise Eight

Between which two logarithms would the following logs be found?

1. log .01345 **2.** log 27.83 **3.** log 1,478 **4.** log 1.899 **5.** log 1111

(43)

Example: Interpolate to find the logarithm of 123.4
Solution: Log 123.4 is found between log 123.0 and log 124.0.

log 123.0 = 2.0899 ←┐
log 123.4 = ? ├difference
log 124.0 = 2.0934 ←┘ is 35

$$\tfrac{4}{10} \cdot 35 = \frac{140}{10}$$
$$= 14$$

add .0014 to 2.0899 to get 2.0913
thus log 123.4 = 2.0913

(44)

Example: Interpolate to find log 62.46
Solution: Log 62.46 is found between log 62.40 and log 62.50

log 62.40 = 1.7952 ←┐
log 62.46 = ? ├difference
log 62.50 = 1.7959 ←┘ is 7

$$\frac{6}{10} \cdot 7 = \frac{42}{10}$$
$$\doteq 4$$

add .0004 to 1.7952 to get 1.7956
thus log 62.46 = 1.7956

(45)

Example: Find log .07623 by interpolation
Solution: log .0762 = 8.8820 − 10 ←┐
log .07623 = ? ├difference
log .0763 = 8.8825 − 10 ←┘ is 5

$$\frac{3}{10} \cdot 5 = \frac{15}{10}$$
$$= 2$$

add .0002 to 8.8820 − 10 to get 8.8822 − 10
thus log .07623 = 8.8822 − 10.

(46)

Rounding

If the digit in the place position to be rounded is 0, 1, 2, 3, or 4, round down; if the digit is 6, 7, 8, or 9, round up.

If the digit is 5, we agree to round to an even number. That is, round up if the digit to the left is odd and round down if the digit to the left is even.

(47)

To the nearest whole number;

6.8 rounds to 7	8.2 rounds to 8	5.5 rounds to 6
6.5 rounds to 6	8.5 rounds to 8	9.5 rounds to 10

(48)

Study Exercise Nine

A. Round to one digit:
1. 5.2 **2.** 3.8 **3.** 2.5 **4.** 7.5 **5.** 8.4 **6.** 4.5

B. Find the following logarithms by interpolation:
7. log .003821 **8.** log 7.777 **9.** log 4736 **10.** log .4142 **11.** log .06854

(49)

Antilogs By Interpolation

Example: Find N such that $\log N = 2.5370$.

Solution:

$\log 344 = 2.5366$
$\log N = 2.5370$ — 4 (difference between smaller log and given log)
$\log 345 = 2.5378$ — difference is 12

$$\frac{x}{10} \cdot 12 = 4$$

x: 4th digit in the number
4: difference between smaller log and given log
12: difference between two known logs

$$12x = 40$$
$$x \doteq 3$$

3 is the 4th digit in the number N.
thus $N = 344.3$

(50)

Example: Find N if $\log N = 0.3062$.

Solution:

$\log 2.02 = 0.3054$
$\log N = 0.3062$ — 8
$\log 2.03 = 0.3075$ — 21

$$\frac{x}{10} \cdot 21 = 8$$
$$21x = 80$$
$$x \doteq 4$$
$$N = 2.024$$

(51)

Study Exercise Ten

Find N in each case:

1. $\log N = 1.6660$ **2.** $\log N = 9.5495 - 10$

3. $\log N = 0.1504$ **4.** $\log N = 2.7863$

(52)

REVIEW EXERCISES

1. Write 847.1 in scientific notation.
2. Without a table, give log 10 and log 100.
3. Given that log 4.13 = 0.6355, fill in the table giving the characteristic and mantissa of the given logarithms.

	Characteristic	*Mantissa*
(a) log 413		
(b) log .0413		
(c) log 41,300		

4. Fill in the table giving the characteristic and mantissa.

	Characteristic	*Mantissa*
(a) $\log N = 1.0000$		
(b) $\log N = -.4362$		
(c) $\log N = -1.1428$		
(d) $\log N = 2.6263$		

5. Using a four place log table, find;
   (a) log 68.48 **(b)** log .4697 **(c)** N if $\log N = 3.3907$
   (d) N if $\log N = .7055$ **(e)** antilog (1.3240)

SOLUTIONS TO REVIEW EXERCISES

1. $(8.471)(10^2)$
2. log 10 = 1 and log 100 = 2
3.

	characteristic	*mantissa*
(a)	2	.6355
(b)	-2	.6355
(c)	4	.6355

4.

(a)	1	.0000
(b)	-1	.5638
(c)	-2	.8572
(d)	2	.6263

5. **(a)** log 68.4 = 1.8351
   log 68.48 = ______ (difference 6) $8/10 \cdot 6 \doteq 5$
   log 68.5 = 1.8357
   thus log 68.48 = 1.8356

   (b) log .469 = 9.6712 − 10
   log .4697 = ______ (difference 9) $7/10 \cdot 9 \doteq 6$
   log .470 = 9.6721 − 10
   thus log .4697 = 9.6718 − 10

   (c) log 2450 = 3.3892
   $\log N$ = 3.3907 (differences 15, 17) $x/10 \cdot 17 = 15$
   log 2460 = 3.3909 $17x = 150$
   $N = 2459$ $x \doteq 9$

   (d) log 5.07 = .7050
   $\log N$ = .7055 (differences 5, 9) $x/10 \cdot 9 = 5$
   log 5.08 = .7059 $9x = 50$
   $N = 5.076$ $x \doteq 6$

SOLUTIONS TO REVIEW EXERCISES, CONTD.

(e) $\log 21.0 = 1.3222$
$\log N = 1.3240$ — 18 — 21
$\log 21.1 = 1.3243$
$N = 21.09$

$x/10 \cdot 21 = 18$
$21x = 180$
$x \doteq 9$

SOLUTIONS TO STUDY EXERCISES

Study Exercise One (Frame 5)

1. 2 (since $10^2 = 100$)
2. -4 (since $10^{-4} = .0001$)
3. 2.13 (since $\log 10^{2.13} = 2.13 \log 10 = (2.13)(1)$
4. π (since $\log 10^{\pi} = \pi \log 10 = \pi (1)$
5. 2 (since $\log 4 + \log 25 = \log (4)(25) = \log 100 = 2$

5A

Study Exercise Two (Frame 8)

1. $(1.35)(10^2)$
2. $(7.2854)(10^4)$
3. $(1.23)(10^{-3})$
4. $(1)(10^3)$
5. $(7.314)(10^0)$
6. $(7.4)(10^{-2})$

8A

Study Exercise Three (Frame 17)

1. characteristic is 3 since $\log (3.5)(10^3)$
$= \log 3.5 + 3 \log 10$
$= 3 + \log 3.5$
2. characteristic is -1 since $\log (1.123)(10^{-1})$
$= \log (1.123) + (-1) \log 10$
$= -1 + \log 1.123$
3. characteristic is -4 since $\log (7)(10^{-4})$
$= \log 7 + (-4) \log 10$
$= -4 + \log 7$
4. characteristic is 2 since $\log (4.513)(10^2)$
$= \log 4.513 + 2 \log 10$
$= 2 + \log 4.513$

17A

Study Exercise Four (Frame 19)

1. 1
2. -1
3. 2
4. -3
5. 3
6. 0

19A

Study Exercise Five (Frame 23)

	characteristic	*mantissa*
1.	3	.4122
2.	0	.5162
3.	-1	.5877
4.	-2	.4682
5.	-2	.7600
6.	4	.0000

23A

Study Exercise Six (Frame 32)

1. 2.1271 **2.** 0.7945 **3.** 8.5809 − 10 **4.** 3.0792

5. 9.2529 − 10 **6.** 2.6990 **7.** 9.9595 − 10

32A

Study Exercise Seven (Frame 36)

1. $N = 332$ **2.** $N = .0543$ **3.** $N = 6250$ **4.** $N = 4.67$ **5.** $N = .246$

36A

Study Exercise Eight (Frame 43)

1. Between log .01340 and log .01350

2. Between log 27.80 and log 27.90

3. Between log 1470 and log 1480

4. Between log 1.890 and log 1.900

5. Between log 1110 and log 1120

43A

Study Exercise Nine (Frame 49)

A. **1.** 5 **2.** 4 **3.** 2 **4.** 8 **5.** 8 **6.** 4

B. **7.**
log .00382 = 7.5821 − 10
log .003821 = ______ (difference 11)
log .00383 = 7.5832 − 10

$1/10 \cdot 11 \doteq 1$

thus log .003821 = 7.5822 − 10

8.
log 7.77 = 0.8904
log 7.777 = ______ (difference 6)
log 7.78 = 0.8910

$\frac{7}{10} \cdot 6 \doteq 4$

thus log 7.777 = 0.8908

9.
log 4730 = 3.6749
log 4736 = ______ (difference 9)
log 4740 = 3.6758

$\frac{6}{10} \cdot 9 \doteq 5$

thus log 4736 = 3.6754

10.
log .414 = 9.6170 − 10
log .4142 = ______
log .415 = 9.6180 − 10

$2/10 \cdot 10 = 2$

thus log .4142 = 9.6172 − 10

11.
log .0685 = 8.8357 − 10
log .06854 = ______ (difference 6)
log .0686 = 8.8363 − 10

$4/10 \cdot 6 \doteq 2$

thus log .06854 = 8.8359 − 10

49A

Study Exercise Ten (Frame 52)

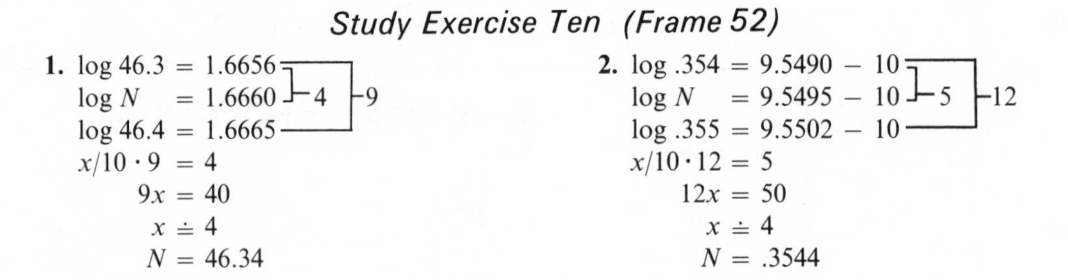

1.
log 46.3 = 1.6656
log N = 1.6660 (difference 4; total 9)
log 46.4 = 1.6665

$x/10 \cdot 9 = 4$

$9x = 40$

$x \doteq 4$

$N = 46.34$

2.
log .354 = 9.5490 − 10
log N = 9.5495 − 10 (difference 5; total 12)
log .355 = 9.5502 − 10

$x/10 \cdot 12 = 5$

$12x = 50$

$x \doteq 4$

$N = .3544$

3. $\log 1.41 = .1492$
$\log N = .1504$ } 12 } 31
$\log 1.42 = .1523$

$$\frac{x}{10} \cdot 31 = 12$$
$$31x = 120$$
$$x \doteq 4$$
$$N = 1.414$$

4. $\log 611 = 2.7860$
$\log N = 2.7863$ } 3 } 8
$\log 612 = 2.7868$

$$\frac{x}{10} \cdot 8 = 3$$
$$8x = 30$$
$$x \doteq 4$$
$$N = 611.4$$

52A

UNIT 19—SUPPLEMENTARY PROBLEMS

A. Fill in the blanks to make a true statement.

1. Logarithms to base 10 are called ____________ logarithms.
2. The logarithm of a postive number is the sum of the characteristic and the ____________.
3. When a number is expressed in scientific notation, the exponent of its power of 10 is the ____________ of its logarithm.

B. True or false:

4. $\log 5x = 5 \log x$
5. If $\log N = -.3413$, the mantissa is .3413.
6. $\log x^n = (\log x)^n$
7. $\log (5 + 6) = \log 5 + \log 6$
8. $\dfrac{\log 5}{\log 3} = \log 2$

C. Give the characteristic of each of these logarithms.

9. log 15, 485
10. log . 867
11. log 100
12. log 3.1416
13. log .00263
14. log 375.1
15. log 37

D. Give the characteristic and mantissa of the following logarithms.

16. $\log N = .1423$
17. $\log N = 3.0020$
18. $\log N = 2.1634$
19. $\log N = 5$
20. $\log N = -1.1346$
21. $\log N = -.6384$

E. Using a four place log table and the method of interpolation, find:

22. log 2.718
23. log .01672
24. log 5802
25. log .007923
26. log 38.34
27. N if $\log N = 2.9184$
28. N if $\log N = 0.0243$
29. N if $\log N = 1.8683$
30. N if $\log N = 7.4862 - 10$
31. N if $\log N = 9.8166 - 10$
32. N if $\log N = 8.4646 - 10$

unit

20

Computation With Logarithms Part 2

Objectives

Be able to compute to four significant figures with the aid of a logarithm table.

①

Properties of Logarithms

m and n are positive real numbers, $b > 0$ and $b \neq 1$

I. $\log_b (m \cdot n) = \log_b m + \log_b n$

II. $\log_b \left(\frac{m}{n}\right) = \log_b m - \log_b n$

III. $\log_b (m)^k = k \cdot \log_b m, k \in R$

②

Notation. $\log_{10} x = \log x$.
Values for $\log_{10} x$ are called logarithms to base 10 or *common logarithms*.

③

Study Exercise One

Using the properties of logarithms express as the sum or difference of simpler logarithmic quantities.

Example: $\log (23)(31) = \log 23 + \log 31$

1. $\log (57)(38)$

2. $\log (19)(45)^3$

3. $\log \frac{\sqrt{14}}{(121)^5}$

4. $\log \sqrt{\frac{(24)^3 \cdot (12)}{45}}$

④

Assumptions

1. If $M = N$, then $\log M = \log N$; $M, N > 0$

2. If $\log M = \log N$, then $M = N$

3. If $M = N$, then $b^M = b^N$

⑤

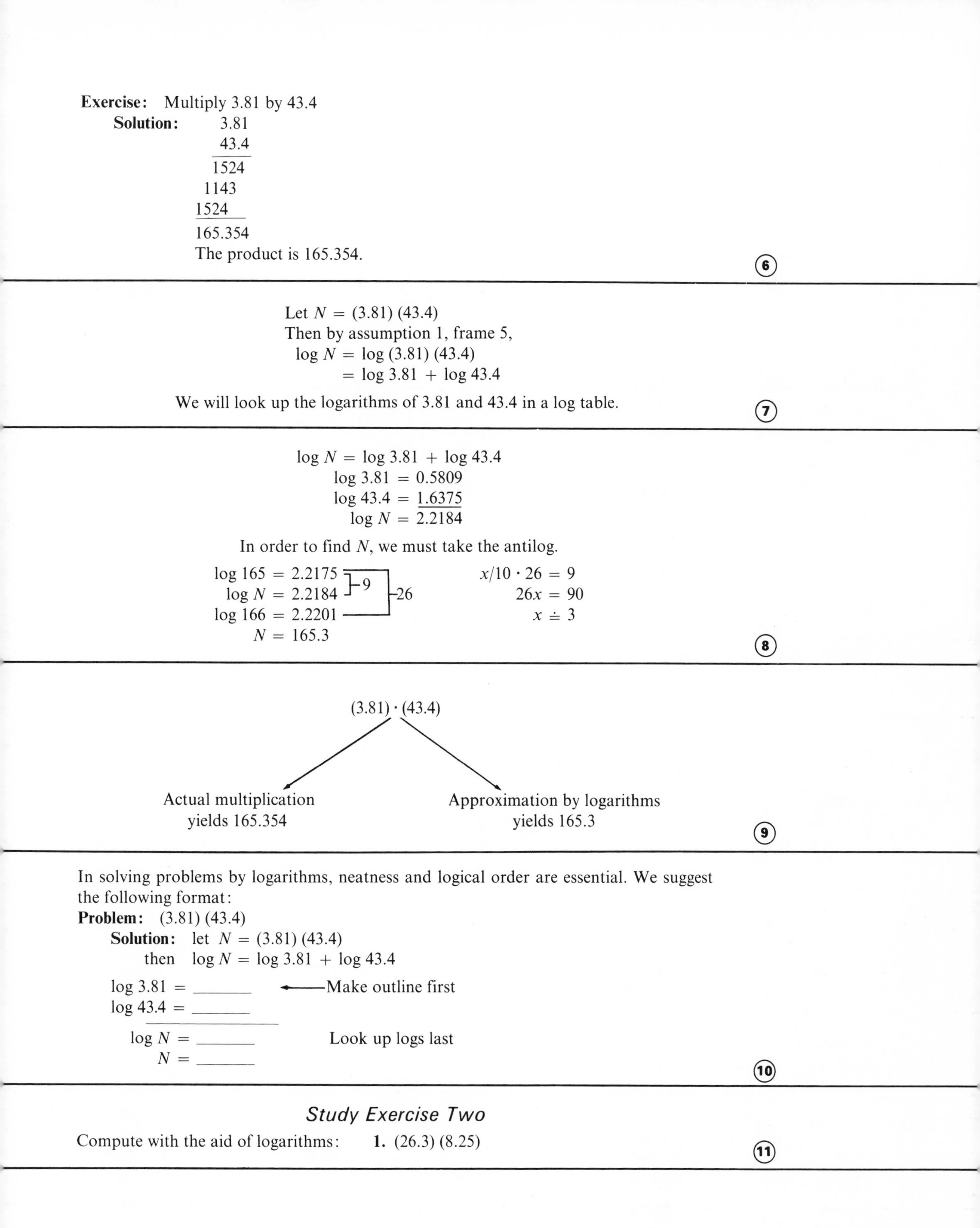

Exercise: Multiply 3.81 by 43.4

Solution:

```
     3.81
     43.4
    -----
     1524
    1143
   1524
   -------
   165.354
```

The product is 165.354.

(6)

Let $N = (3.81)\,(43.4)$

Then by assumption 1, frame 5,

$$\log N = \log (3.81)\,(43.4)$$
$$= \log 3.81 + \log 43.4$$

We will look up the logarithms of 3.81 and 43.4 in a log table.

(7)

$$\log N = \log 3.81 + \log 43.4$$
$$\log 3.81 = 0.5809$$
$$\log 43.4 = \underline{1.6375}$$
$$\log N = 2.2184$$

In order to find N, we must take the antilog.

$\log 165 = 2.2175$ ⎤ 9 ⎤ 26
$\log N = 2.2184$ ⎦
$\log 166 = 2.2201$ ⎦
$N = 165.3$

$x/10 \cdot 26 = 9$
$26x = 90$
$x \doteq 3$

(8)

$(3.81) \cdot (43.4)$

(9)

In solving problems by logarithms, neatness and logical order are essential. We suggest the following format:

Problem: (3.81) (43.4)

Solution: let $N = (3.81)\,(43.4)$

then $\log N = \log 3.81 + \log 43.4$

$\log 3.81 =$ ______ ←——Make outline first
$\log 43.4 =$ ______

$\log N =$ ______ Look up logs last
$N =$ ______

(10)

Study Exercise Two

Compute with the aid of logarithms: **1.** (26.3) (8.25)

(11)

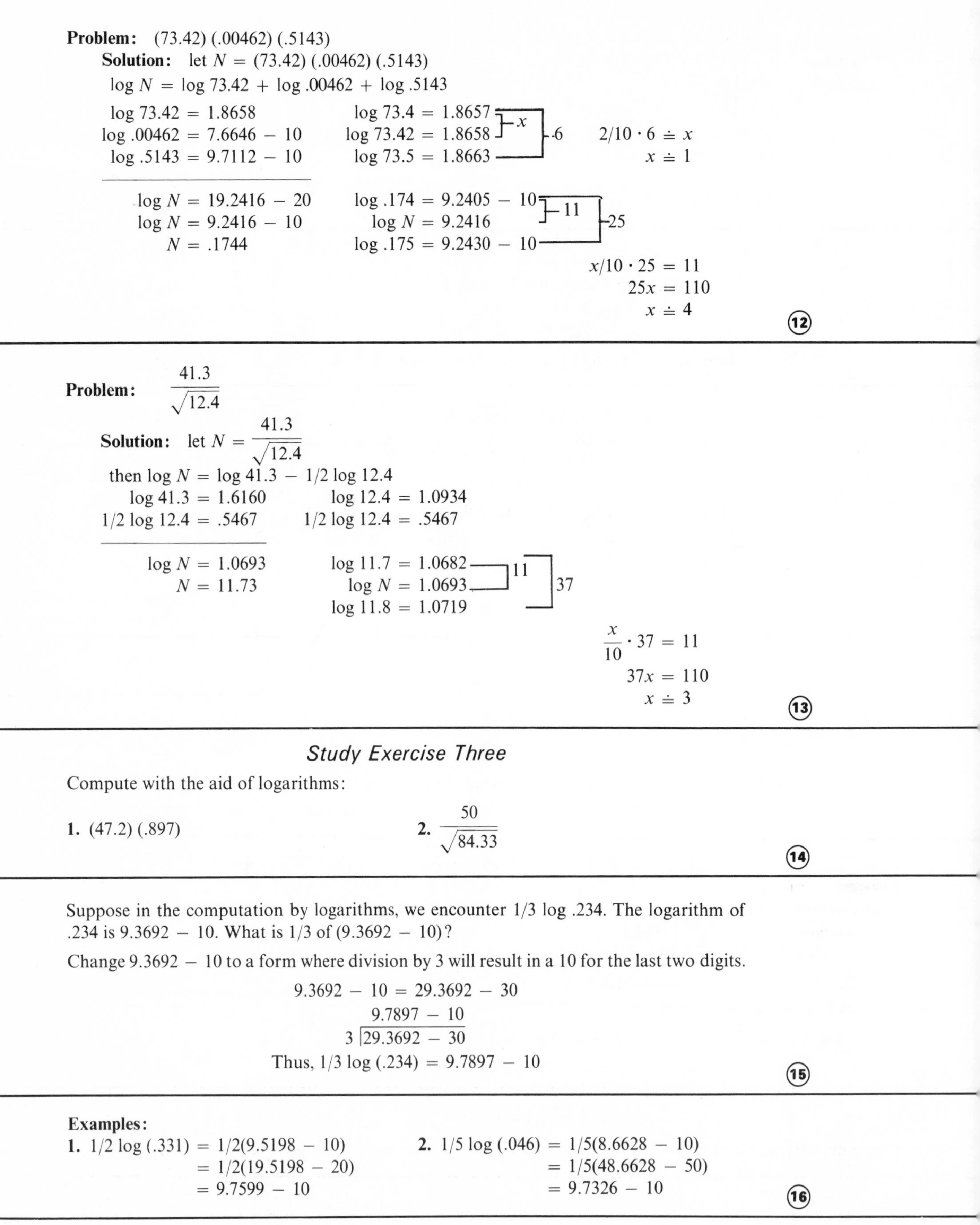

Problem: (73.42) (.00462) (.5143)

Solution: let $N = (73.42)(.00462)(.5143)$

$\log N = \log 73.42 + \log .00462 + \log .5143$

$$\begin{aligned}
\log 73.42 &= 1.8658\\
\log .00462 &= 7.6646 - 10\\
\log .5143 &= 9.7112 - 10\\
\hline
\log N &= 19.2416 - 20\\
\log N &= 9.2416 - 10\\
N &= .1744
\end{aligned}$$

$$\begin{aligned}
\log 73.4 &= 1.8657\\
\log 73.42 &= 1.8658\\
\log 73.5 &= 1.8663
\end{aligned}
\quad \left.\begin{matrix} x \\ \end{matrix}\right\} \quad .6$$

$$2/10 \cdot 6 \doteq x$$
$$x \doteq 1$$

$$\begin{aligned}
\log .174 &= 9.2405 - 10\\
\log N &= 9.2416\\
\log .175 &= 9.2430 - 10
\end{aligned}
\quad 11 \quad 25$$

$$x/10 \cdot 25 = 11$$
$$25x = 110$$
$$x \doteq 4$$

(12)

Problem: $\dfrac{41.3}{\sqrt{12.4}}$

Solution: let $N = \dfrac{41.3}{\sqrt{12.4}}$

then $\log N = \log 41.3 - 1/2 \log 12.4$

$$\begin{aligned}
\log 41.3 &= 1.6160\\
1/2 \log 12.4 &= .5467\\
\hline
\log N &= 1.0693\\
N &= 11.73
\end{aligned}$$

$$\begin{aligned}
\log 12.4 &= 1.0934\\
1/2 \log 12.4 &= .5467
\end{aligned}$$

$$\begin{aligned}
\log 11.7 &= 1.0682\\
\log N &= 1.0693\\
\log 11.8 &= 1.0719
\end{aligned}
\quad 11 \quad 37$$

$$\frac{x}{10} \cdot 37 = 11$$
$$37x = 110$$
$$x \doteq 3$$

(13)

Study Exercise Three

Compute with the aid of logarithms:

1. (47.2) (.897)

2. $\dfrac{50}{\sqrt{84.33}}$

(14)

Suppose in the computation by logarithms, we encounter 1/3 log .234. The logarithm of .234 is 9.3692 − 10. What is 1/3 of (9.3692 − 10)?

Change 9.3692 − 10 to a form where division by 3 will result in a 10 for the last two digits.

$$9.3692 - 10 = 29.3692 - 30$$

$$3\,\overline{)29.3692 - 30} = 9.7897 - 10$$

Thus, $1/3 \log (.234) = 9.7897 - 10$

(15)

Examples:

1. $\begin{aligned}[t] 1/2 \log (.331) &= 1/2(9.5198 - 10)\\ &= 1/2(19.5198 - 20)\\ &= 9.7599 - 10 \end{aligned}$

2. $\begin{aligned}[t] 1/5 \log (.046) &= 1/5(8.6628 - 10)\\ &= 1/5(48.6628 - 50)\\ &= 9.7326 - 10 \end{aligned}$

(16)

Example: Find $\sqrt[3]{.234}$ by using logarithms.

Solution: let $N = \sqrt[3]{.234}$

Then $\log N = 1/3 \log (.234)$

$1/3\,(.234) = 9.7897 - 10$

$\log N = 9.7897 - 10$

$N = .6161$

$\log (.234) = 9.3692 - 10$

$1/3 \log (.234) = 1/3(29.3692 - 30)$

$= 9.7897 - 10$

$\log .616 = 9.7896 - 10$ ⎤ 1 (difference to $\log N$); 7 (difference to $\log .617$)

$\log N = 9.7897 - 10$

$\log .617 = 9.7903 - 10$

$$\frac{x}{10} \cdot 7 = 1$$

$$7x = 10$$

$$x \doteq 1$$

⑰

Another type of problem that can be worked by logarithms is $(.561)^{-1}$.

Let $N = (.561)^{-1}$, then $\log N = \log \frac{1}{.561}$

$\log N = \log 1 - \log .561$

$\log 1 = 0.0000$

$\log .561 = 9.7490 - 10$

To avoid a negative logarithm, change $\log 1 = 0.0000$ to $\log 1 = 10.000 - 10$

$\log 1 = 10.0000 - 10$

$\log .561 = 9.7490 - 10$

$\log N = .2510$

$N = 1.782$

⑱

Do you think $(-42.4)^{-3}$ can be found by the use of logarithms?

$$(-42.4)^{-3} = \frac{1}{(-42.4)^3}$$

Since the cube of a negative number is negative, the result will be negative. To apply logarithms we disregard the negative sign and attach it to the end result.

⑲

Example: Find $(-42.4)^{-3}$ by logarithms.

Solution: Let $N = (+42.4)^{-3}$

$$\log N = \log \frac{1}{(42.4)^3} \quad \left(\begin{array}{c}\text{disregard the} \\ \text{negative sign}\end{array}\right)$$

$\log N = \log 1 - 3 \log 42.4$

$\log 1 = 0.0000$

$3 \log 42.4 = 4.8822$

$\log 1 = 10.0000 - 10$

$3 \log 42.4 = 4.8822$

$\log N = 5.1178 - 10$

$N = .00001312$

Remember that the end result is negative.

$(-42.4)^{-3} = -.00001312$

⑳

REVIEW EXERCISES

A. **1.** Express $\log \frac{\sqrt{20}}{(25)(1.2)}$ as the sum and/or difference of logarithms.

2. If $\log N = 9.7621 - 10$, find $1/5 \log N$.

B. Evaluate by logarithms:

3. $(3.81)(43.4)$

4. $\frac{32.7}{\sqrt{.892}}$

5. $(23)^8$

6. $(7.284)^5 (1.31)^2$

7. $(-3.47)^{-3} \sqrt[4]{8.2}$

(21)

SOLUTIONS TO REVIEW EXERCISES

A. **1.** $1/2 \log 20 - \log 25 - \log 1.2$

2. $1/5(9.7621 - 10) = 1/5(49.7621 - 50)$
$= 9.9524 - 10$

B. **3.**
$\log 3.81 = .5809$
$\log 43.4 = 1.6375$
$\log N = 2.2184$
$N = 165.3$

4.
$\log 32.7 = 11.5145 - 10$
$1/2 \log .892 = 9.9752 - 10$
$\log N = 1.5393$
$N = 34.62$

5.
$\log 23 = 1.3617$
$8 \log 23 = 10.8936$
$\log N = 10.8936$
$N = 78{,}270{,}000{,}000$

6.
$5 \log 7.284 = 4.3115$
$2 \log 1.31 = .2346$
$\log N = 4.5461$
$N = 35{,}170$

7.
$1/4 \log 8.2 = 10.22845 - 10$
$3 \log 3.47 = 1.6209$
$\log N = 8.60755 - 10$
$\log N = 8.6076 - 10$ (rounded)
$N = .04051$

The answer is $-.04051$.

(21A)

SOLUTIONS TO STUDY EXERCISES

Study Exercise One (Frame 4)

1. $\log (57)(38) = \log 57 + \log 38$

2. $\log (19)(45)^3 = \log 19 + \log (45)^3$
$= \log 19 + 3 \log 45$

3. $\log \frac{\sqrt{14}}{(121)^5} = \log \sqrt{14} - \log (121)^5$
$= 1/2 \log 14 - 5 \log 121$

4. $\log \frac{(24)^3 \cdot (12)}{45} = 1/2 [\log (24)^3 + \log 12 - \log 45]$
$= 1/2[3 \log 24 + \log 12 - \log 45]$
$= 3/2 \log 24 + 1/2 \log 12 - 1/2 \log 45$

(4A)

Study Exercise Two (Frame 11)

1. Let $N = (26.3)(8.25)$

then $\log N = \log 26.3 + \log 8.25$

$$\begin{aligned}
\log 26.3 &= 1.4200 \\
\log 8.25 &= 0.9165 \\
\log N &= 2.3365 \\
N &= 217.0
\end{aligned}$$

(11A)

Study Exercise Three (Frame 14)

1. Let $N = (47.2)(.897)$

then $\log N = \log 47.2 + \log .897$

$$\begin{aligned}
\log 47.2 &= 1.6739 \\
\log .897 &= 9.9528 - 10 \\
\log N &= 11.6267 - 10 \\
\log N &= 1.6267 \\
N &= 42.34
\end{aligned}$$

$$\left.\begin{aligned}
&\left.\begin{aligned}\log 42.3 &= 1.6263 \\ \log N &= 1.6267\end{aligned}\right\} 4 \\
&\log 42.4 = 1.6274
\end{aligned}\right\} 11$$

$$\frac{x}{10} \cdot 11 = 4$$

$$11x = 40$$

$$x = 4$$

2. Let $N = \dfrac{50}{\sqrt{84.33}}$

$\log N = \log 50 - 1/2 \log 84.33$ $\quad (\log 84.33 = 1.9260)$

$$\begin{aligned}
\log 50 &= 1.6990 \\
1/2 \log 84.33 &= .9630 \\
\log N &= .7360 \\
N &= 5.445
\end{aligned}$$

$$\left.\begin{aligned}
&\left.\begin{aligned}\log 5.44 &= .7356 \\ \log N &= .7360\end{aligned}\right\} 4 \\
&\log 5.45 = .7364
\end{aligned}\right\} 8$$

$$x/10 \cdot 8 = 4$$

$$8x = 40$$

$$x = 5$$

(14A)

UNIT 20—SUPPLEMENTARY PROBLEMS

A. Find the indicated logarithms. Use a 4 place log table for problems 3 and 4.

1. If $\log N = 9.4876 - 10$:

a) find $1/2 \log N$ **b)** find $3 \log N$ **c)** find $2/3 \log N$

2. If $\log M = .4286$ and $\log N = .8672$, find $\log M - \log N$.

3. Find log (log 2)

4. Find $\dfrac{\log (12)}{\log (3)}$

B. Perform the indicated computations using a 4 place log table.

5. $(12.3)(.268)$ **6.** $\dfrac{29.34}{(6.8)^2}$ **7.** $(16.8)^{3/4}$

8. $\sqrt{136}$ **9.** $\sqrt{(4.73)(38)}$ **10.** $(.451)^{-2}$

11. $\sqrt[5]{51.6}$ **12.** $\dfrac{(0.352)(1.74)^2}{\sqrt{.00526}}$ **13.** $(45.5)^{-1}(\sqrt[3]{9200})$

14. $(-1.034)^7$ **15.** $\sqrt{\dfrac{(4.17)(61.3)^2}{.0352}}$

More On Solution Sets of Equations and Inequalities

unit

21

Objectives

1. Find solution sets of:
   (a) exponential equations
   (b) logarithm equations
   (c) radical equations
   (d) quadratic in form equations
   (e) inequalities
2. Be able to graph equations involving absolute value.

(1)

An equation of the form $a^x = b$ where $a > 0$ will be called an *exponential equation.*

Before proceeding to the solution of an exponential equation, we will review some ideas about logarithms.

(2)

From a table of logarithms, we find;

$$\log 8 = .9031$$
$$\log 4 = .6021$$
$$\log 2 = .3010$$

$$\begin{aligned} \log 8/2 &= \log 8 - \log 2 \\ &= .9031 - .3010 \\ &= .6021 \end{aligned}$$

$$\begin{aligned} \frac{\log 8}{\log 2} &= \frac{.9031}{.3010} \\ &\doteq 3.0003 \end{aligned}$$

(3)

Thus we see from the previous frame that $\log 8/2 \neq \dfrac{\log 8}{\log 2}$.

We conclude that for $a, b > 0, b \neq 1$:

$$\frac{\log a}{\log b} \neq \log a/b$$

$$\frac{\log a}{\log b} \neq \log a - \log b$$

(4)

To find the value of $\frac{\log 7}{\log 2}$, we first look up the logarithms.

Thus $\frac{\log 7}{\log 2} = \frac{.8451}{.3010}$. To evaluate $\frac{.8451}{.3010}$, we could use long division; then $\frac{.8451}{.3010} \doteq 2.808$.

We could also perform the division by logarithms. Let $N = \frac{.8451}{.3010}$.

$$\begin{aligned} \text{Then } \log N &= \log \frac{.8451}{.3010} \\ &= \log .8451 - \log .3010 \\ \log .8451 &= 9.9269 - 10 \\ \log .3010 &= 9.4786 - 10 \\ \log N &= .4483 \\ N &\doteq 2.807 \end{aligned}$$

⑤

To find the solution set of an exponential equation of the form $a^x = b, a > 0$, we will make use of the property:

If $a > 0$ and $a = b$, then $\log a = \log b$.

⑥

Example: Solve $2^x = 7$

Solution:

Line (1) $\log 2^x = \log 7$

Line (2) $x \log 2 = \log 7$

Line (3) $x = \frac{\log 7}{\log 2}$

Line (4) $x = \frac{.8451}{.3010}$

Line (5) $x \doteq 2.808$

The solution set is $\{2.808\}$.

⑦

Study Exercise One

Find the solution sets:

1. $5^x = 17$ **2.** $3^x = 5$ **3.** $4^x = 100$

⑧

We next focus our attention on equations involving logarithms.

Remember that $\log x$ exists only when $x > 0$. Thus in the equation $\log (x + 3) - \log x = 2$, $\log (x + 3)$ exists only when $(x + 3) > 0$ or $x > -3$ and $\log x$ exists only when $x > 0$. Thus both logs exist when $x > 0$. The replacement set is $\{x|x > 0\}$.

⑨

Example: Solve $\log (x + 3) - \log x = 2$

Solution:

Line (1) $\log (x + 3) - \log x = 2$

Line (2) $\log \frac{x + 3}{x} = 2$

Line (3) $10^2 = \frac{x + 3}{x}$

(Frame 10, contd.)

Line (4) $100x = x + 3$
Line (5) $99x = 3$
Line (6) $x = 1/33$

The solution set is $\{1/33\}$.

(10)

Example: Find the numerical value for x if $x = \log_5 3$.

Solution:

Line (1) $x = \log_5 3$
Line (2) $5^x = 3$
Line (3) $\log 5^x = \log 3$
Line (4) $x \log 5 = \log 3$
Line (5) $x = \frac{\log 3}{\log 5}$
Line (6) $x = \frac{.4771}{.6990}$
Line (7) $x \doteq .6825$

(11)

Study Exercise Two

A. Find the solution sets:

1. $\log (3x + 2) - \log (x - 4) = 1$ **2.** $\log x - 2 \log 4 = \log 32$

3. $\log (x + 2) - \log x = \log 12$

B. Find the numerical value of x:

4. $x = \log_4 6$

(12)

Consider the sentence $x = 2$. The solution set is $\{2\}$.

Consider the sentence $x^2 = 4$. The solution set is $\{2, -2\}$.

Thus the two equations are not equivalent. *Squaring both sides of an equation does not yield an equivalent equation.*

(13)

Squaring both sides of an equation may introduce extraneous roots.

If any operation capable of changing the solution set has been used, it is necessary to check the answers.

(14)

The solution set of an equation obtained by squaring both sides will contain the solution set of the original. That is, the solution set of the original is a subset of the solution set of the squared equation.

(15)

Procedure For Solving Radical Equations

I. The equation contains one radical:

1. Isolate the term containing the radical.
2. Put parenthesis around each side and square each side.
3. Check all roots obtained in the original equation.

II. The equation contains two radicals:

1. Isolate one of the terms containing one of the radicals.
2. Put parenthesis around each side and square each side.

(Frame 16, contd.)

3. If a radical remains, proceed as above.
4. Check all roots obtained in the original equation.

(16)

Example: Find the solution set of $\sqrt{2x-3} - 1 = 0$.

Solution:

Line (1) $\sqrt{2x-3} - 1 = 0$

Line (2) $\sqrt{2x-3} = 1$

Line (3) $(\sqrt{2x-3})^2 = (1)^2$

Line (4) $2x - 3 = 1$

Line (5) $2x = 4$

Line (6) $x = 2$

check $x = 2$. $\sqrt{2(2)-3} - 1 = 0$

$\sqrt{1} - 1 = 0$

$0 = 0$

The solution set is $\{2\}$.

(17)

Example: Solve $\sqrt{5x-1} = \sqrt{x} + 1$.

Solution:

Line (1) $(\sqrt{5x-1}) = (\sqrt{x} + 1)$

Line (2) $(\sqrt{5x-1})^2 = (\sqrt{x} + 1)^2$

Line (3) $5x - 1 = x + 2\sqrt{x} + 1$

Line (4) $4x - 2 = 2\sqrt{x}$

Line (5) $2x - 1 = \sqrt{x}$

Line (6) $(2x-1)^2 = (\sqrt{x})^2$

Line (7) $4x^2 - 4x + 1 = x$

Line (8) $4x^2 - 5x + 1 = 0$

Line (9) $(4x-1)(x-1) = 0$

Line (10) $x = 1/4, x = 1$

check $x = 1/4$ $\sqrt{\frac{5}{4} - 1} = \sqrt{\frac{1}{4}} + 1$

$1/2 \neq 1/2 + 1$

does not check

check $x = 1$ $\sqrt{5(1) - 1} = \sqrt{1} + 1$

$\sqrt{4} = 1 + 1$

$2 = 2$

The solution set is $\{1\}$.

(18)

Study Exercise Three

Find the solution sets:

1. $\sqrt{2x + 5} = 4$

2. $\sqrt{3x + 1} + 1 = x$

3. $\sqrt{3x + 1} = \sqrt{x} + 3$

4. $\sqrt{x - 3} = x - 5$

5. $\sqrt{4x - 3} = \sqrt{8x + 1} - 2$

(19)

Equations in Quadratic Form

Certain equations, although not quadratic, may be solved by methods learned for quadratic equations. One such method is by *factoring*.

(20)

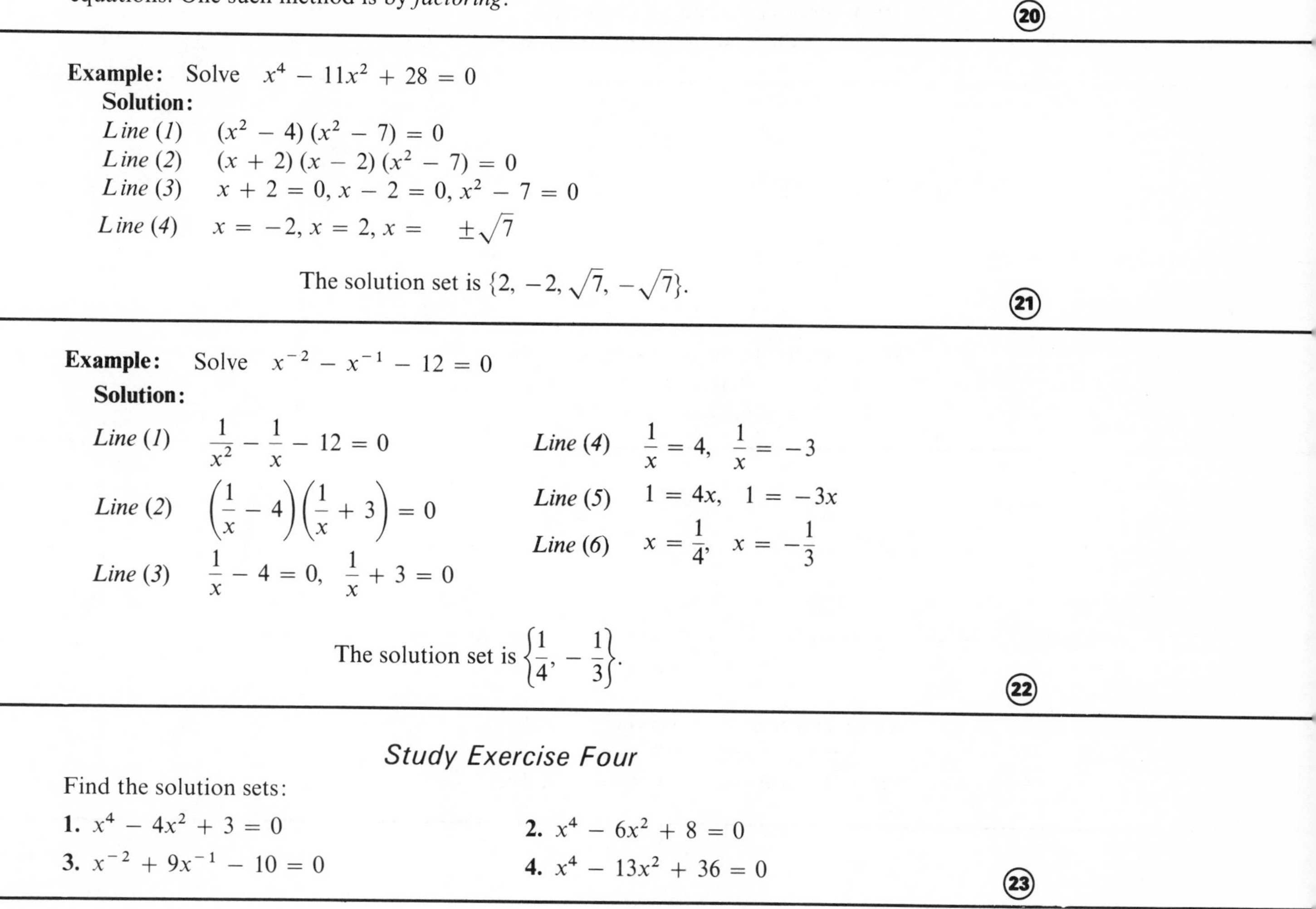

Example: Solve $x^4 - 11x^2 + 28 = 0$

Solution:

Line (1) $(x^2 - 4)(x^2 - 7) = 0$

Line (2) $(x + 2)(x - 2)(x^2 - 7) = 0$

Line (3) $x + 2 = 0, x - 2 = 0, x^2 - 7 = 0$

Line (4) $x = -2, x = 2, x = \pm\sqrt{7}$

The solution set is $\{2, -2, \sqrt{7}, -\sqrt{7}\}$.

(21)

Example: Solve $x^{-2} - x^{-1} - 12 = 0$

Solution:

Line (1) $\frac{1}{x^2} - \frac{1}{x} - 12 = 0$

Line (2) $\left(\frac{1}{x} - 4\right)\left(\frac{1}{x} + 3\right) = 0$

Line (3) $\frac{1}{x} - 4 = 0, \quad \frac{1}{x} + 3 = 0$

Line (4) $\frac{1}{x} = 4, \quad \frac{1}{x} = -3$

Line (5) $1 = 4x, \quad 1 = -3x$

Line (6) $x = \frac{1}{4}, \quad x = -\frac{1}{3}$

The solution set is $\left\{\frac{1}{4}, -\frac{1}{3}\right\}$.

(22)

Study Exercise Four

Find the solution sets:

1. $x^4 - 4x^2 + 3 = 0$

2. $x^4 - 6x^2 + 8 = 0$

3. $x^{-2} + 9x^{-1} - 10 = 0$

4. $x^4 - 13x^2 + 36 = 0$

(23)

Order

The number line motivates us to make additional assumptions about real numbers. The origin separates the line into three disjoint sets, namely the origin, the points to the left of the origin, and the points to the right of the origin.

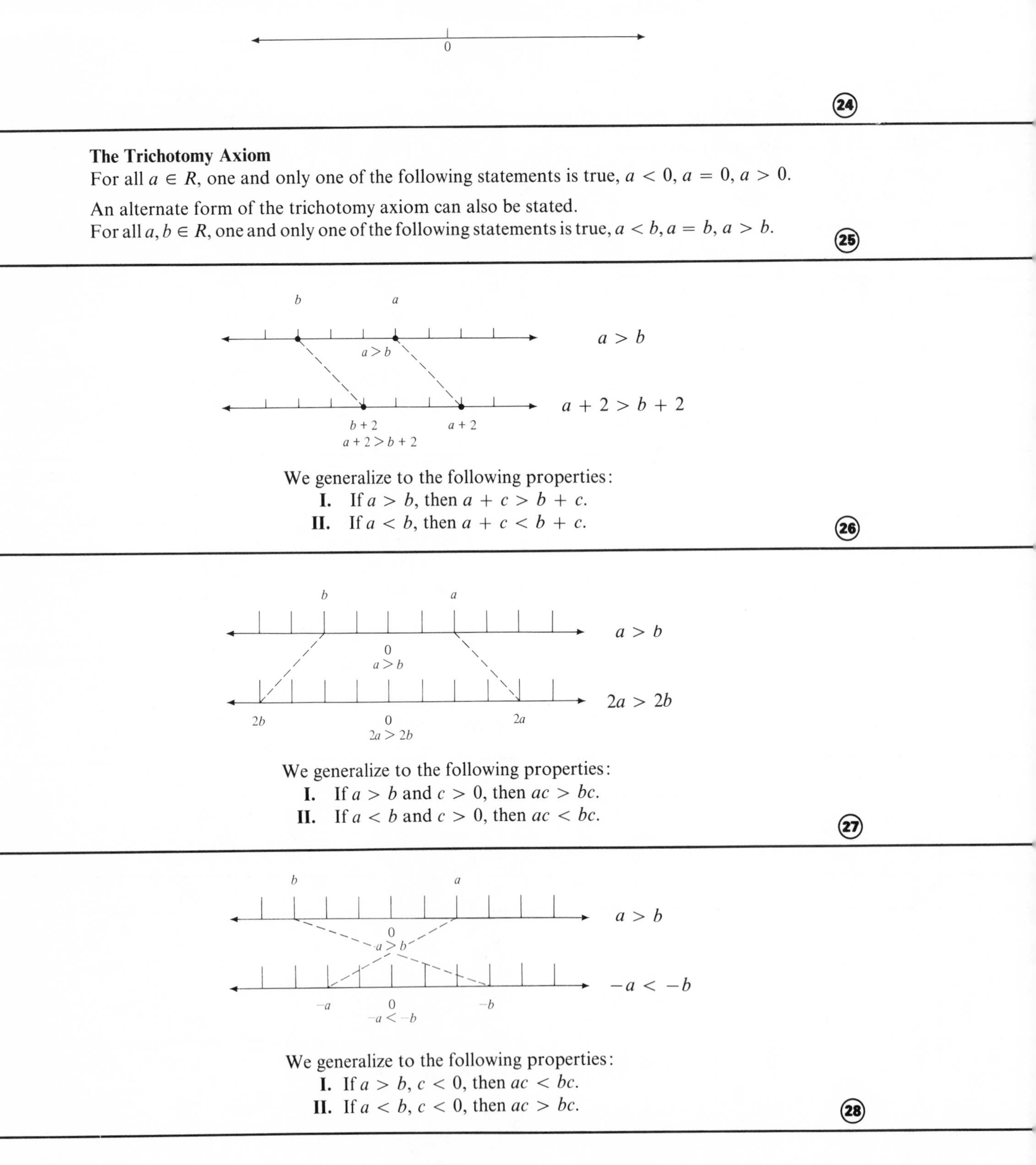

The Trichotomy Axiom
For all $a \in R$, one and only one of the following statements is true, $a < 0$, $a = 0$, $a > 0$.

An alternate form of the trichotomy axiom can also be stated.
For all $a, b \in R$, one and only one of the following statements is true, $a < b$, $a = b$, $a > b$.

(25)

We generalize to the following properties:

I. If $a > b$, then $a + c > b + c$.
II. If $a < b$, then $a + c < b + c$.

(26)

We generalize to the following properties:

I. If $a > b$ and $c > 0$, then $ac > bc$.
II. If $a < b$ and $c > 0$, then $ac < bc$.

(27)

We generalize to the following properties:

I. If $a > b$, $c < 0$, then $ac < bc$.
II. If $a < b$, $c < 0$, then $ac > bc$.

(28)

The important thing to remember concerning inequality statements is that *multiplication of an inequality by a negative number reverses the sign of the inequality* while multiplication by a positive number keeps the inequality sign fixed.

(29)

Study Exercise Five

Fill in with either $<$, $>$, or $=$ to make a true statement.

1. If $x + 1 > y + 2$, then x ___ $y + 1$.
2. If $-x < 0$, then x ___ 0.
3. If $a \nleq 0$, then a ___ 0.
4. If $x < 3$, then $2x$ ___ 6.
5. If $-2x < 8$, then x ___ -4.
6. If $a \nless y$ and $a \ngtr y$, then a ___ y.

(30)

Two inequality statements are equivalent if they have the same solution set over the same replacement set.

Example: The solution set of $2x > 4$ is $\{x|x > 2\}$
The solution set of $x - 2 > 0$ is $\{x|x > 2\}$
Therefore, $2x > 4$ and $x - 2 > 0$ are equivalent.

(31)

Principles of Equivalence

I. If $P(x) > Q(x)$, then $P(x) + R(x) > Q(x) + R(x)$ is an equivalent statement.

II. If $P(x) > Q(x)$, and if $R(x) > 0$, then
$P(x) \cdot R(x) > Q(x) \cdot R(x)$ is an equivalent statement.

III. If $P(x) > Q(x)$ and if $R(x) < 0$, then
$P(x) \cdot R(x) < Q(x) \cdot R(x)$ is an equivalent statement.

(32)

Example 1: Find the solution set of $3x - 27 > 0$.

Solution:

Line (1) $3x - 27 > 0$
Line (2) $3x > 27$
Line (3) $x > 9$

The solution set is $\{x|x > 9\}$.

Example 2: Find the solution set of $2x + 5 > 4x - 9$.

Solution:

Line (1) $2x + 5 > 4x - 9$
Line (2) $2x > 4x - 14$
Line (3) $-2x > -14$ — remember to reverse the inequality
Line (4) $x < 7$

The solution set is $\{x|x < 7\}$.

(33)

Study Exercise Six

Find the solution sets:

1. $5x - 3 < 8x - 12$
2. $x \leqslant 5(x - 2)$
3. $5x - 4 > 7x + 9$
4. $2x + 6 \leqslant 4 - 6x$
5. $\frac{x}{2} - 1 > 3 - x$
6. $\frac{2x - 5}{2} > \frac{5x + 4}{5}$

(34)

Absolute Value

The definition of absolute value states:

$$|x| = \begin{cases} x \text{ if } x > 0 \\ 0 \text{ if } x = 0 \\ -x \text{ if } x < 0 \end{cases}$$

Does $y = |x|$ represent a function?

35

Below is the graph of $\{(x, y) | y = |x|\}$.

x	y
0	0
1	1
2	2
3	3
4	4
5	5
-1	1
-2	2
-3	3
-4	4
-5	5

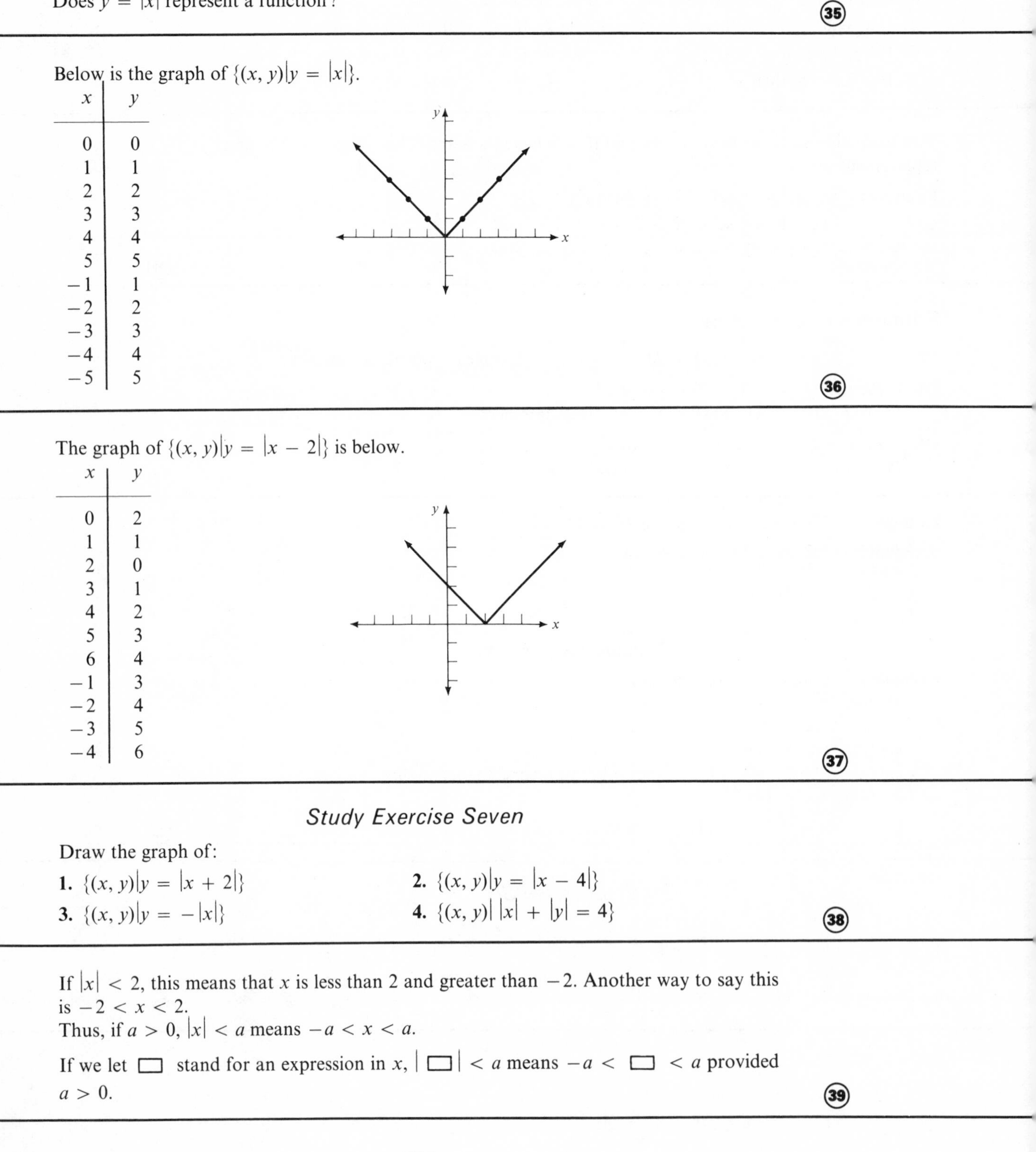

36

The graph of $\{(x, y) | y = |x - 2|\}$ is below.

x	y
0	2
1	1
2	0
3	1
4	2
5	3
6	4
-1	3
-2	4
-3	5
-4	6

37

Study Exercise Seven

Draw the graph of:

1. $\{(x, y) | y = |x + 2|\}$

2. $\{(x, y) | y = |x - 4|\}$

3. $\{(x, y) | y = -|x|\}$

4. $\{(x, y) | \, |x| + |y| = 4\}$

38

If $|x| < 2$, this means that x is less than 2 and greater than -2. Another way to say this is $-2 < x < 2$.
Thus, if $a > 0$, $|x| < a$ means $-a < x < a$.

If we let $\square$ stand for an expression in x, $|\square| < a$ means $-a < \square < a$ provided $a > 0$.

39

Since $|\square| < a$ means $-a < \square < a$, $|(x - 2)| < 4$ means $-4 < (x - 2) < 4$.

Also $|2x - 3/2| < 10$ means $-10 < 2x - 3/2 < 10$.

(40)

Since $-a < x < a$ also means $-a < x$ and $x < a$,
$-4 < (x - 2) < 4$ also means
$-4 < (x - 2)$ and $(x - 2) < 4$
$-4 < (x - 2)$ is equivalent to $-2 < x$
$(x - 2) < 4$ is equivalent to $x < 6$
$-2 < x$ and $x < 6$ can be written $-2 < x < 6$

This can be shortened to

$-4 < x - 2 < 4$
$-2 < x < 6$ by adding 2 to each part.

(41)

Example: Find the solution set of $|x - 3| < 6$

Solution:

Line (1) $|x - 3| < 6$
Line (2) $-6 < (x - 3) < 6$
Line (3) $-3 < x < 9$

The solution set is $\{x | -3 < x < 9\}$.

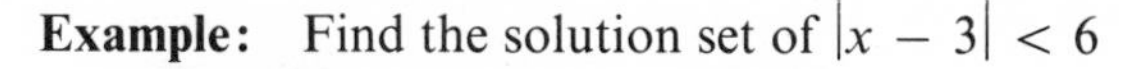

(42)

Example: Find the solution set of $|2x - 3| \leqslant 5$

Solution:

Line (1) $|2x - 3| \leqslant 5$
Line (2) $-5 \leqslant 2x - 3 \leqslant 5$
Line (3) $-2 \leqslant 2x \leqslant 8$
Line (4) $-1 \leqslant x \leqslant 4$

The solution set is $\{x | -1 \leqslant x \leqslant 4\}$.

−3 −2 −1 0 1 2 3 4 5

(43)

If $|x| > 2$, this means that x is greater than 2 or that x is less than -2. Thus if $a > 0$, $|x| > a$ means $x > a$ or $x < -a$.

If we let $\square$ stand for an expression in x, $|\square| > a$ means $\square > a$ or $\square < -a$ provided $a > 0$. Thus $|x - 2| > 4$ means $(x - 2) > 4$ or $(x - 2) < -4$.

(44)

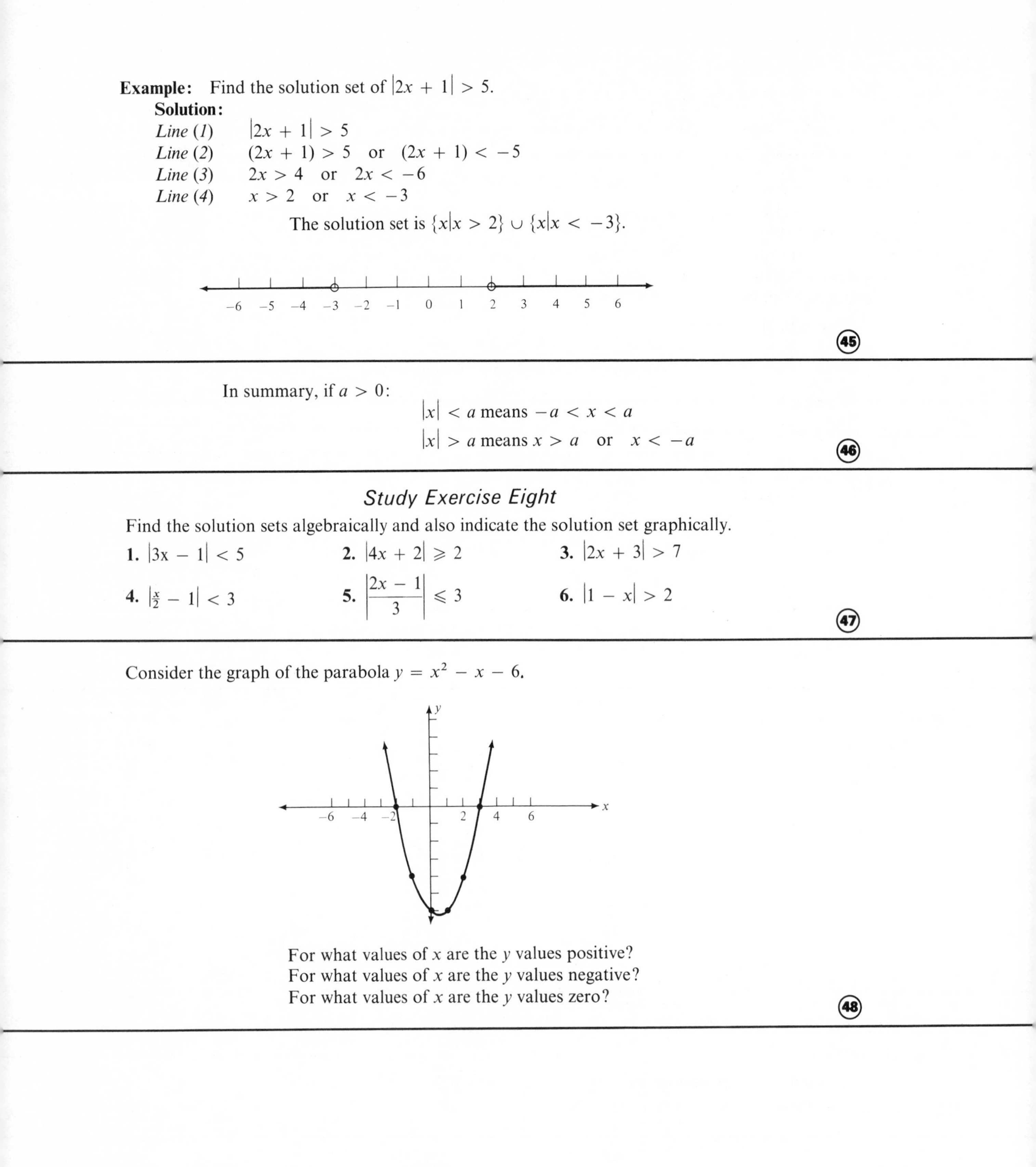

Example: Find the solution set of $|2x + 1| > 5$.

Solution:

Line (1) $|2x + 1| > 5$

Line (2) $(2x + 1) > 5$ or $(2x + 1) < -5$

Line (3) $2x > 4$ or $2x < -6$

Line (4) $x > 2$ or $x < -3$

The solution set is $\{x|x > 2\} \cup \{x|x < -3\}$.

(45)

In summary, if $a > 0$:

$|x| < a$ means $-a < x < a$

$|x| > a$ means $x > a$ or $x < -a$

(46)

Study Exercise Eight

Find the solution sets algebraically and also indicate the solution set graphically.

1. $|3x - 1| < 5$ **2.** $|4x + 2| \geqslant 2$ **3.** $|2x + 3| > 7$

4. $|\frac{x}{2} - 1| < 3$ **5.** $\left|\frac{2x - 1}{3}\right| \leqslant 3$ **6.** $|1 - x| > 2$

(47)

Consider the graph of the parabola $y = x^2 - x - 6$.

For what values of x are the y values positive?
For what values of x are the y values negative?
For what values of x are the y values zero?

(48)

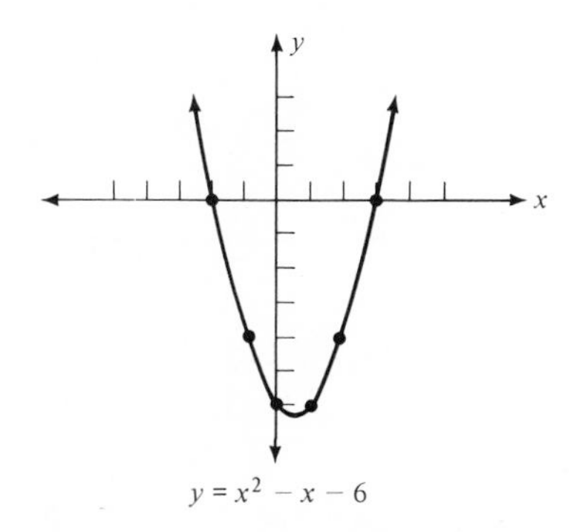

$y = x^2 - x - 6$

The y values are positive for all x greater than 3 or less than -2.

The y values are negative for all values of x between -2 and 3.

The y values are zero for x equal to -2 or 3.

(49)

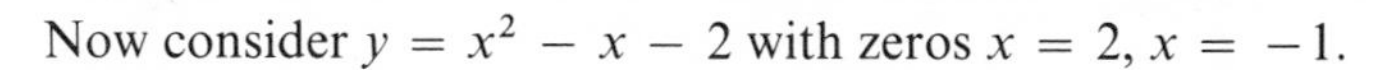

Now consider $y = x^2 - x - 2$ with zeros $x = 2, x = -1$.

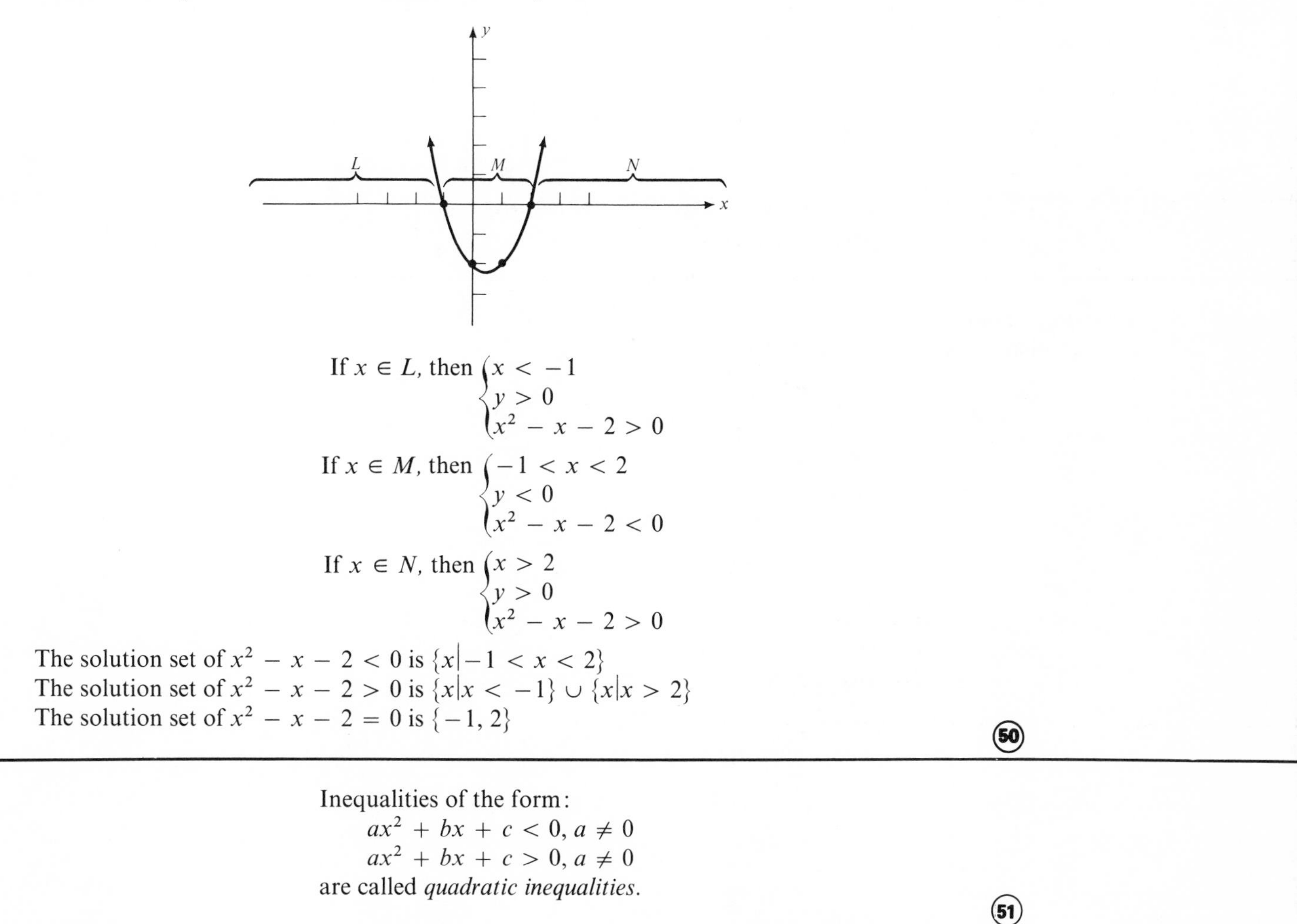

If $x \in L$, then $\begin{cases} x < -1 \\ y > 0 \\ x^2 - x - 2 > 0 \end{cases}$

If $x \in M$, then $\begin{cases} -1 < x < 2 \\ y < 0 \\ x^2 - x - 2 < 0 \end{cases}$

If $x \in N$, then $\begin{cases} x > 2 \\ y > 0 \\ x^2 - x - 2 > 0 \end{cases}$

The solution set of $x^2 - x - 2 < 0$ is $\{x \mid -1 < x < 2\}$
The solution set of $x^2 - x - 2 > 0$ is $\{x \mid x < -1\} \cup \{x \mid x > 2\}$
The solution set of $x^2 - x - 2 = 0$ is $\{-1, 2\}$

(50)

Inequalities of the form:

$ax^2 + bx + c < 0, a \neq 0$

$ax^2 + bx + c > 0, a \neq 0$

are called *quadratic inequalities*.

(51)

To find the solution set of a quadratic inequality:

1. If $a < 0$, multiply both sides by -1.
2. Graph $y = ax^2 + bx + c$.
3. Note where the graph is above or below the x axis.
4. Write the solution set.

(52)

Example: Find the solution set of $x^2 - 2x - 8 < 0$

Solution: Graph $y = x^2 - 2x - 8$ by first finding the x-intercepts.

$$x^2 - 2x - 8 = 0$$
$$(x - 4)(x + 2) = 0$$
$$x = 4, \quad x = -2$$

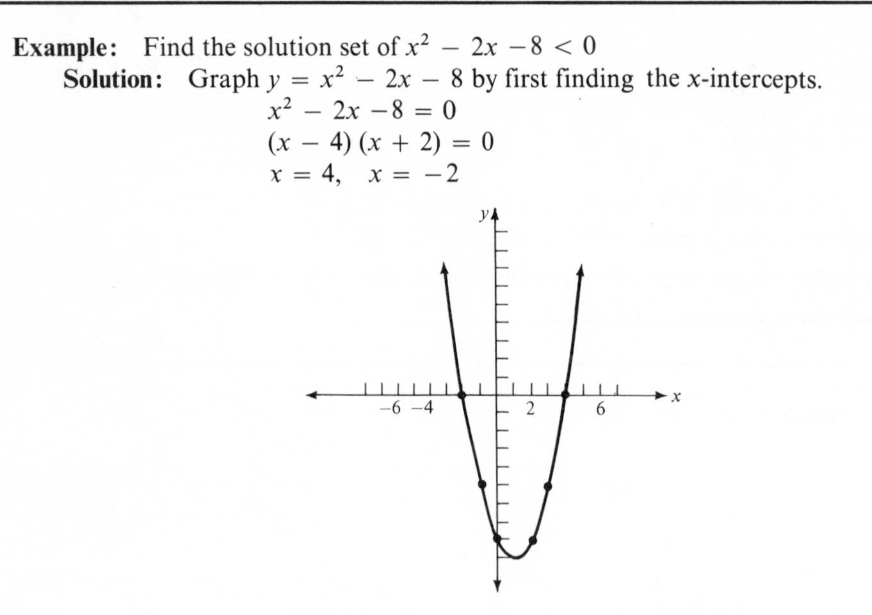

Observe where the y values are negative.
The solution set is $\{x| -2 < x < 4\}$.

(53)

Example: Find the solution set of $-x^2 + x - 1 < 0$.

Solution: First multiply both sides by -1.

$$x^2 - x + 1 > 0$$

Then graph $y = x^2 - x + 1$.
Find x intercepts.

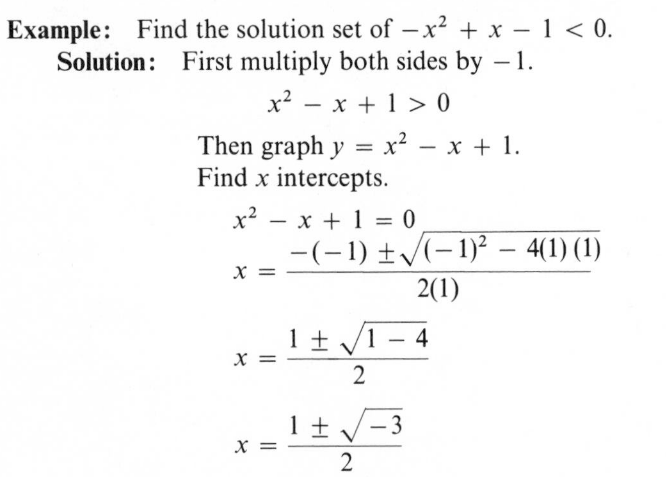

$$x^2 - x + 1 = 0$$
$$x = \frac{-(-1) \pm \sqrt{(-1)^2 - 4(1)(1)}}{2(1)}$$
$$x = \frac{1 \pm \sqrt{1 - 4}}{2}$$
$$x = \frac{1 \pm \sqrt{-3}}{2}$$

There are no x intercepts.

(Frame 54 contd.)

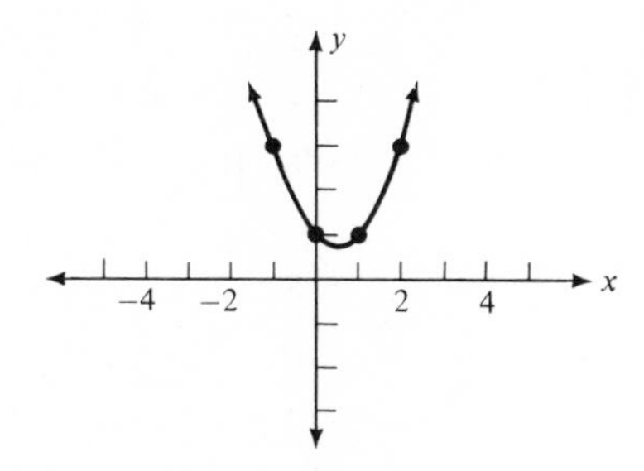

To solve $x^2 - x + 1 > 0$, we observe where the y values are positive. The solution set is $\{x | x \in R\}$.

(54)

Study Exercise Nine

Find the solution sets:

1. $x^2 - x - 6 < 0$

2. $x^2 - 5x + 4 > 0$

3. $-x^2 + 3x - 2 > 0$

4. $x^2 - 2x + 2 < 0$

(55)

REVIEW EXERCISES

A. Fill in with either $<$ or $>$ to make a true statement.

1. If $1 < 10$, then $1(-3)$ ___ $10(-3)$.

2. If $-3 < 3$, then $-3(2)$ ___ $3(2)$.

3. If $a > b$ and $c < 0$, then ac ___ bc.

4. If $x < 0$, then $-x$ ___ 0.

5. If $a < 0, b > 0$, then ab ___ 0.

6. If $a > b$, then $a - c$ ___ $b - c$.

B. Find the solution sets.

7. $3^x = 15$

8. $\log(x^2 - 1) - \log(x + 1) = 2$

9. $\sqrt{x - 2} - 3 = 0$

10. $-2x > x$

11. $|2x - 1| \geqslant 3$

12. $|3x + 2| < 2$

13. $x^4 - 3x^2 - 4 = 0$

14. $x^{-4} - 5x^{-2} + 4 = 0$

15. $7(3x - 1) \geqslant 4 + 5(2x + 1)$

C. Draw the graph.

16. $y = |2x|$

D. By graphing, find the solution set.

17. $x^2 + 3x - 4 > 0$

18. $x^2 + 3x - 4 < 0$

SOLUTIONS TO REVIEW EXERCISES

A. **1.** $>$ **2.** $<$ **3.** $<$ **4.** $>$ **5.** $<$ **6.** $>$

B. **7.** $3^x = 15$

$x \log 3 = \log 15$

$x = \dfrac{\log 15}{\log 3}$

$x = \dfrac{1.1761}{.4771}$

The solution set is $\{2.465\}$.

8. $\log(x^2 - 1) - \log(x + 1) = 2$

$\log \dfrac{x^2 - 1}{x + 1} = 2$

$\log(x - 1) = 2$

$10^2 = x - 1$

The solution set is $\{101\}$.

9. $\sqrt{x - 2} = 3$

$x - 2 = 9$

$x = 11$

The solution set is $\{11\}$.

Check: $x = 11$

$\sqrt{11 - 2} - 3 = 0$

$\sqrt{9} - 3 = 0$

$3 - 3 = 0$

$0 = 0$

10. $-2x > x$

$-3x > 0$

$x < 0$

The solution set is $\{x|x < 0\}$.

11. $|2x - 1| \geqslant 3$

$2x - \geqslant 3$ or $2x - 1 \leqslant -3$

$2x \geqslant 4$ or $2x \leqslant -2$

$x \geqslant 2$ or $x \leqslant -1$

The solution set is $\{x|x \geqslant 2\} \cup \{x|x \leqslant -1\}$.

(Frame **57** contd.)

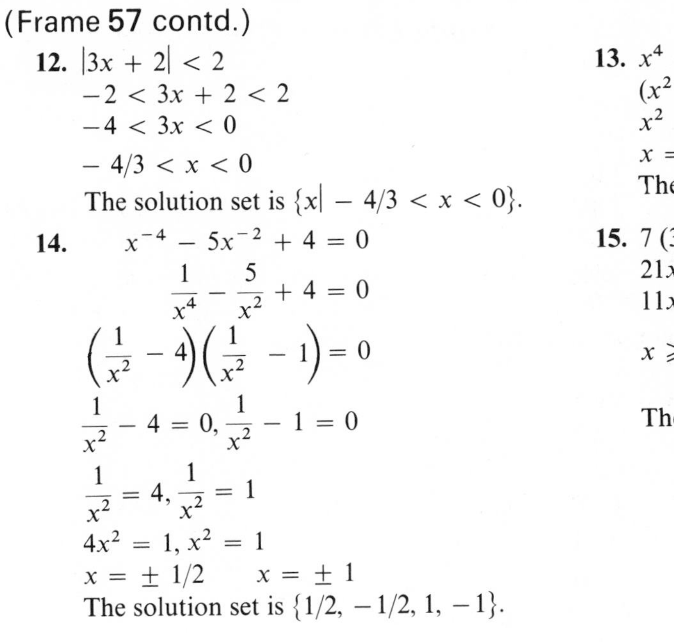

12. $|3x + 2| < 2$

$-2 < 3x + 2 < 2$

$-4 < 3x < 0$

$-4/3 < x < 0$

The solution set is $\{x| -4/3 < x < 0\}$.

13. $x^4 - 3x^2 - 4 = 0$

$(x^2 - 4)(x^2 + 1) = 0$

$x^2 = 4, \quad x^2 = -1$

$x = \pm 2, \quad$ no solution

The solution set is $\{2, -2\}$.

14. $x^{-4} - 5x^{-2} + 4 = 0$

$$\frac{1}{x^4} - \frac{5}{x^2} + 4 = 0$$

$$\left(\frac{1}{x^2} - 4\right)\left(\frac{1}{x^2} - 1\right) = 0$$

$$\frac{1}{x^2} - 4 = 0, \frac{1}{x^2} - 1 = 0$$

$$\frac{1}{x^2} = 4, \frac{1}{x^2} = 1$$

$4x^2 = 1, x^2 = 1$

$x = \pm 1/2 \qquad x = \pm 1$

The solution set is $\{1/2, -1/2, 1, -1\}$.

15. $7(3x - 1) \geqslant 4 + 5(2x + 1)$

$21x - 7 \geqslant 4 + 10x + 5$

$11x \geqslant 16$

$$x \geqslant \frac{16}{11}$$

The solution set is $\left\{x | x \geqslant \frac{16}{11}\right\}$.

C. 16.

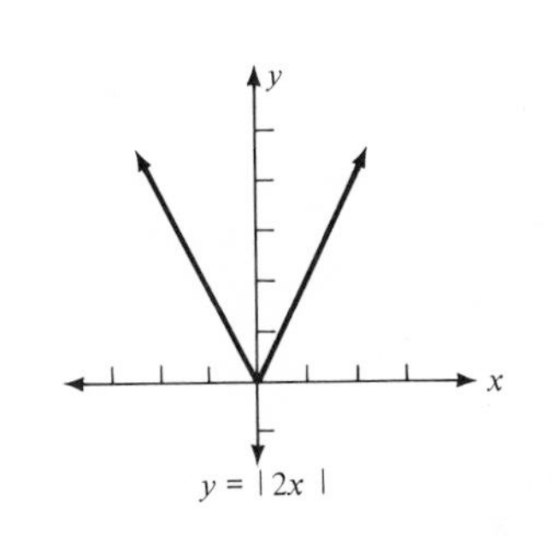

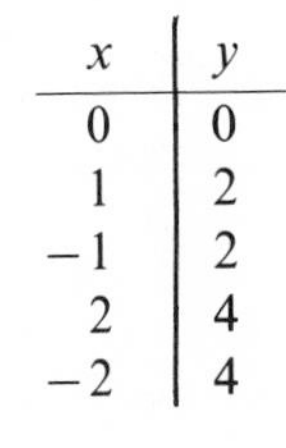

x	y
0	0
1	2
−1	2
2	4
−2	4

D. 17 – 18

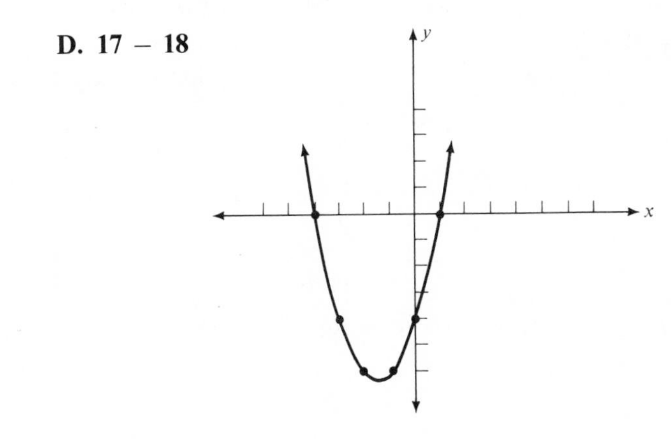

17. $\{x|x > 1\} \cup \{x|x < -4\}$

18. $\{x| -4 < x < 1\}$

SOLUTIONS TO STUDY EXERCISES

Study Exercise One (Frame 8)

1.
$$5^x = 17$$
$$\log 5^x = \log 17$$
$$x \log 5 = \log 17$$
$$x = \frac{\log 17}{\log 5}$$
$$x \doteq \frac{1.2304}{.6990}$$
$$x \doteq 1.76$$

$$\begin{aligned} \log 1.2304 &= 10.0900 - 10 \\ \log\ .6990 &= \ \ 9.8445 - 10 \\ \hline \text{log difference} &= \ \ .2455 \\ x &\doteq 1.76 \end{aligned}$$

The solution set is $\{1.76\}$.

2.
$$3^x = 5$$
$$x \log 3 = \log 5$$
$$x = \frac{\log 5}{\log 3}$$
$$x = \frac{.6990}{.4771}$$
$$x \doteq 1.465$$

The solution set is $\{1.465\}$.

3.
$$4^x = 100$$
$$x \log 4 = \log 100$$
$$x = \frac{\log 100}{\log 4}$$
$$x = \frac{2}{.6021}$$
$$x \doteq 3.3217$$

The solution set is $\{3.3217\}$.

8A

Study Exercise Two (Frame 12)

A. 1.
$$\log(3x + 2) - \log(x - 4) = 1$$
$$\log \frac{3x + 2}{x - 4} = 1$$
$$10^1 = \frac{3x + 2}{x - 4}$$
$$10x - 40 = 3x + 2$$
$$7x = 42$$
$$x = 6$$

The solution set is $\{6\}$.

2.
$$\log x - 2 \log 4 = \log 32$$
$$\log x - \log 4^2 = \log 32$$
$$\log \frac{x}{16} = \log 32$$
$$\frac{x}{16} = 32$$
$$x = 512$$

The solution set is $\{512\}$.

Study Exercise Two (Frame 12, contd.)

3. $\log(x+2) - \log x = \log 12$

$$\log \frac{x+2}{x} = \log 12$$

$$\frac{x+2}{x} = 12$$

$$12x = x + 2$$

$$11x = 2$$

$$x = \frac{2}{11}$$

The solution set is $\left\{\frac{2}{11}\right\}$.

4. $x = \log_4 6$

$$4^x = 6$$

$$\log 4^x = \log 6$$

$$x \log 4 = \log 6$$

$$x = \frac{\log 6}{\log 4}$$

$$x = \frac{.7782}{.6021}$$

$$x \doteq 1.2925$$

12A

Study Exercise Three (Frame 19)

1. $(\sqrt{2x+5})^2 = (4)^2$

$$2x + 5 = 16$$

$$2x = 11$$

$$x = \frac{11}{2}$$

The solution set is $\left\{\frac{11}{2}\right\}$.

check:

$$\sqrt{2\left(\frac{11}{2}\right) + 5} = 4$$

$$\sqrt{11 + 5} = 4$$

$$4 = 4$$

2. $(\sqrt{3x+1})^2 = (x-1)^2$

$$3x + 1 = x^2 - 2x + 1$$

$$0 = x^2 - 5x$$

$$0 = x(x-5)$$

$x = 0, x = 5$

The solution set is $\{5\}$.

check: $x = 0$

$$\sqrt{3 \cdot 0 + 1} + 1 = 0$$

$$\sqrt{1} + 1 \neq 0$$

check: $x = 5$

$$\sqrt{3 \cdot 5 + 1} + 1 = 5$$

$$4 + 1 = 5$$

3. $(\sqrt{3x+1})^2 = (\sqrt{x} + 3)^2$

$$3x + 1 = x + 6\sqrt{x} + 9$$

$$2x - 8 = 6\sqrt{x}$$

$$(x-4)^2 = (3\sqrt{x})^2$$

$$x^2 - 8x + 16 = 9x$$

$$x^2 - 17x + 16 = 0$$

$$(x-16)(x-1) = 0$$

$$x = 16, x = 1$$

The solution set is $\{16\}$.

check: $x = 16$

$$\sqrt{3(16) + 1} = \sqrt{16} + 3$$

$$\sqrt{49} = 4 + 3$$

$$7 = 7$$

check: $x = 1$

$$\sqrt{3(1) + 1} = \sqrt{1} + 3$$

$$\sqrt{4} = 1 + 3$$

$$2 \neq 4$$

4. $(\sqrt{x-3})^2 = (x-5)^2$

$$x - 3 = x^2 - 10x + 25$$

$$0 = x^2 - 11x + 28$$

$$0 = (x-4)(x-7)$$

$$x = 4, x = 7$$

The solution set is $\{7\}$.

check: $x = 4$

$$\sqrt{4-3} = 4 - 5$$

$$1 \neq -1$$

check: $x = 7$

$$\sqrt{7-3} = 7 - 5$$

$$2 = 2$$

Study Exercise Three (Frame 19, contd.)

5.

$$(\sqrt{4x-3})^2 = (\sqrt{8x+1}-2)^2$$
$$4x-3 = 8x+1-4\sqrt{8x+1}+4$$
$$-4x-8 = -4\sqrt{8x+1}$$
$$x+2 = \sqrt{8x+1}$$
$$(x+2)^2 = (\sqrt{8x+1})^2$$
$$x^2+4x+4 = 8x+1$$
$$x^2-4x+3 = 0$$
$$(x-3)(x-1) = 0$$
$$x = 3, x = 1$$

The solution set is $\{1, 3\}$.

check: $x = 3$

$$\sqrt{4(3)-3} = \sqrt{8(3)+1}-2$$
$$\sqrt{12-3} = \sqrt{24+1}-2$$
$$\sqrt{9} = \sqrt{25}-2$$
$$3 = 3$$

check: $x = 1$

$$\sqrt{4(1)-3} = \sqrt{8(1)+1}-2$$
$$\sqrt{1} = \sqrt{9}-2$$
$$1 = 1$$

19A

Study Exercise Four (Frame 23)

1.
$$x^4-4x^2+3 = 0$$
$$(x^2-3)(x^2-1) = 0$$
$$x = \pm\sqrt{3}, x = \pm 1$$

The solution set is $\{\sqrt{3}, -\sqrt{3}, 1, -1\}$.

2.
$$x^4-6x^2+8 = 0$$
$$(x^2-4)(x^2-2) = 0$$
$$x = \pm 2, x = \pm\sqrt{2}$$

The solution set is $\{2, -2, \sqrt{2}, -\sqrt{2}\}$.

3.
$$x^{-2}+9x^{-1}-10 = 0$$
$$\frac{1}{x^2}+\frac{9}{x}-10 = 0$$
$$\left(\frac{1}{x}+10\right)\left(\frac{1}{x}-1\right) = 0$$
$$\frac{1}{x}+10 = 0, \frac{1}{x}-1 = 0$$
$$\frac{1}{x} = -10, \quad \frac{1}{x} = 1$$
$$-10x = 1, \quad x = 1$$
$$x = -\frac{1}{10}$$

The solution set is $\left\{-\frac{1}{10}, 1\right\}$.

4.
$$x^4-13x^2+36 = 0$$
$$(x^2-9)(x^2-4) = 0$$
$$x = \pm 3, x = \pm 2$$

The solution set is $\{3, -3, 2, -2\}$.

23A

Study Exercise Five (Frame 30)

1. $>$ **2.** $>$ **3.** $>$ **4.** $<$ **5.** $>$ **6.** $=$

30A

Study Exercise Six (Frame 34)

1.
$$5x-3 < 8x-12$$
$$5x < 8x-9$$
$$-3x < -9$$
$$x > 3$$

The solution set is $\{x|x > 3\}$.

2.
$$x \leqslant 5(x-2)$$
$$x \leqslant 5x-10$$
$$-4x \leqslant -10$$
$$x \geqslant \frac{5}{2}$$

The solution set is $\{x|x \geqslant 5/2\}$.

SOLUTIONS TO STUDY EXERCISES, CONTD.

Study Exercise Six (Frame 34, contd.)

3. $5x - 4 > 7x + 9$
$5x > 7x + 13$
$-2x > 13$
$x < -13/2$
The solution set is $\{x|x < -13/2\}$.

4. $2x + 6 \leqslant 4 - 6x$
$2x \leqslant -2 - 6x$
$8x \leqslant -2$
$x \leqslant -1/4$
The solution set is $\{x|x \leqslant -1/4\}$.

5. $x/2 - 1 > 3 - x$
$x - 2 > 6 - 2x$
$x > 8 - 2x$
$3x > 8$
$x > \frac{8}{3}$
The solution set is $\{x|x > 8/3\}$.

6. $\frac{2x - 5}{2} > \frac{5x + 4}{5}$
$10x - 25 > 10x + 8$
$10x > 10x + 33$
The solution set is empty.

34A

Study Exercise Seven (Frame 38)

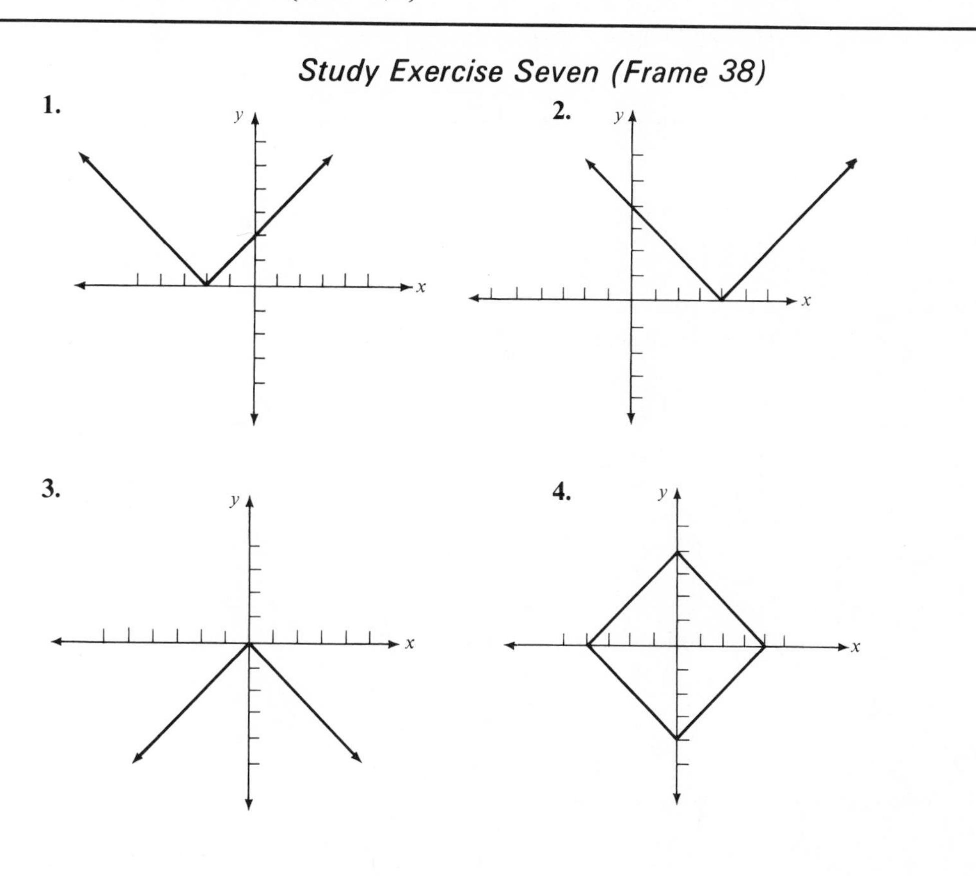

38A

Study Exercise Eight (Frame 47)

1. $|3x - 1| < 5$
$-5 < 3x - 1 < 5$
$-4 < 3x < 6$
$-4/3 < x < 2$
The solution set is $\{x| -4/3 < x < 2\}$.

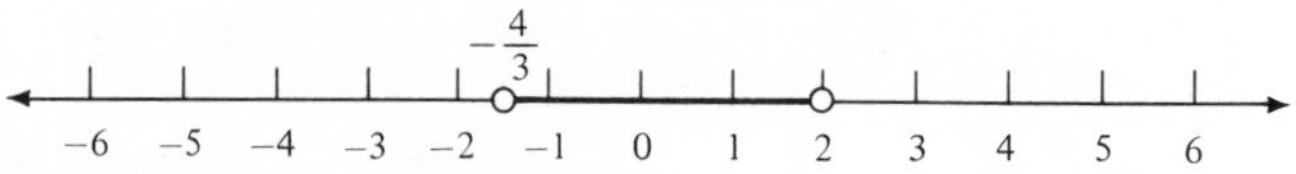

2. $|4x + 2| \geqslant 2$
$4x + 2 \geqslant 2$ or $4x + 2 \leqslant -2$
$4x \geqslant 0$ or $4x \leqslant -4$
$x \geqslant 0$ or $x \leqslant -1$
The solution set is $\{x|x \geqslant 0\} \cup \{x|x \leqslant -1\}$.

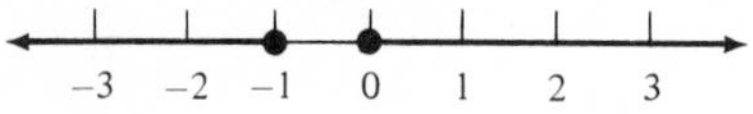

3. $|2x + 3| > 7$
$2x + 3 > 7$ or $2x + 3 < -7$
$2x > 4$ or $2x < -10$
$x > 2$ or $x < -5$
The solution set is $\{x|x > 2\} \cup \{x|x < -5\}$.

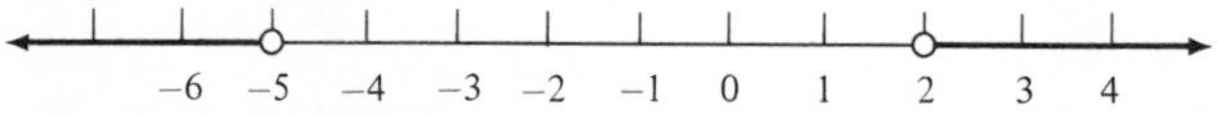

4. $|x/2 - 1| < 3$
$-3 < x/2 - 1 < 3$
$-6 < x - 2 < 6$
$-4 < x < 8$
The solution set is $\{x| -4 < x < 8\}$.

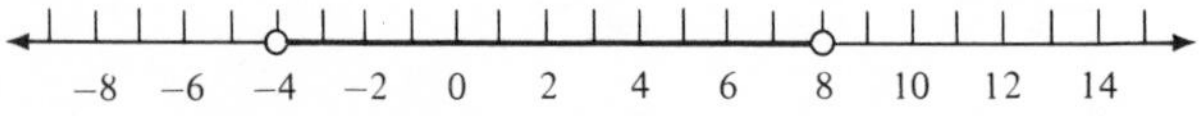

5. $\left|\dfrac{2x - 1}{3}\right| \leqslant 3$

$-3 \leqslant \dfrac{2x - 1}{3} \leqslant 3$
$-9 \leqslant 2x - 1 \leqslant 9$
$-8 \leqslant 2x \leqslant 10$
$-4 \leqslant x \leqslant 5$

The solution set is $\{x| -4 \leqslant x \leqslant 5\}$.

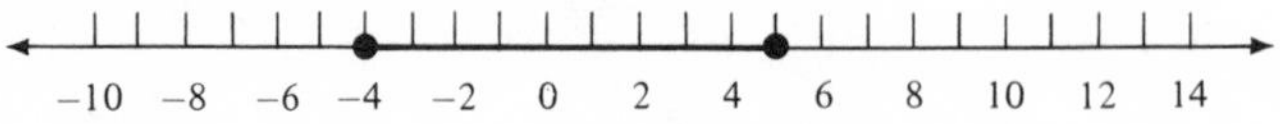

SOLUTIONS TO STUDY EXERCISES, CONTD.

Study Exercise Eight (Frame 47, contd.)

6. $|1 - x| > 2$

$1 - x > 2$ or $1 - x < -2$

$-x > 1$ or $-x < -3$

$x < -1$ or $x > 3$

The solution set is $\{x|x < -1\} \cup \{x|x > 3\}$.

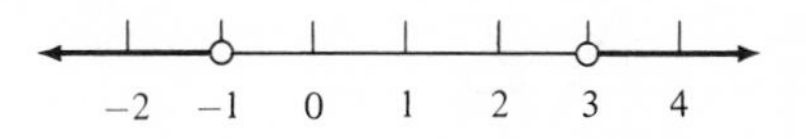

47A

Study Exercise Nine (Frame 55)

1.

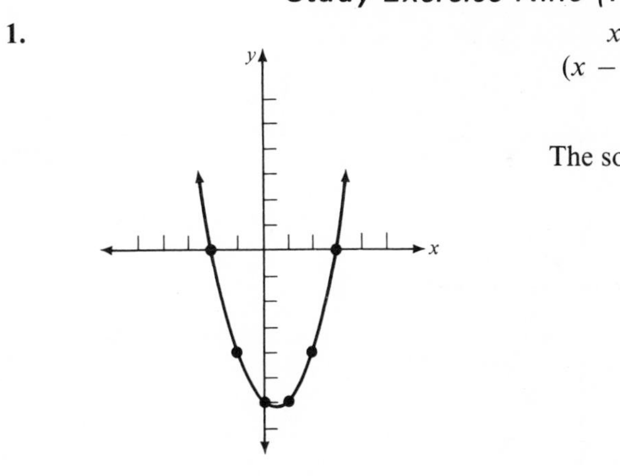

$$x^2 - x - 6 = 0$$
$$(x - 3)(x + 2) = 0$$

The solution set is $\{x|-2 < x < 3\}$.

2.

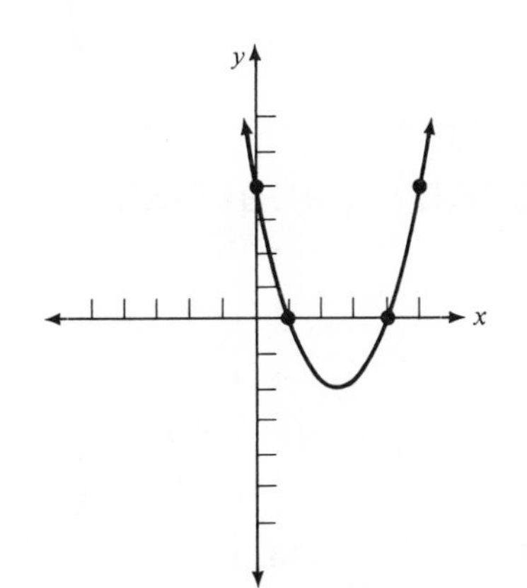

$$x^2 - 5x + 4 = 0$$
$$(x - 4)(x - 1) = 0$$
$$x = 4, \quad x = 1$$

The solution set is
$\{x|x > 4\} \cup \{x|x < 1\}$.

SOLUTIONS TO STUDY EXERCISES, CONTD.

Study Exercise Nine (Frame 55, contd.)

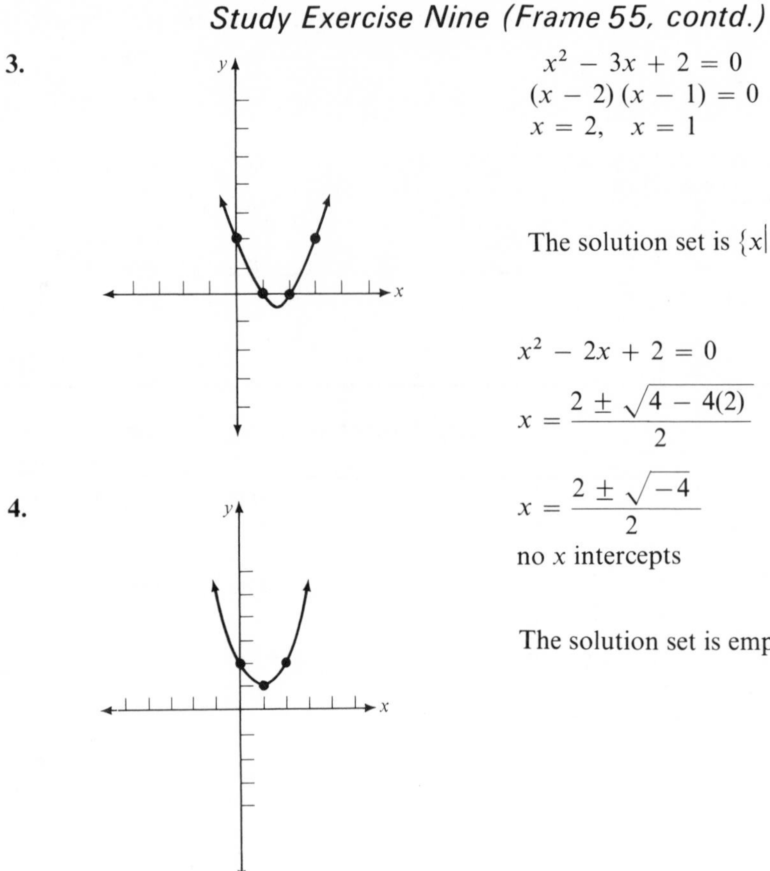

3.

$x^2 - 3x + 2 = 0$

$(x - 2)(x - 1) = 0$

$x = 2, \quad x = 1$

The solution set is $\{x | 1 < x < 2\}$.

4.

$x^2 - 2x + 2 = 0$

$$x = \frac{2 \pm \sqrt{4 - 4(2)}}{2}$$

$$x = \frac{2 \pm \sqrt{-4}}{2}$$

no x intercepts

The solution set is empty.

55A

UNIT 21—SUPPLEMENTARY PROBLEMS

Find the solution sets.

1. $2^x = 7$

2. $6^x = 25.3$

3. $3^{2x} = 15$

4. $\log x + \log 2x = \log 50$

5. $\log x + 2 \log 2 = 3 \log 4$

6. $\log x + \log 2 = 3$

7. $\sqrt{x - 1} = 7$

8. $\sqrt{x^2 + 3x} - 2 = 0$

9. $\sqrt{2x + 7} - \sqrt{x + 3} = 5$

10. $\sqrt{2x + 6} - \sqrt{x + 4} = 1$

11. $2\sqrt{2x} = 12$

12. $\sqrt{x + 1} = \sqrt{x + 5}$

13. $2x^4 - 17x^2 - 9 = 0$

14. $3x^4 - 4x^2 - 7 = 0$

15. $x^4 - 3x^2 - 4 = 0$

16. $x^4 - 5x^2 + 4 = 0$

17. $x^{-2} + 9x^{-1} - 10 = 0$

18. $2x^4 + 17x^2 - 9 = 0$

19. $x^{-2} - x^{-1} - 12 = 0$

20. $7x - 2(3 - x) < 4(1 + x)$

21. $3 - 2x < 11$

22. $\frac{7 + 5x}{3} - \frac{2 - x}{2} \geq 0$

23. $3x - 2 < 4x$

24. $5x \leq 6x$

UNIT 21—SUPPLEMENTARY PROBLEMS, CONTD.

25. $6 < 3x + 2 < 10$

26. $-1 < 3 - x < 1$

27. $-1/2 < 3x - 1/3 < 1/2$

28. $|3x - 1| \geq 4$

29. $|2x + 1| < 5$

30. $|5x| < 10$

31. $|2 - x| > 4$

32. $|4 - x| > 6$

33. $|2 - x| < 4$

34. $x^2 - 4 < 0$

35. $2x^2 + 5x > 12$

36. $3x^2 + 14x - 5 < 0$

37. $2x^2 - 9x + 4 > 0$

unit

System of Linear Equations and Determinants 22

Objectives

1. Know what is meant by a system of equations.
2. Be able to find solution sets of a system of two equations, two unknowns, and three equations, three unknowns both algebraically and by Cramer's Method.
3. Be able to evaluate second and third order determinants.

Terms

determinant	inconsistent
consistent	dependent

①

Solution Set of a System

We have seen that an equation in x and y has a solution set which consists of all number pairs (x, y) which satisfy the equation.

When we deal with two or more equations in x and y at the same time, we shall speak of the equations as a *system*. The solution set of such a system is the set of all ordered pairs (x, y) which satisfy all equations in the system.

②

Consider the system $\begin{cases} 2x - y = 4 \\ x + y = 5 \end{cases}$

The ordered pair (3, 2) is a solution since (3, 2) satisfies both equations

$$\begin{aligned} \text{check:}\quad 2(3) - 2 &= 4 \\ 6 - 2 &= 4 \\ 4 &= 4 \\ 3 + 2 &= 5 \\ 5 &= 5 \end{aligned}$$

③

The solution of the system $\begin{cases} 2x - y = 4 \\ x + y = 5 \end{cases}$ is $\{(x, y)|2x - y = 4\} \cap \{(x, y)|x + y = 5\}$.

Since each equation in the system has a graph forming a straight line, the ordered pair corresponding to the intersection of the lines is the solution set.

④

The solution set to a system of two equations in two variables x and y consists of all pairs of numbers (x, y) which satisfy both equations.

In set language, the solution set of a system of two equations in two variables is the intersection of their respective solution sets.

⑤

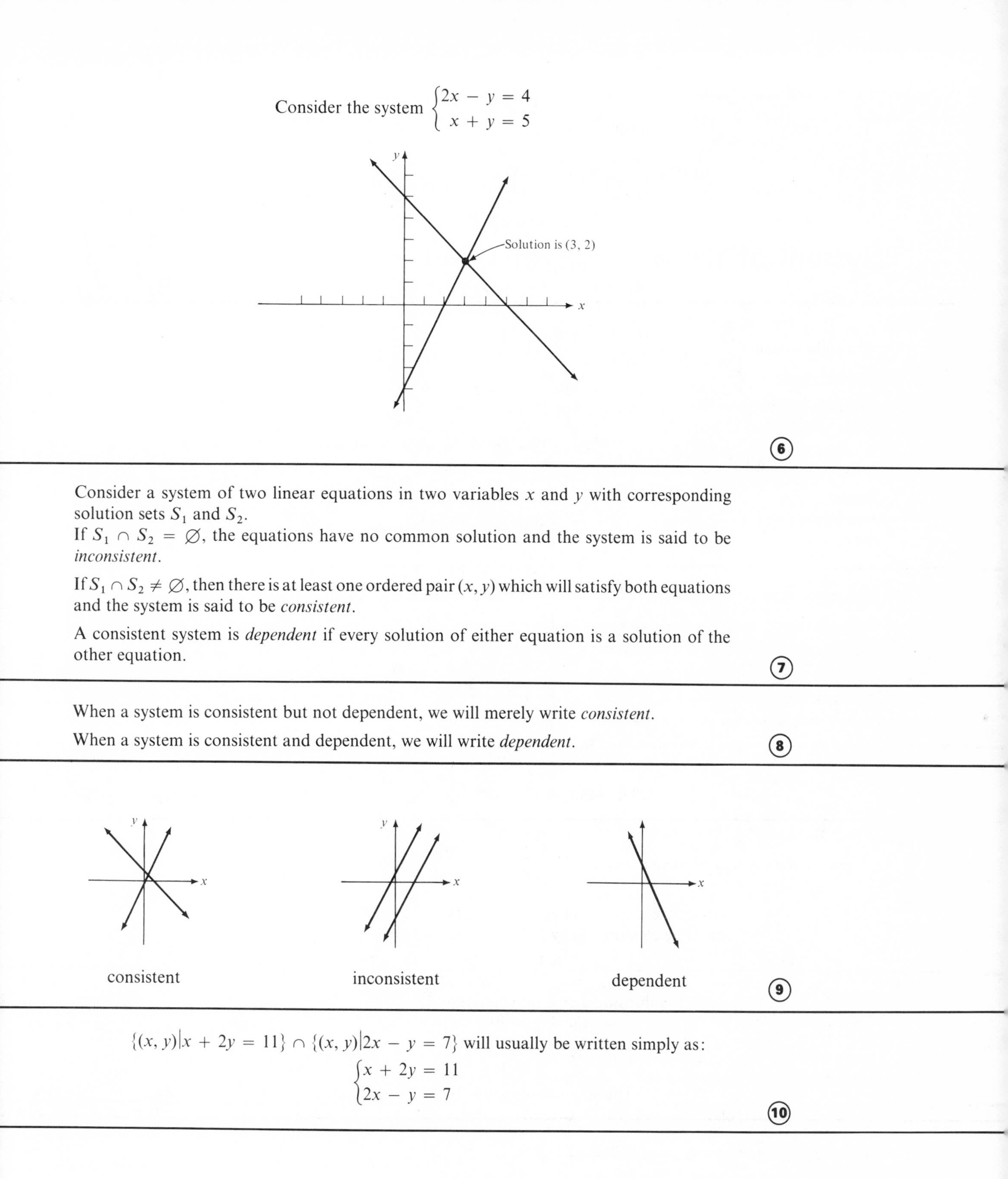

Consider the system $\begin{cases} 2x - y = 4 \\ x + y = 5 \end{cases}$

(6)

Consider a system of two linear equations in two variables x and y with corresponding solution sets S_1 and S_2.
If $S_1 \cap S_2 = \varnothing$, the equations have no common solution and the system is said to be *inconsistent*.

If $S_1 \cap S_2 \neq \varnothing$, then there is at least one ordered pair (x, y) which will satisfy both equations and the system is said to be *consistent*.

A consistent system is *dependent* if every solution of either equation is a solution of the other equation.

(7)

When a system is consistent but not dependent, we will merely write *consistent*.

When a system is consistent and dependent, we will write *dependent*.

(8)

consistent | inconsistent | dependent

(9)

$\{(x, y)|x + 2y = 11\} \cap \{(x, y)|2x - y = 7\}$ will usually be written simply as:

$$\begin{cases} x + 2y = 11 \\ 2x - y = 7 \end{cases}$$

(10)

Let $f(x, y)$ and $g(x, y)$ denote expressions involving x and y.

The system $\begin{cases} f(x, y) = 0 \\ g(x, y) = 0 \end{cases}$ is equivalent to:

$$\begin{cases} a \cdot f(x, y) + b \cdot g(x, y) = 0 \\ f(x, y = 0 \end{cases} \quad \text{where } b \neq 0$$

or is equivalent to

$$\begin{cases} a \cdot f(x, y) + b \cdot g(x, y) = 0 \\ g(x, y) = 0 \end{cases} \quad \text{where } a \neq o$$

(11)

Thus, the system $\begin{cases} 2x - y - 10 = 0 \\ 5x + 2y - 34 = 0 \end{cases}$

is equivalent to $\begin{cases} 2[2x - y - 10] + 1 \cdot [5x + 2y - 34] = 0 \\ 2x - y - 10 = 0 \end{cases}$

which is equivalent to $\begin{cases} 4x - 2y - 20 + 5x + 2y - 34 = 0 \\ 2x - y - 10 = 0 \end{cases}$

$$\begin{cases} 9x - 54 = 0 \\ 2x - y - 10 = 0 \end{cases}$$

$$\begin{cases} x - 6 = 0 \\ 2x - y - 10 = 0 \end{cases}$$

$$\begin{cases} -2(x - 6) + 1(2x - y - 10) = 0 \\ x - 6 = 0 \end{cases}$$

$$\begin{cases} -2x + 12 + 2x - y - 10 = 0 \\ x - 6 = 0 \end{cases}$$

$$\begin{cases} -y + 2 = 0 \\ x - 6 = 0 \end{cases}$$

$$\begin{cases} y = 2 \\ x = 6 \end{cases}$$

The solution set is $\{(6, 2)\}$.

(12)

A more compact way to solve the system $\begin{matrix} ① \\ ② \end{matrix}\begin{cases} 2x - y = 10 \\ 5x + 2y = 34 \end{cases}$ is shown below.

multiply equation ① by 2. $\quad 4x - 2y = 20$

multiply equation ② by 1. $\quad 5x + 2y = 34$

add. $\quad 9x = 54$

$x = 6$

substitute $x = 6$ into equation ①

$$2(6) - y = 10$$
$$12 - y = 10$$
$$2 = y$$

The solution set is $\{(6, 2)\}$.

(13)

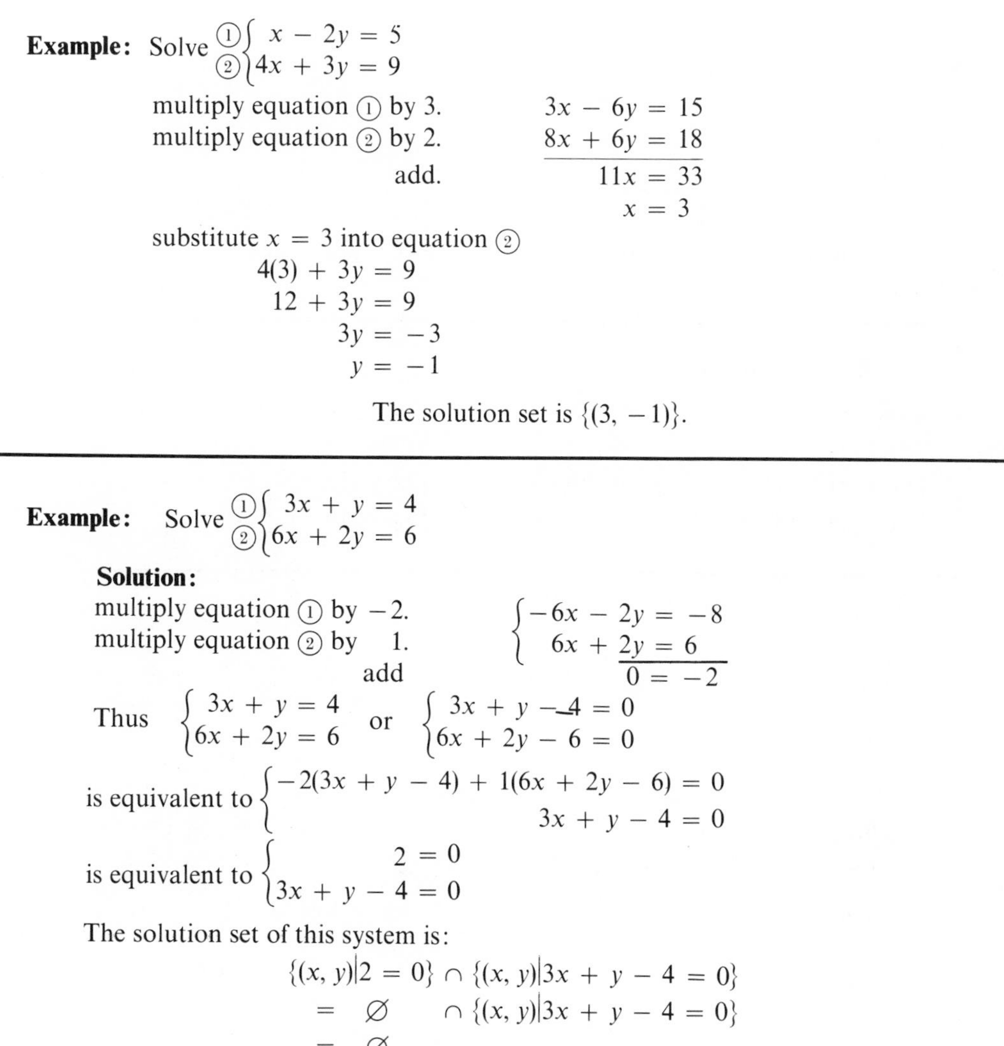

Example: Solve ①$\begin{cases} x - 2y = 5 \\ 4x + 3y = 9 \end{cases}$②

multiply equation ① by 3. $\quad 3x - 6y = 15$

multiply equation ② by 2. $\quad 8x + 6y = 18$

add. $\quad 11x = 33$

$x = 3$

substitute $x = 3$ into equation ②

$$4(3) + 3y = 9$$
$$12 + 3y = 9$$
$$3y = -3$$
$$y = -1$$

The solution set is $\{(3, -1)\}$.

⑭

Example: Solve ①$\begin{cases} 3x + y = 4 \\ 6x + 2y = 6 \end{cases}$②

Solution:

multiply equation ① by -2.

multiply equation ② by 1.

add

$$\begin{cases} -6x - 2y = -8 \\ 6x + 2y = 6 \end{cases}$$
$$0 = -2$$

Thus $\begin{cases} 3x + y = 4 \\ 6x + 2y = 6 \end{cases}$ or $\begin{cases} 3x + y - 4 = 0 \\ 6x + 2y - 6 = 0 \end{cases}$

is equivalent to $\begin{cases} -2(3x + y - 4) + 1(6x + 2y - 6) = 0 \\ 3x + y - 4 = 0 \end{cases}$

is equivalent to $\begin{cases} 2 = 0 \\ 3x + y - 4 = 0 \end{cases}$

The solution set of this system is:

$$\{(x, y) | 2 = 0\} \cap \{(x, y) | 3x + y - 4 = 0\}$$
$$= \varnothing \cap \{(x, y) | 3x + y - 4 = 0\}$$
$$= \varnothing$$

The system $\begin{Bmatrix} 3x + y = 4 \\ 6x + 2y = 6 \end{Bmatrix}$ is inconsistent (parallel lines).

⑮

Example: Solve ①$\begin{cases} 3x - y = 4 \\ 6x - 2y = 8 \end{cases}$②

Solution:

multiply equation ① by -2.

multiply equation ② by 1.

add.

$$\begin{cases} -6x + 2y = -8 \\ 6x - 2y = 8 \end{cases}$$
$$0 = 0$$

Thus $\begin{cases} 3x - y = 4 \\ 6x - 2y = 8 \end{cases}$ or $\begin{cases} 3x - y - 4 = 0 \\ 6x - 2y - 8 = 0 \end{cases}$

is equivalent to $\begin{cases} -2(3x - y - 4) + 1(6x - 2y - 8) = 0 \\ 3x - y - 4 = 0 \end{cases}$

is equivalent to $\begin{cases} 0 = 0 \\ 3x - y - 4 = 0 \end{cases}$

(Frame 16, contd.)

The solution set of this system is:

$\{(x, y)|0 = 0\} \cap \{(x, y)|3x - y - 4 = 0\} =$
$\{(x, y)|3x - y - 4 = 0\}$

Similarly it can be shown the solution set is:

$\{(x, y)|6x - 2y - 8 = 0\}$

Any ordered pair which satisfies one equation will satisfy the other. The equations each graph into the same line. The system is dependent.

(16)

Study Exercise One

Find the solution sets of the following systems. Also state if the system is consistent, inconsistent, or depedent.

1. $\begin{cases} x + y = 8 \\ x - y = 4 \end{cases}$ **2.** $\begin{cases} 7x + 5y = 11 \\ 3x - 2y = 13 \end{cases}$ **3.** $\begin{cases} 2x + y = 3 \\ 4x + 2y = 5 \end{cases}$ **4.** $\begin{cases} 2x - 5y = 3 \\ 4x - 10y = 6 \end{cases}$

(17)

Solve the system $\begin{cases} a_1x + b_1y = c_1 \\ a_2x + b_2y = c_2 \end{cases}$

Solution:

1. First we eliminate y.

$$\begin{aligned} a_1b_2x + b_1b_2y &= b_2c_1 \\ -a_2b_1x - b_1b_2y &= -b_1c_2 \\ \hline a_1b_2x - a_2b_1x &= b_2c_1 - b_1c_2 \\ (a_1b_2 - a_2b_1)x &= b_2c_1 - b_1c_2 \\ x &= \frac{b_2c_1 - b_1c_2}{a_1b_2 - a_2b_1}, \quad a_1b_2 - a_2b_1 \neq 0 \end{aligned}$$

2. Next we eliminate x.

$$\begin{aligned} a_1a_2x + a_2b_1y &= a_2c_1 \\ -a_1a_2x - a_1b_2y &= -a_1c_2 \\ \hline a_2b_1y - a_1b_2y &= a_2c_1 - a_1c_2 \\ -a_2b_1y + a_1b_2y &= -a_2c_1 + a_1c_2 \\ a_1b_2y - a_2b_1y &= a_1c_2 - a_2c_1 \\ (a_1b_2 - a_2b_1)y &= a_1c_2 - a_2c_1 \\ y &= \frac{a_1c_2 - a_2c_1}{a_1b_2 - a_2b_1}, \quad a_1b_2 - a_2b_1 \neq 0 \end{aligned}$$

(18)

Since the solution of $\begin{cases} a_1x + b_1y = c_1 \\ a_2x + b_2y = c_2 \end{cases}$ is given

by $x = \frac{b_2c_1 - b_1c_2}{a_1b_2 - a_2b_1}$, $y = \frac{a_1c_2 - a_2c_1}{a_1b_2 - a_2b_1}$, $a, b_2 - a_2b, \neq 0$.

We introduce the symbol $\begin{vmatrix} a_1 & b_1 \\ a_2 & b_2 \end{vmatrix}$ which is called the *determinant* and define it as another form for $(a_1b_2 - a_2b_1)$.

(19)

Second Order Determinant

By definition $\begin{vmatrix} a_1 & b_1 \\ a_2 & b_2 \end{vmatrix} = a_1b_2 - a_2b_1.$

The symbol $\begin{vmatrix} a_1 & b_1 \\ a_2 & b_2 \end{vmatrix}$ is called a *second order determinant* or a *two by two determinant* since it has two rows and two columns.

(20)

Example 1: Evaluate $\begin{vmatrix} 2 & 1 \\ 4 & 5 \end{vmatrix}$

Solution: $\begin{vmatrix} 2 & 1 \\ 4 & 5 \end{vmatrix} = (2)(5) - (1)(4)$
$= 10 - 4$
$= 6$

Example 2: Evaluate $\begin{vmatrix} 4 & 3 \\ -1 & -1 \end{vmatrix}$

Solution: $\begin{vmatrix} 4 & 3 \\ -1 & -1 \end{vmatrix} = 4(-1) - (3)(-1)$
$= (-4) - (-3)$
$= -1$

(21)

Study Exercise Two

Evaluate:

1. $\begin{vmatrix} 3 & 0 \\ 2 & 3 \end{vmatrix}$ **2.** $\begin{vmatrix} -5 & -1 \\ 2 & 3 \end{vmatrix}$ **3.** $\begin{vmatrix} 1 & -2 \\ -1 & 2 \end{vmatrix}$ **4.** $\begin{vmatrix} 3/4 & 1/2 \\ -1/2 & 1/4 \end{vmatrix}$

(22)

The solution of $\begin{cases} a_1x + b_1y = c_1 \\ a_2x + b_2y = c_2 \end{cases}$ is $x = \dfrac{b_2c_1 - b_1c_2}{a_1b_2 - a_2b_1}$, $y = \dfrac{a_1c_2 - a_2c_1}{a_1b_2 - a_2b_1}$ and can be written:

$$x = \frac{\begin{vmatrix} c_1 & b_1 \\ c_2 & b_2 \end{vmatrix}}{\begin{vmatrix} a_1 & b_1 \\ a_2 & b_2 \end{vmatrix}}, \quad y = \frac{\begin{vmatrix} a_1 & c_1 \\ a_2 & c_2 \end{vmatrix}}{\begin{vmatrix} a_1 & b_1 \\ a_2 & b_2 \end{vmatrix}}$$

(23)

When determinants are used to solve the system $\begin{cases} a_1x + b_1y = c_1 \\ a_2x + b_2y = c_2 \end{cases}$, the procedure is known as *Cramer's Method.*

$$x = \frac{\begin{vmatrix} c_1 & b_1 \\ c_2 & b_2 \end{vmatrix}}{\begin{vmatrix} a_1 & b_1 \\ a_2 & b_2 \end{vmatrix}}, \quad y = \frac{\begin{vmatrix} a_1 & c_1 \\ a_2 & c_2 \end{vmatrix}}{\begin{vmatrix} a_1 & b_1 \\ a_2 & b_2 \end{vmatrix}}$$

provided the determinant of the denominator is not equal to zero.

(24)

Example:

$$\text{Solve } \begin{cases} 3x + 2y = 10 \\ x - 3y = -7 \end{cases} \text{ using determinants.}$$

Solution:

$$x = \frac{\begin{vmatrix} 10 & 2 \\ -7 & -3 \end{vmatrix}}{\begin{vmatrix} 3 & 2 \\ 1 & -3 \end{vmatrix}} \qquad y = \frac{\begin{vmatrix} 3 & 10 \\ 1 & -7 \end{vmatrix}}{\begin{vmatrix} 3 & 2 \\ 1 & -3 \end{vmatrix}}$$

$$x = \frac{(-30)-(-14)}{(-9)-(2)} \qquad y = \frac{(-21)-(10)}{(-9)-2}$$

$$x = \frac{-30 + 14}{-11} \qquad y = \frac{-31}{-11}$$

$$x = \frac{-16}{-11}$$

The solution set is $\left\{\left(\frac{16}{11}, \frac{31}{11}\right)\right\}$.

(25)

Study Exercise Three

Solve by Cramer's Method:

1. $\begin{cases} 3x + 2y = 4 \\ x - 2y = 8 \end{cases}$

2. $\begin{cases} x - 3y = -5 \\ 2x + 3y = 5 \end{cases}$

3. $\begin{cases} 2x + y - 1 = 0 \\ 3x - 2y + 9 = 0 \end{cases}$

(26)

Cramer's Method can be extended to three equations in three unknowns.

$$\begin{cases} x + 2y - z = 6 \\ 2x - y + 3z = -13 \\ 3x - 2y + 3z = -16 \end{cases}$$

$$x = \frac{\begin{vmatrix} 6 & 2 & -1 \\ -13 & -1 & 3 \\ -16 & -2 & 3 \end{vmatrix}}{\begin{vmatrix} 1 & 2 & -1 \\ 2 & -1 & 3 \\ 3 & -2 & 3 \end{vmatrix}} \qquad y = \frac{\begin{vmatrix} 1 & 6 & -1 \\ 2 & -13 & 3 \\ 3 & -16 & 3 \end{vmatrix}}{\begin{vmatrix} 1 & 2 & -1 \\ 2 & -1 & 3 \\ 3 & -2 & 3 \end{vmatrix}} \qquad z = \frac{\begin{vmatrix} 1 & 2 & 6 \\ 2 & -1 & -13 \\ 3 & -2 & -16 \end{vmatrix}}{\begin{vmatrix} 1 & 2 & -1 \\ 2 & -1 & 3 \\ 3 & -2 & 3 \end{vmatrix}}$$

But how do you evaluate these determinants?

(27)

Third Order Determinants

A determinant with three rows and three columns is called a *third order determinant*. A third order determinant may be expanded as follows:

1. Rewrite the first and second columns to the right of the determinant.

$$\begin{vmatrix} a_1 & b_1 & c_1 \\ a_2 & b_2 & c_2 \\ a_3 & b_3 & c_3 \end{vmatrix} \begin{matrix} a_1 & b_1 \\ a_2 & b_2 \\ a_3 & b_3 \end{matrix}$$

(Frame 28, contd.)

2. The products of the elements in each of the three diagonals running down from left to right are given positive signs, and the products of the elements in each of the three diagonals running down from right to left are given negative signs; the algebraic sum of these six products is the value of the determinant.

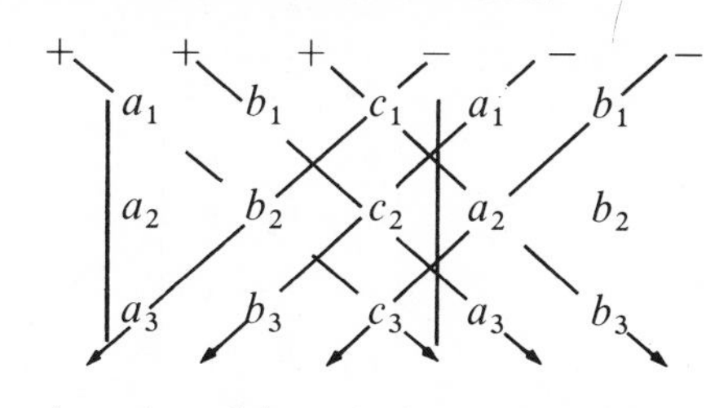

$$= (a_1b_2c_3) + (b_1c_2a_3) + (c_1a_2b_3) - (c_1b_2a_3) - (a_1c_2b_3) - (b_1a_2c_3).$$

(28)

Example: Evaluate $\begin{vmatrix} 1 & 2 & -1 \\ 2 & -1 & 3 \\ -3 & -4 & 5 \end{vmatrix}$

Solution:

$$\begin{vmatrix} 1 & 2 & -1 \\ 2 & -1 & 3 \\ -3 & -4 & 5 \end{vmatrix} \begin{matrix} 1 & 2 \\ 2 & -1 \\ -3 & -4 \end{matrix}$$

$$\begin{aligned} &= (1)(-1)(5) + (2)(3)(-3) + (-1)(2)(-4) \\ &\quad - (-1)(-1)(-3) - (1)(3)(-4) \\ &\quad - (2)(2)(5) \\ &= (-5) + (-18) + (8) - (-3) - (-12) \\ &\quad - (20) \\ &= -5 - 18 + 8 + 3 + 12 - 20 \\ &= -20 \end{aligned}$$

(29)

Study Exercise Four

Evaluate:

1. $\begin{vmatrix} 1 & 2 & -1 \\ 2 & -1 & 3 \\ 3 & -2 & 3 \end{vmatrix}$

2. $\begin{vmatrix} 1 & 0 & 3 \\ 0 & 2 & 1 \\ 1 & 2 & -1 \end{vmatrix}$

3. $\begin{vmatrix} 1 & 6 & -1 \\ 2 & -13 & 3 \\ 3 & -16 & 3 \end{vmatrix}$

4. $\begin{vmatrix} 1 & 2 & 3 \\ 3 & 2 & 1 \\ 1 & 1 & 1 \end{vmatrix}$

(30)

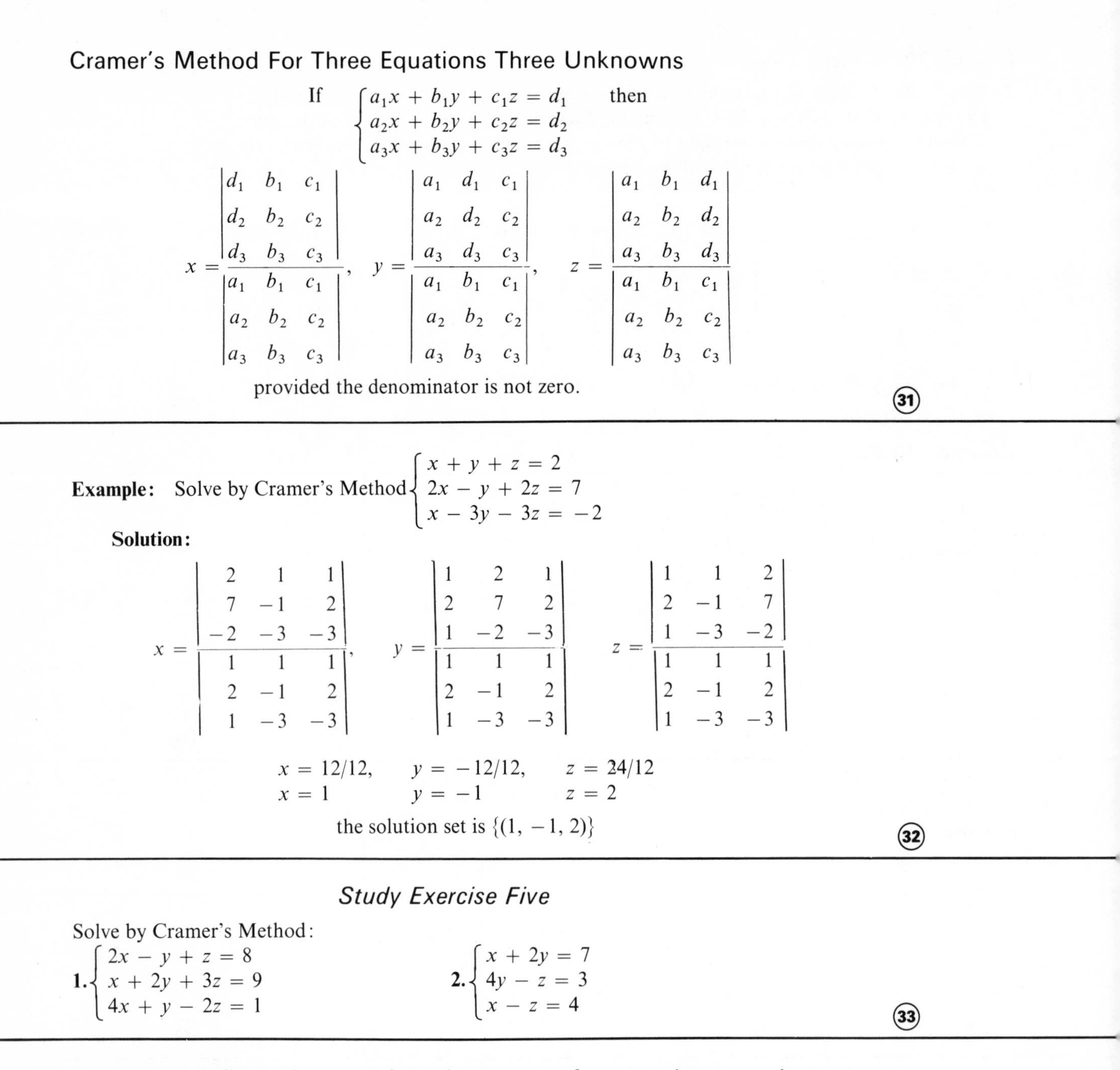

Cramer's Method For Three Equations Three Unknowns

If $\begin{cases} a_1x + b_1y + c_1z = d_1 \\ a_2x + b_2y + c_2z = d_2 \\ a_3x + b_3y + c_3z = d_3 \end{cases}$ then

$$x = \frac{\begin{vmatrix} d_1 & b_1 & c_1 \\ d_2 & b_2 & c_2 \\ d_3 & b_3 & c_3 \end{vmatrix}}{\begin{vmatrix} a_1 & b_1 & c_1 \\ a_2 & b_2 & c_2 \\ a_3 & b_3 & c_3 \end{vmatrix}}, \quad y = \frac{\begin{vmatrix} a_1 & d_1 & c_1 \\ a_2 & d_2 & c_2 \\ a_3 & d_3 & c_3 \end{vmatrix}}{\begin{vmatrix} a_1 & b_1 & c_1 \\ a_2 & b_2 & c_2 \\ a_3 & b_3 & c_3 \end{vmatrix}}, \quad z = \frac{\begin{vmatrix} a_1 & b_1 & d_1 \\ a_2 & b_2 & d_2 \\ a_3 & b_3 & d_3 \end{vmatrix}}{\begin{vmatrix} a_1 & b_1 & c_1 \\ a_2 & b_2 & c_2 \\ a_3 & b_3 & c_3 \end{vmatrix}}$$

provided the denominator is not zero.

(31)

Example: Solve by Cramer's Method $\begin{cases} x + y + z = 2 \\ 2x - y + 2z = 7 \\ x - 3y - 3z = -2 \end{cases}$

Solution:

$$x = \frac{\begin{vmatrix} 2 & 1 & 1 \\ 7 & -1 & 2 \\ -2 & -3 & -3 \end{vmatrix}}{\begin{vmatrix} 1 & 1 & 1 \\ 2 & -1 & 2 \\ 1 & -3 & -3 \end{vmatrix}}, \quad y = \frac{\begin{vmatrix} 1 & 2 & 1 \\ 2 & 7 & 2 \\ 1 & -2 & -3 \end{vmatrix}}{\begin{vmatrix} 1 & 1 & 1 \\ 2 & -1 & 2 \\ 1 & -3 & -3 \end{vmatrix}}, \quad z = \frac{\begin{vmatrix} 1 & 1 & 2 \\ 2 & -1 & 7 \\ 1 & -3 & -2 \end{vmatrix}}{\begin{vmatrix} 1 & 1 & 1 \\ 2 & -1 & 2 \\ 1 & -3 & -3 \end{vmatrix}}$$

$x = 12/12, \quad y = -12/12, \quad z = 24/12$

$x = 1 \quad\quad y = -1 \quad\quad z = 2$

the solution set is $\{(1, -1, 2)\}$

(32)

Study Exercise Five

Solve by Cramer's Method:

1. $\begin{cases} 2x - y + z = 8 \\ x + 2y + 3z = 9 \\ 4x + y - 2z = 1 \end{cases}$

2. $\begin{cases} x + 2y = 7 \\ 4y - z = 3 \\ x - z = 4 \end{cases}$

(33)

A procedure similar to the one used to solve a system of two equations two unknowns algebraically can be used to solve three equations three unknowns. The first step is to select one of the unknowns to eliminate.

(34)

Example:

Solve
$$\begin{cases} ① & x + 2y - z = 6 \\ ② & 2x - y + 3z = -13 \\ ③ & 3x - 2y + 3z = -16 \end{cases}$$

Solution: eliminate z in ① and ②

multiply ① by 3 $\quad 3x + 6y - 3z = 18$

multiply ② by 1 $\quad 2x - y + 3z = -13$

$$5x + 5y = 5$$
$$x + y = 1 \qquad \text{equation ④}$$

eliminate z in ② and ③

multiply ② by 1 $\quad 2x - y + 3z = -13$

multiply ③ by -1 $\quad -3x + 2y - 3z = 16$

$$-x + y = 3 \qquad \text{equation ⑤}$$

now consider equations ④ and ⑤

$$x + y = 1$$
$$-x + y = 3$$
$$sy = 4$$
$$y = 2$$

substitute $y = 2$ into ④

$$x + 2 = 1$$
$$x = -1$$

substitute $y = 2, x = -1$ into ①

$$-1 + 4 - z = 6$$
$$z = -3$$

The solution set is $\{(-1, 2, -3)\}$.

㉟

Study Exercise Six

Find the solution sets:

1. $\begin{cases} x + 5y - z = 2 \\ 2x + y + z = 7 \\ x - y + 2z = 11 \end{cases}$

2. $\begin{cases} x + y + z = 3 \\ x - z = 1 \\ y - z = -4 \end{cases}$

㊱

REVIEW EXERCISES

A. State whether each system is consistent, inconsistent, or dependent.

1. $\begin{cases} 2x - y = 4 \\ 2x + y = 4 \end{cases}$

2. $\begin{cases} 2x - 3y = 5 \\ 2x - 3y = 1 \end{cases}$

3. $\begin{cases} 2x + 3y = 5 \\ 4x + 6y = 10 \end{cases}$

4. $\begin{cases} x + y = 4 \\ x - 2y = 4 \end{cases}$

B. Find the solution sets (*do not use determinants*).

5. $\begin{cases} 2x + 3y = 1 \\ 3x + 5y = -4 \end{cases}$

6. $\begin{cases} 2x + y - z = -5 \\ -5x - 3y + 2z = 7 \\ x + 4y - 3z = 0 \end{cases}$

C. Evaluate

7. $\begin{vmatrix} -4 & -2 \\ 3 & 2 \end{vmatrix}$

8. $\begin{vmatrix} 2 & 1 & -1 \\ -5 & -3 & 2 \\ 1 & 4 & -3 \end{vmatrix}$

D. Find the solution sets by *Cramer's Method.*

9. $\begin{cases} 2x + 3y = 1 \\ 3x + 5y = -2 \end{cases}$

10. $\begin{cases} 2x + y - z = -5 \\ -5x - 3y + 2z = 7 \\ x + 4y - 3z = 0 \end{cases}$

(37)

SOLUTIONS TO REVIEW EXERCISES

A. 1. consistent **2.** inconsistent **3.** dependent **4.** consistent

B. 5. multiply equation ① by -5
multiply equation ② by 3

$$\begin{aligned} -10x - 15y &= -5 \\ 9x + 15y &= -12 \\ \hline -x &= -17 \\ x &= 17 \end{aligned}$$

substitute $x = 17$ into equation ①

$$\begin{aligned} 34 + 3y &= 1 \\ 3y &= -33 \\ y &= -11 \end{aligned}$$

The solution set is $\{(17, -11)\}$.

6. eliminate z in equations ① and ②

$$\begin{aligned} 4x + 2y - 2z &= -10 \\ -5x - 3y + 2z &= 7 \\ \hline -x - y &= -3 \\ \text{or} \quad x + y &= 3 \end{aligned}$$

eliminate z in equations ① and ③

$$\begin{aligned} -6x - 3y + 3z &= 15 \\ x + 4y - 3z &= 0 \\ \hline -5x + y &= 15 \\ \text{or} \quad 5x - y &= -15 \end{aligned}$$

solve for x and y

$$\begin{aligned} x + y &= 3 \\ 5x - y &= -15 \\ \hline 6x &= -12 \\ x &= -2 \\ y &= 5 \end{aligned}$$

$$\begin{aligned} 2x + y - z &= -5 \\ -4 + 5 - z &= -5 \\ -z &= -6 \\ z &= 6 \end{aligned}$$

The solution set is $\{(-2, 5, 6)\}$.

SOLUTIONS TO REVIEW EXERCISES, CONTD.

(Frame 37, contd.)

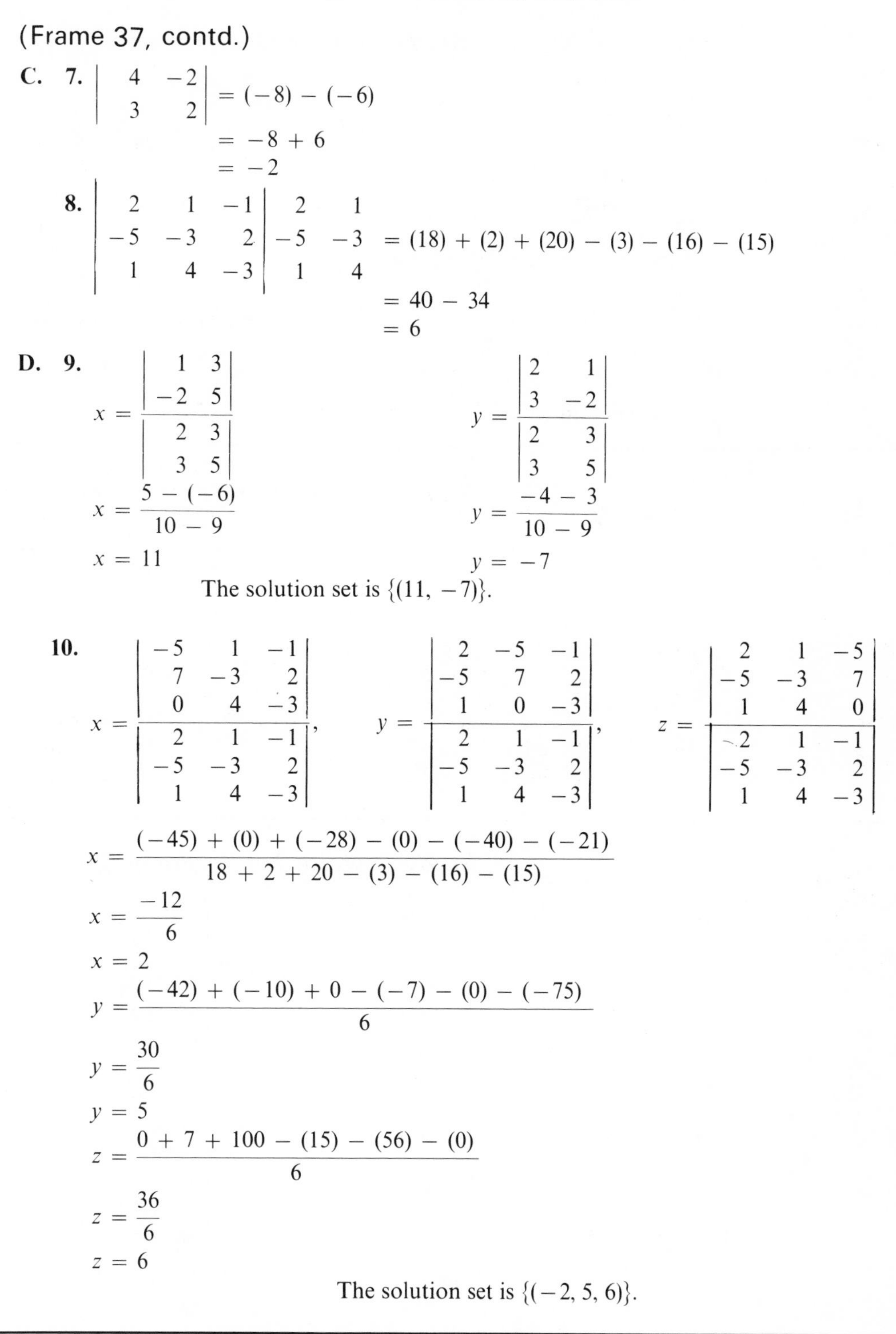

C. 7.

$$\begin{vmatrix} 4 & -2 \\ 3 & 2 \end{vmatrix} = (-8) - (-6)$$
$$= -8 + 6$$
$$= -2$$

8.

$$\left|\begin{array}{ccc|cc} 2 & 1 & -1 & 2 & 1 \\ -5 & -3 & 2 & -5 & -3 \\ 1 & 4 & -3 & 1 & 4 \end{array}\right| = (18) + (2) + (20) - (3) - (16) - (15)$$
$$= 40 - 34$$
$$= 6$$

D. 9.

$$x = \frac{\begin{vmatrix} 1 & 3 \\ -2 & 5 \end{vmatrix}}{\begin{vmatrix} 2 & 3 \\ 3 & 5 \end{vmatrix}} \qquad y = \frac{\begin{vmatrix} 2 & 1 \\ 3 & -2 \end{vmatrix}}{\begin{vmatrix} 2 & 3 \\ 3 & 5 \end{vmatrix}}$$

$$x = \frac{5 - (-6)}{10 - 9} \qquad y = \frac{-4 - 3}{10 - 9}$$

$$x = 11 \qquad y = -7$$

The solution set is $\{(11, -7)\}$.

10.

$$x = \frac{\begin{vmatrix} -5 & 1 & -1 \\ 7 & -3 & 2 \\ 0 & 4 & -3 \end{vmatrix}}{\begin{vmatrix} 2 & 1 & -1 \\ -5 & -3 & 2 \\ 1 & 4 & -3 \end{vmatrix}}, \qquad y = \frac{\begin{vmatrix} 2 & -5 & -1 \\ -5 & 7 & 2 \\ 1 & 0 & -3 \end{vmatrix}}{\begin{vmatrix} 2 & 1 & -1 \\ -5 & -3 & 2 \\ 1 & 4 & -3 \end{vmatrix}}, \qquad z = \frac{\begin{vmatrix} 2 & 1 & -5 \\ -5 & -3 & 7 \\ 1 & 4 & 0 \end{vmatrix}}{\begin{vmatrix} 2 & 1 & -1 \\ -5 & -3 & 2 \\ 1 & 4 & -3 \end{vmatrix}}$$

$$x = \frac{(-45) + (0) + (-28) - (0) - (-40) - (-21)}{18 + 2 + 20 - (3) - (16) - (15)}$$

$$x = \frac{-12}{6}$$

$$x = 2$$

$$y = \frac{(-42) + (-10) + 0 - (-7) - (0) - (-75)}{6}$$

$$y = \frac{30}{6}$$

$$y = 5$$

$$z = \frac{0 + 7 + 100 - (15) - (56) - (0)}{6}$$

$$z = \frac{36}{6}$$

$$z = 6$$

The solution set is $\{(-2, 5, 6)\}$.

37A

SOLUTIONS TO STUDY EXERCISES

Study Exercise One (Frame 17)

1.
$$\begin{aligned} x + y &= 8 \\ x - y &= 4 \\ \hline 2x &= 12 \\ x &= 6 \\ 6 + y &= 11 \\ y &= 2 \end{aligned}$$
the solution set is $\{(6, 2)\}$; consistent

2.
$$\begin{aligned} 14x + 10y &= 22 \\ 15x - 10y &= 65 \\ \hline 29x &= 87 \\ x &= 3 \\ 21 + 5y &= 11 \\ 5y &= -10 \\ y &= -2 \end{aligned}$$
the solution set is $\{(3, -2)\}$; consistent

3.
$$\begin{aligned} -4x - 2y &= -6 \\ 4x + 2y &= 5 \\ \hline 0 &\neq -1 \end{aligned}$$
the solution set is $\{\quad\}$; inconsistent

4. same lines, any point on either line is a solution; dependent

17A

Study Exercise Two (Frame 22)

1. $\begin{vmatrix} 3 & 0 \\ 2 & 3 \end{vmatrix} = (3)(3) - (2)(0)$
$= 9 - 0$
$= 9$

2. $\begin{vmatrix} -5 & -1 \\ 2 & 3 \end{vmatrix} = (-5)(3) - (2)(-1)$
$= -15 + 2$
$= -13$

3. $\begin{vmatrix} 1 & -2 \\ -1 & 2 \end{vmatrix} = 1(2) - (-2)(-1)$
$= 2 - (2)$
$= 0$

4. $\begin{vmatrix} 3/4 & 1/2 \\ -1/2 & 1/4 \end{vmatrix} = (3/4)(1/4) - (1/2)(-1/2)$
$= 3/16 - (-1/4)$
$= 3/16 + 4/16$
$= 7/16$

22A

Study Exercise Three (Frame 26)

1.
$$x = \frac{\begin{vmatrix} 4 & 2 \\ 8 & -2 \end{vmatrix}}{\begin{vmatrix} 3 & 2 \\ 1 & -2 \end{vmatrix}}, \qquad y = \frac{\begin{vmatrix} 3 & 4 \\ 1 & 8 \end{vmatrix}}{\begin{vmatrix} 3 & 2 \\ 1 & -2 \end{vmatrix}}$$
$x = -24/-8$ $y = 20/-8$
$x = 3$ $y = -5/2$

The solution set is $\{(3, -5/2)\}$.

2.
$$x = \frac{\begin{vmatrix} -5 & -3 \\ 5 & 3 \end{vmatrix}}{\begin{vmatrix} 1 & -3 \\ 2 & 3 \end{vmatrix}}, \qquad y = \frac{\begin{vmatrix} 1 & -5 \\ 2 & 5 \end{vmatrix}}{\begin{vmatrix} 1 & -3 \\ 2 & 3 \end{vmatrix}}$$
$x = 0/9$ $y = 15/9$
$x = 0$ $y = 5/3$

The solution set is $\{(0, 5/3)\}$.

Study Exercise Three (Frame 26, contd.)

3.

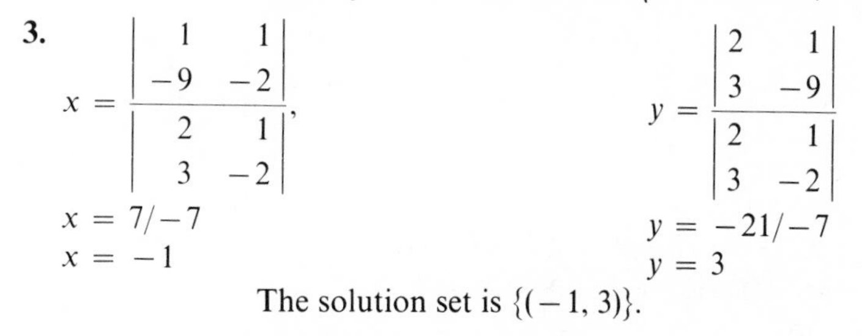

$$x = \frac{\begin{vmatrix} 1 & 1 \\ -9 & -2 \end{vmatrix}}{\begin{vmatrix} 2 & 1 \\ 3 & -2 \end{vmatrix}}, \qquad y = \frac{\begin{vmatrix} 2 & 1 \\ 3 & -9 \end{vmatrix}}{\begin{vmatrix} 2 & 1 \\ 3 & -2 \end{vmatrix}}$$

$x = 7/-7$ $\qquad$ $y = -21/-7$

$x = -1$ $\qquad$ $y = 3$

The solution set is $\{(-1, 3)\}$.

26A

Study Exercise Four (Frame 30)

1. $\begin{vmatrix} 1 & 2 & -1 \\ 2 & -1 & 3 \\ 3 & -2 & 3 \end{vmatrix} = 1(-1)(3) + 2(3)(3) + (-1)(2)(-2) - (-1)(-1)(3) - (-2)(3)(1) - 3(2)(2)$

$= -3 + 18 + 4 - 3 + 6 - 12$

$= 10$

2. $\begin{vmatrix} 1 & 0 & 3 \\ 0 & 2 & 1 \\ 1 & 2 & -1 \end{vmatrix} = 1(2)(-1) + 0(1)(1) + 3(2)(0) - 1(2)(3) - 2(1)(1) - (-1)(0)(0)$

$= -2 + 0 + 0 - 6 - 2 - 0$

$= -10$

3. $\begin{vmatrix} 1 & 6 & -1 \\ 2 & -13 & 3 \\ 3 & -16 & 3 \end{vmatrix} = 1(-13)(3) + 6(3)(3) + (-1)(2)(-16) - (3)(-13)(-1) - (-16)(3)(1) - (3)(6)(2)$

$= -39 + 54 + 32 - 39 + 48 - 36$

$= 20$

4. $\begin{vmatrix} 1 & 2 & 3 \\ 3 & 2 & 1 \\ 1 & 1 & 1 \end{vmatrix} = 1(2)(1) + 2(1)(1) + 3(3)(1) - 1(2)(3) - (1)(1)(1) - 1(2)(3)$

$= 2 + 2 + 9 - 6 - 1 - 6$

$= 0$

30A

Study Exercise Five (Frame 33)

1.

$$x = \frac{\begin{vmatrix} 8 & -1 & 1 \\ 9 & 2 & 3 \\ 1 & 1 & -2 \end{vmatrix}}{\begin{vmatrix} 2 & -1 & 1 \\ 1 & 2 & 3 \\ 4 & 1 & -2 \end{vmatrix}}, \qquad y = \frac{\begin{vmatrix} 2 & 8 & 1 \\ 1 & 9 & 3 \\ 4 & 1 & -2 \end{vmatrix}}{\begin{vmatrix} 2 & -1 & 1 \\ 1 & 2 & 3 \\ 4 & 1 & -2 \end{vmatrix}}$$

$$x = \frac{-32 + (-3) + 9 - 2 - 24 - 18}{-8 + (-12) + 1 - 8 - 6 - 2}, \qquad y = \frac{-36 + 96 + 1 - 36 - 6 - (-16)}{-35}$$

$$x = \frac{-70}{-35} \qquad y = \frac{35}{-35}$$

$$x = 2 \qquad y = -1$$

$$z = \frac{\begin{vmatrix} 2 & -1 & 8 \\ 1 & 2 & 9 \\ 4 & 1 & 1 \end{vmatrix}}{\begin{vmatrix} 2 & -1 & 1 \\ 1 & 2 & 3 \\ 4 & 1 & -2 \end{vmatrix}}$$

$$z = \frac{4 + (-36) + 8 - 64 - 18 - (-1)}{-35}$$

$$z = \frac{-105}{-35}$$

$$z = 3$$

The solution set is $\{(2, -1, 3)\}$.

2.

$$x = \frac{\begin{vmatrix} 7 & 2 & 0 \\ 3 & 4 & -1 \\ 4 & 0 & -1 \end{vmatrix}}{\begin{vmatrix} 1 & 2 & 0 \\ 0 & 4 & -1 \\ 1 & 0 & -1 \end{vmatrix}}, \qquad y = \frac{\begin{vmatrix} 1 & 7 & 0 \\ 0 & 3 & -1 \\ 1 & 4 & -1 \end{vmatrix}}{\begin{vmatrix} 1 & 2 & 0 \\ 0 & 4 & -1 \\ 1 & 0 & -1 \end{vmatrix}}$$

$$x = \frac{-28 + (-8) + 0 - 0 - 0 - (-6)}{(-4) + (-2) + 0 - 0 - 0 - 0} \qquad y = \frac{(-3) + (-7) + 0 - 0 - (-4) - 0}{-6}$$

$$x = -30/-6 \qquad y = -6/-6$$

$$x = 5 \qquad y = 1$$

$$z = \frac{\begin{vmatrix} 1 & 2 & 7 \\ 0 & 4 & 3 \\ 1 & 0 & 4 \end{vmatrix}}{\begin{vmatrix} 1 & 2 & 0 \\ 0 & 4 & -1 \\ 1 & 0 & -1 \end{vmatrix}}$$

$$z = \frac{16 + 6 + 0 - 28 - 0 - 0}{-6}$$

$$z = -6/-6$$

$$z = 1$$

The solution set is $\{(5, 1, 1)\}$.

33A

SOLUTIONS TO STUDY EXERCISES, CONTD.

Study Exercise Six (Frame 36)

1. Add equations ① and ②
multiply ② by -2
multiply ③ by 1

$$\begin{array}{r} 3x + 6y = 9 \\ -4x - 2y - 2z = -14 \\ x - y + +2z = 11 \\ \hline -3x - 3y = -3 \end{array}$$

$$\begin{array}{r} 3x + 6y = 9 \\ -3x - 3y = -3 \\ \hline 3y = 6 \\ y = 2 \\ x = -1 \\ z = 7 \end{array}$$

The solution set is $\{(-1, 2, 7)\}$.

2. Add equations ① and ②
multiply ② by 1
multiply ③ by -1

$$\begin{array}{r} 2x + y = 4 \\ x - z = 1 \\ -y + z = 4 \\ \hline x - y = 5 \end{array}$$

$$\begin{array}{r} 2x + y = 4 \\ x - y = 5 \\ \hline 3x = 9 \\ x = 3 \\ y = -2 \\ z = 2 \end{array}$$

The solution set is $\{(3, -2, 2)\}$.

36A

UNIT 22—SUPPLEMENTARY PROBLEMS

1. Evaluate $\begin{vmatrix} 3 & 4 \\ 15 & -2 \end{vmatrix}$

2. Evaluate $\begin{vmatrix} 1 & 3 \\ -4 & 7 \end{vmatrix} + \begin{vmatrix} 3 & 1 \\ 7 & -4 \end{vmatrix}$

3. Find the value of $\begin{vmatrix} a & b \\ c & d \end{vmatrix} + \begin{vmatrix} b & a \\ d & c \end{vmatrix}$

4. If $a \neq 0$, $b \neq 0$, find $\begin{vmatrix} a & b \\ 0 & 1 \end{vmatrix} \cdot \begin{vmatrix} 1/a & 1/b \\ 0 & 1 \end{vmatrix}$

5. Evaluate $\begin{vmatrix} 1 & 1 & 4 \\ 3 & 2 & 6 \\ 1 & -1 & 2 \end{vmatrix}$

6. Evaluate $\begin{vmatrix} 1 & -1 & 1 \\ -1 & 3 & -1 \\ 2 & 4 & 2 \end{vmatrix}$

7. Determine if the following systems are consistent, inconsistent, or dependent:

a) $\begin{cases} 2x + 5y = 8 \\ 4x + 10y = 12 \end{cases}$ **b)** $\begin{cases} 2x + 5y = 8 \\ 2x - y = 8 \end{cases}$ **c)** $\begin{cases} 2x - 3y = 4 \\ x + 2y = 7 \end{cases}$ **d)** $\begin{cases} x = 3y \\ y = 4 \end{cases}$

8. Solve $\begin{cases} 5x + 3y = 6 \\ 7x - y = 11 \end{cases}$

9. Solve $\begin{cases} 4x - y = 11 \\ 2x + 3y = -5 \end{cases}$

10. Solve by Cramer's Method: $\begin{cases} 5x + 4y = 3 \\ 2x + 3y = 4 \end{cases}$

11. Solve by Cramer's Method: $\begin{cases} 2x - 3y = -4 \\ 5x + 7y = 1 \end{cases}$

12. Solve $\begin{cases} x - 2y + 3z = 2 \\ 2x - 3y + z = 1 \\ 3x - y + 2z = 9 \end{cases}$

13. Solve $\begin{cases} x + y + 2z = 0 \\ 2x - 2y + z = 8 \\ 3x + 2y + z = 2 \end{cases}$

14. Solve by Cramer's Method: $\begin{cases} 4x - 2y + 3z = 2 \\ 5x - 6y + 2z = -1 \\ 3x - 4y - 5z = 7 \end{cases}$

15. Solve by Cramer's Method: $\begin{cases} x - 2y = 3 \\ -y + 3z = 1 \\ 2x + 5z = 0 \end{cases}$

unit

23

Systems of Equations and Inequalities

Objectives

1. Be able to find solution sets of certain systems such as one linear and one conic, and two conics.
2. Be able to graph solution sets of linear and certain quadratic inequalities.

Terms

linear inequality | open half plane | closed half plane

①

Find graphically: $\{(x, y)|x^2 + y^2 = 13\} \cap \{(x, y)|x - y = 1\}$.

Solution: $\{(x, y)|x^2 + y^2 = 13\}$ represents a circle and $\{(x, y)|x - y = 1\}$ represents a straight line.

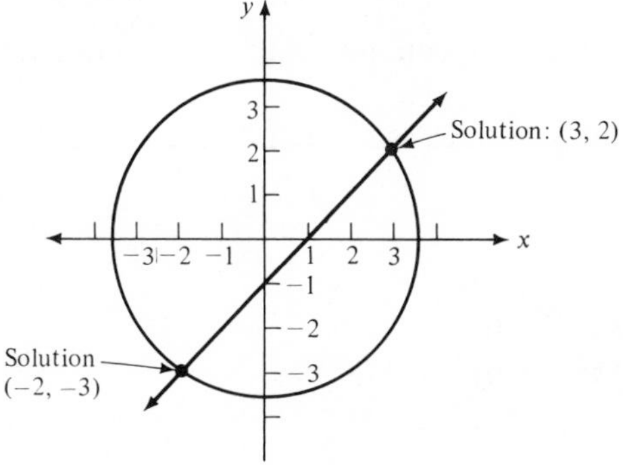

②

With one linear equation and one equation whose graph is a conic section (these are sometimes called quadratic in x and y), solve for one of the variables in the linear and substitute into the quadratic.

③

Example: Solve the system $\begin{cases} x^2 + y^2 = 13 \\ x - y = 1 \end{cases}$

Solution:

Line (1) $x - y = 1$

Line (2) $x = y + 1$

Line (3) $(y + 1)^2 + y^2 = 13$

Line (4) $y^2 + 2y + 1 + y^2 = 13$

Line (5) $2y^2 + 2y - 12 = 0$

Line (6) $y^2 + y - 6 = 0$

Line (7) $(y + 3)(y - 2) = 0$

Line (8) $y = -3, y = 2$

Line (9) Since $x = y + 1$ (line (2) above)
if $y = -3$, $x = -3 + 1$ or -2
if $y = 2$, $x = 2 + 1$ or 3

The solution set is $\{(-2, -3), (3, 2)\}$.

④

Study Exercise One

Find the solution sets:

1. $\begin{cases} x^2 - y^2 = 25 \\ x - y = 5 \end{cases}$
2. $\begin{cases} x^2 + 4y + 2x = 11 \\ y = x + 5 \end{cases}$
3. $\begin{cases} 4x^2 + 8xy - y^2 = 11 \\ 2x + 3y = 3 \end{cases}$

⑤

If the two equations in a system are of the form $\begin{cases} Ax^2 + By^2 = C \\ Dx^2 + Ey^2 = F \end{cases}$ where $A, B, C, D, E, F \in R$, eliminate one of the variables by techniques used for solving a system of two equations, two unknowns.

⑥

Example: Solve $\begin{cases} ① \ 4x^2 - 5y^2 = 16 \\ ② \ 3x^2 + 2y^2 = 35 \end{cases}$

Solution:

multiply ① by 2 $\quad 8x^2 - 10y^2 = 32$
multiply ② by 5 $\quad 15x^2 + 10y^2 = 175$

$$23x^2 = 207$$
$$x^2 = 9$$
$$x = 3 \quad or \quad -3$$

$x = 3$
substitute $x = 3$ into ②
$3(3)^2 + 2y^2 = 35$
$2y^2 = 8$
$y^2 = 4$
$y = 2 \quad or \quad -2$
$(3, 2)\ (3, -2)$

$x = -3$
substitute $x = -3$ into ②
$3(-3)^2 + 2y^2 = 35$
$2y^2 = 8$
$y^2 = 4$
$y = 2 \quad or \quad -2$
$(-3, 2)\ (-3, -2)$

The solution set is $\{(3, 2), (3, -2), (-3, 2), (-3, -2)\}$.

(Frame 7, contd.)

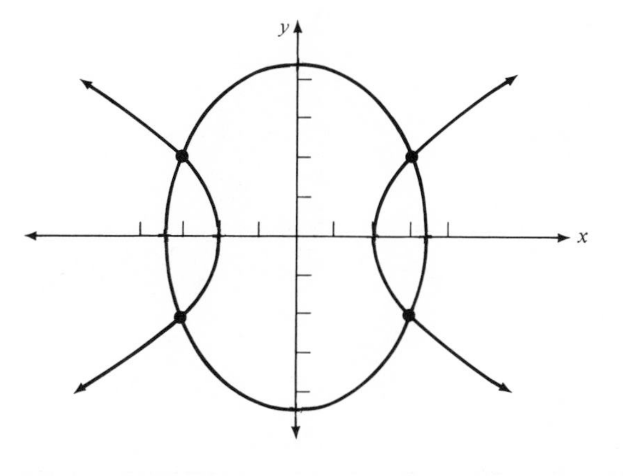

(7)

Study Exercise Two

Find the solution sets algebraically:

1. $\begin{cases} x^2 + y^2 = 25 \\ xy = -12 \end{cases}$ **2.** $\begin{cases} x^2 - y^2 = 1 \\ x^2 + y^2 = 3 \end{cases}$

(8)

Graphical Representation of Linear Inequalities

We have already studied linear relations of the form $\{(x, y) | Ax + By + C = 0\}$. If the equal sign is changed to $>$, $<$, $\geqslant$, or $\leqslant 0$, the resulting relation will be called a *linear inequality*.

(9)

Let us consider $\{(x, y) | x + y - 5 > 0\}$.

We plot this relation by looking at $x + y - 5 = 0$ or $y = 5 - x$ and then find where $y > 5 - x$.

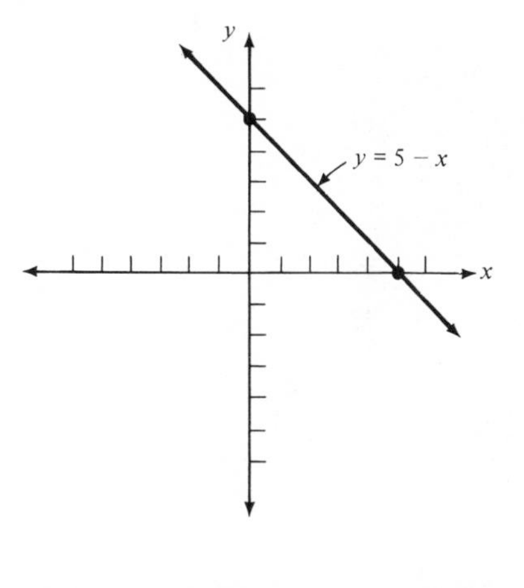

(10)

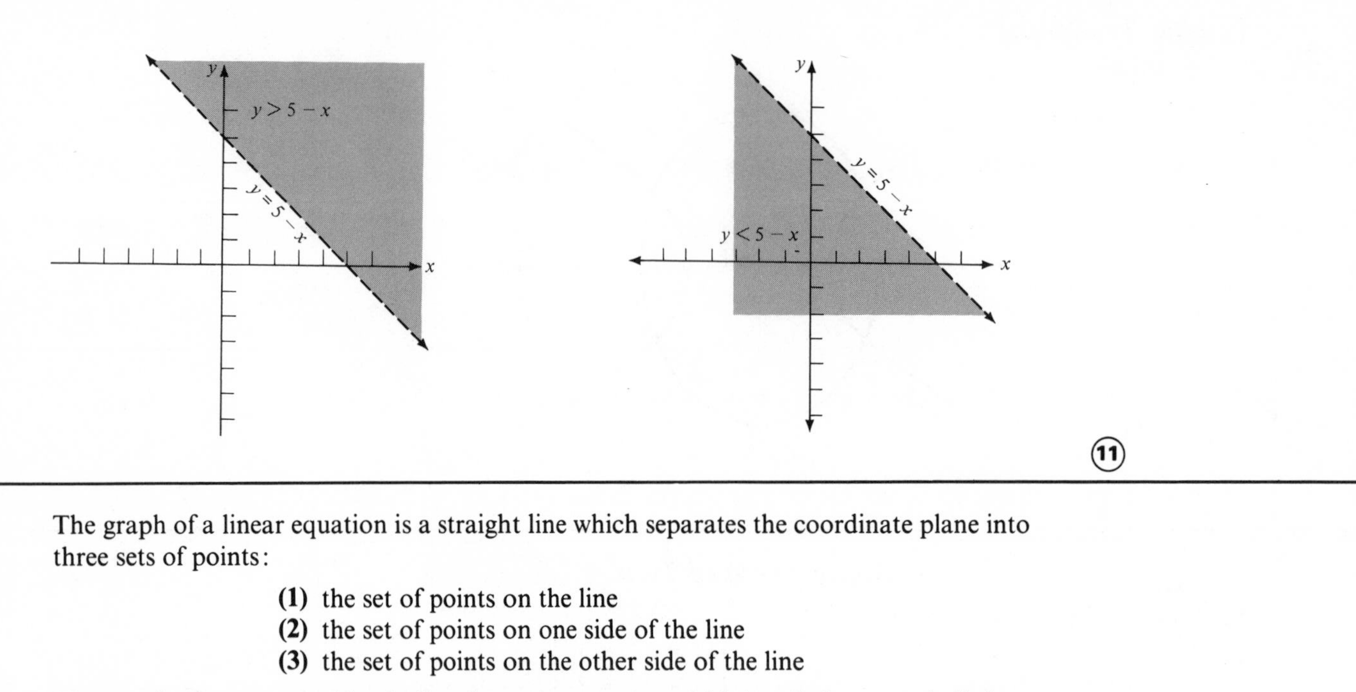

(11)

The graph of a linear equation is a straight line which separates the coordinate plane into three sets of points:

(1) the set of points on the line
(2) the set of points on one side of the line
(3) the set of points on the other side of the line

The set of points on one side of a line determine a region which is called an *open half plane.* If the set of points of the line are combined with the points of an open half plane determined by the line, the resulting region is called a *closed half plane.*

(12)

To graph $\{(x, y)|x + y > 5\}$:

1. first draw the graph of the line $x + y = 5$ (points on this line are not points of $x + y > 5$)
2. choose a point in each open half plane and test the truth of the given inequality
3. shade the half plane corresponding to points in which the given inequality is true.

(13)

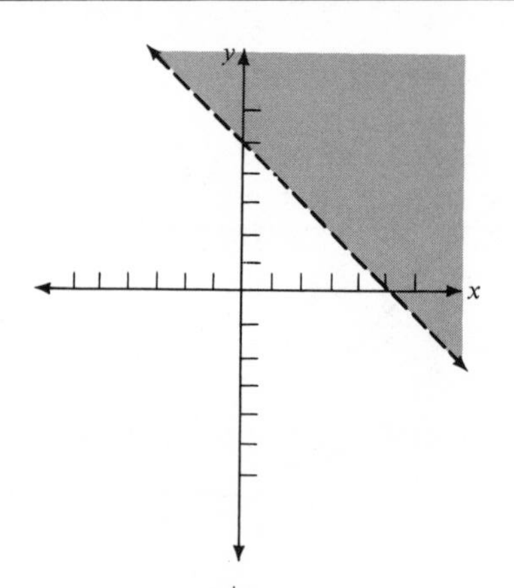

$\{(x, y)|x + y > 5\}$

(0, 0) is below the line $x + y = 5$. Check (0, 0) in $x + y > 5$
$0 + 0 > 5$; false.

$(7 + 2)$ is above the line. Check (7, 2) in $x + y > 5$.
$7 + 2 > 5$, true. Shade all points around (7, 2).

(14)

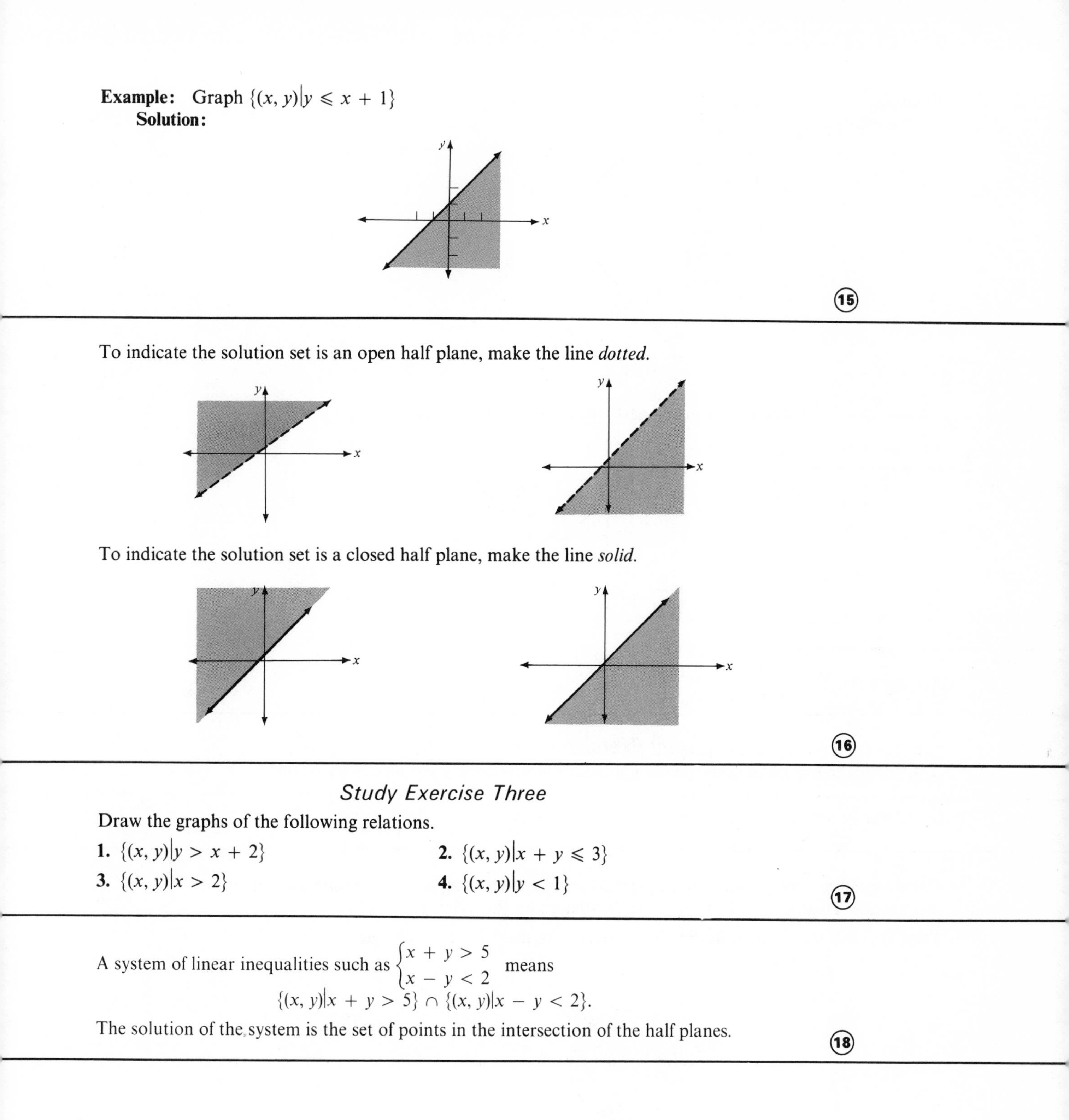

Example: Graph $\{(x, y) | y \leqslant x + 1\}$

Solution:

(15)

To indicate the solution set is an open half plane, make the line *dotted*.

To indicate the solution set is a closed half plane, make the line *solid*.

(16)

Study Exercise Three

Draw the graphs of the following relations.

1. $\{(x, y) | y > x + 2\}$

2. $\{(x, y) | x + y \leqslant 3\}$

3. $\{(x, y) | x > 2\}$

4. $\{(x, y) | y < 1\}$

(17)

A system of linear inequalities such as $\begin{cases} x + y > 5 \\ x - y < 2 \end{cases}$ means

$$\{(x, y) | x + y > 5\} \cap \{(x, y) | x - y < 2\}.$$

The solution of the system is the set of points in the intersection of the half planes.

(18)

Example: Graph $\{(x, y)|x + y > 5\} \cap \{(x, y)|x - y < 2\}$
Solution:

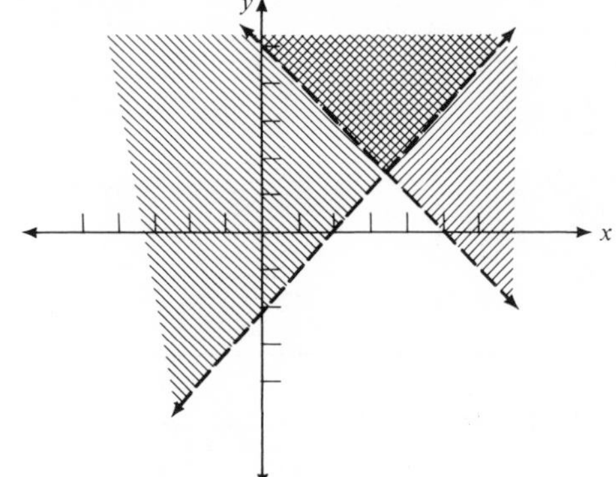

The cross hatched part represents the solution set.

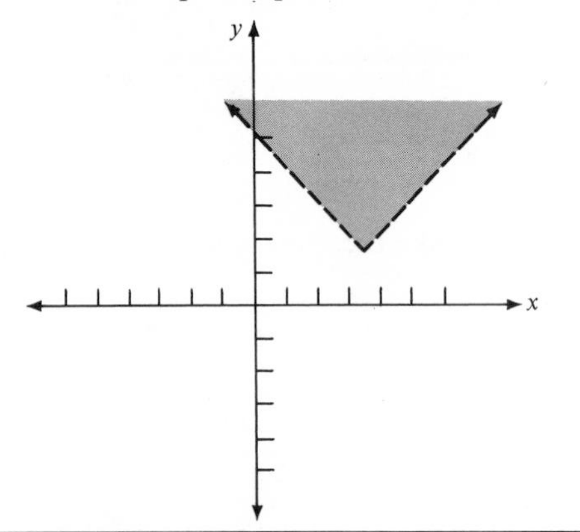

⑲

Study Exercise Four

Find the solution sets graphically:

1. $\{(x, y)|y < x + 3\} \cap \{(x, y)|y \geqslant -2x + 1\}$
2. $\begin{cases} x > 3 \\ y > 2 \end{cases}$
3. $\begin{cases} y \geqslant x + 2 \\ x + y \geqslant 0 \end{cases}$

⑳

Certain quadratic inequalities can be graphed using methods employed in graphing linear inequalities. To graph $\{(x, y)|x^2 + y^2 < 9\}$, first graph $\{(x, y)|x^2 + y^2 = 9\}$.

Then, decide if the points in the interior or exterior of the circle are such that $x^2 + y^2 < 9$.

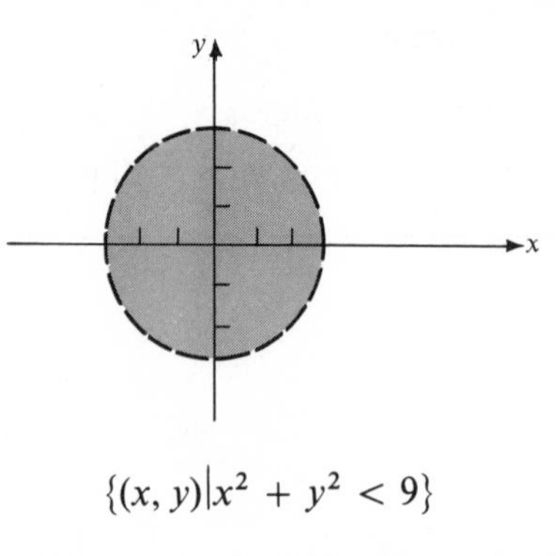

$\{(x, y)|x^2 + y^2 < 9\}$

㉑

Study Exercise Five

Draw the graphs:

1. $\{(x, y) | 4x^2 + 9y^2 > 36\}$ **2.** $\{(x, y) | y \leqslant x^2\}$ **3.** $\{(x, y) | x^2 + y^2 \geqslant 4\}$

(22)

REVIEW EXERCISES

1. Solve algebraically $\begin{cases} x + y = 7 \\ x^2 + y^2 = 25 \end{cases}$

2. Solve algebraically $\begin{cases} x^2 - y^2 = -7 \\ 2x^2 + 3y^2 = 66 \end{cases}$

3. Graph $\{(x, y) | y \leqslant x + 2\}$

4. Graph $\{(x, y) | y > x - 1\} \cap \{(x, y) | y < x + 2\}$

5. Graph $\{(x, y) | y > x^2\}$

SOLUTIONS TO REVIEW EXERCISES

1. $\begin{cases} x + y = 7 \\ x^2 + y^2 = 25 \end{cases}$

$$\begin{aligned} x &= 7 - y \\ (7 - y)^2 + y^2 &= 25 \\ 49 - 14y + y^2 + y^2 &= 25 \\ 2y^2 - 14y + 24 &= 0 \\ y^2 - 7y + 12 &= 0 \\ (y - 4)(y - 3) &= 0 \end{aligned}$$

$$y = 4, \quad y = 3$$
$$x = 7 - y, \; x = 7 - y$$
$$x = 3 \qquad x = 4$$

The solution set is $\{(3, 4), (4, 3)\}$.

2. $\begin{cases} x^2 - y^2 = -7 \\ 2x^2 + 3y^2 = 66 \end{cases}$

$$\begin{aligned} 3x^2 - 3y^2 &= -21 \\ 2x^2 + 3y^2 &= 66 \\ \hline 5x^2 &= 45 \\ x^2 &= 9 \end{aligned}$$

$$x = 3, \; x = -3$$

if $x = 3$, $x^2 - y^2 = -7$; $9 - y^2 = -7$; $-y^2 = -16$; $y^2 = 16$; $y = \pm 4$

if $x = -3$, $x^2 - y^2 = -7$; $9 - y^2 = -7$; $-y^2 = -16$; $y^2 = 16$; $y = \pm 4$

The solution set is $\{(3, 4), (3, -4), (-3, 4), (-3, -4)\}$.

3.

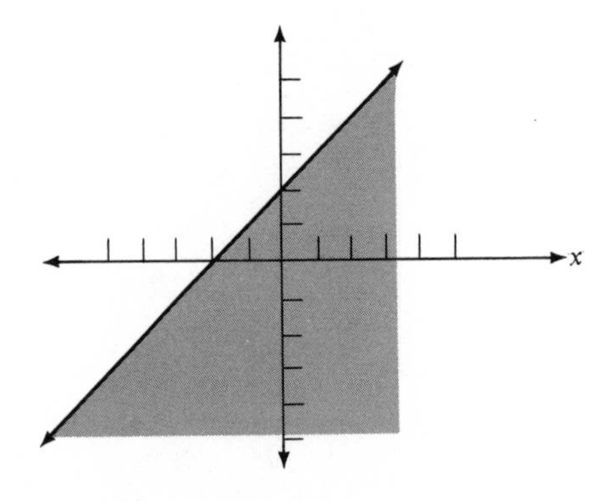

SOLUTIONS TO REVIEW EXERCISES, CONTD.

(Frame 24, contd.)

4.

5.

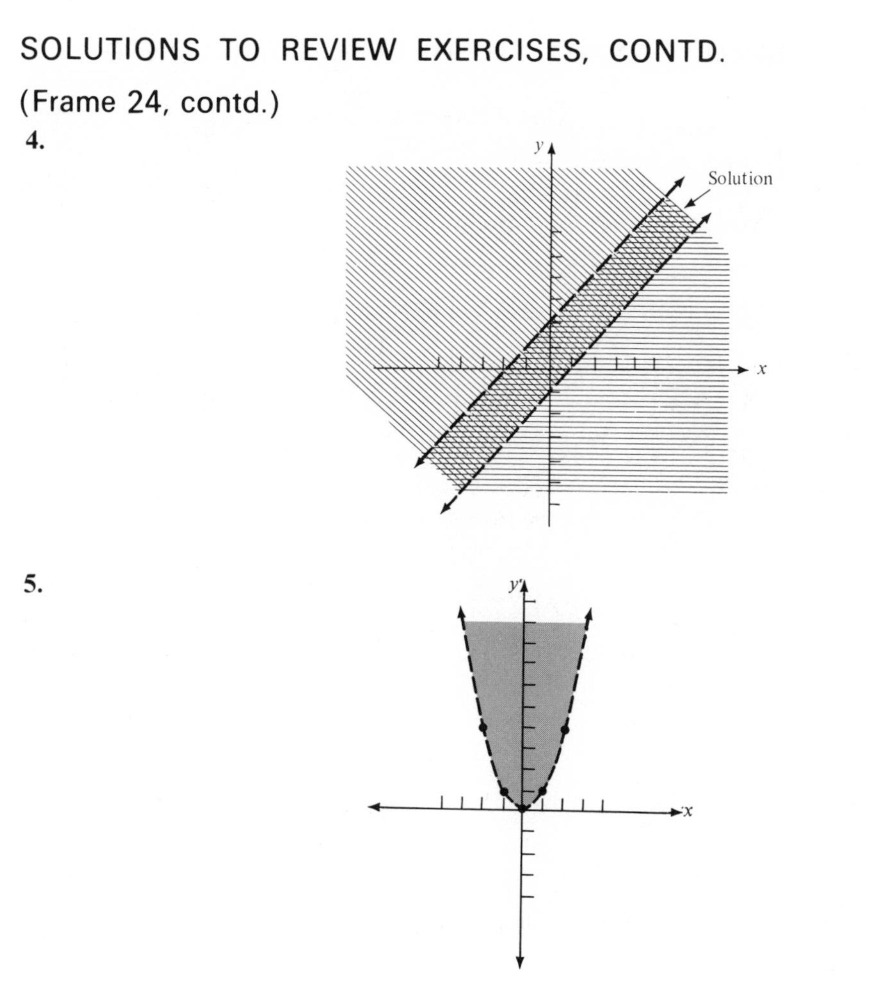

(24)

SOLUTIONS TO STUDY EXERCISES

Study Exercise One (Frame 5)

1. $\begin{cases} x^2 - y^2 = 25 \\ x - y = 5 \end{cases}$

$$\begin{aligned} x &= 5 + y \\ (5 + y)^2 - y^2 &= 25 \\ 25 + 10y + y^2 - y^2 &= 25 \\ 10y &= 0 \\ y &= 0 \\ x &= 5 \end{aligned}$$

The solution set is $\{(5, 0)\}$.

2. $\begin{cases} x^2 + 4y + 2x = 11 \\ y = x + 5 \end{cases}$

$$\begin{aligned} x^2 + 4(x + 5) + 2x &= 11 \\ x^2 + 6x + 9 &= 0 \\ (x + 3)(x + 3) &= 0 \\ x &= -3 \\ y &= x + 5 \\ y &= -3 + 5 \\ y &= 2 \end{aligned}$$

The solution set is $\{(-3, 2)\}$.

Study Exercise One (Frame 5, contd.)

3. $\begin{cases} 4x^2 + 8xy - y^2 = 11 \\ 2x + 3y = 3 \end{cases}$

$$3y = 3 - 2x$$
$$y = \frac{3 - 2x}{3}$$
$$4x^2 + 8x\left(\frac{3 - 2x}{3}\right) - \left[\left(\frac{3 - 2x}{3}\right)^2\right] = 11$$
$$4x^2 + \frac{24x - 16x^2}{3} - \left[\frac{9 - 12x + 4x^2}{9}\right] = 11$$
$$36x^2 + 72x - 48x^2 - (9 - 12x + 4x^2) = 99$$
$$-12x^2 + 72x - 9 + 12x - 4x^2 = 99$$
$$-16x^2 + 84x - 108 = 0$$
$$4x^2 - 21x + 27 = 0$$
$$(4x - 9)(x - 3) = 0$$
$$x = \frac{9}{4}, \quad x = 3$$

$x = 9/4$
$$y = \frac{3 - 2x}{3}$$
$$y = \frac{3 - \frac{9}{2}}{3}$$
$$= -1/2$$

$x = 3$
$$y = \frac{3 - 2x}{3}$$
$$y = \frac{3 - 6}{3}$$
$$y = -3/3$$
$$y = -1$$

The solution set is $\{(9/4, -1/2), (3, -1)\}$.

5A

Study Exercise Two (Frame 8)

1. $\begin{cases} x^2 + y^2 = 25 \\ xy = -12 \end{cases}$

$$y = \frac{-12}{x}$$
$$x^2 + \left(\frac{-12}{x}\right)^2 = 25$$
$$x^2 + \frac{144}{x^2} = 25$$
$$x^4 + 144 = 25x^2$$
$$x^4 - 25x^2 + 144 = 0$$
$$(x^2 - 16)(x^2 - 9) = 0$$
$$x = 4, x = -4, x = 3, x = -3$$
$$y = \frac{-12}{x}$$
$$y = -3, y = 3, y = -4, y = 4$$

The solution set is $\{(4, -3), (-4, 3), (3, -4), (-3, 4)\}$.

SOLUTIONS TO STUDY EXERCISES, CONTD.

Study Exercise Two (Frame 8, contd.)

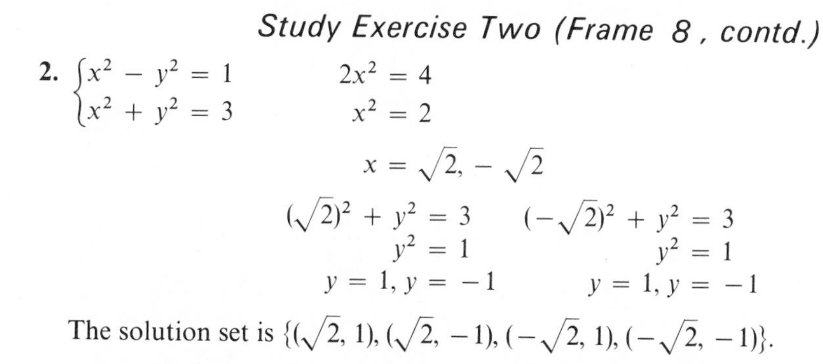

2. $\begin{cases} x^2 - y^2 = 1 \\ x^2 + y^2 = 3 \end{cases}$ $\quad 2x^2 = 4$

$$x^2 = 2$$

$$x = \sqrt{2}, -\sqrt{2}$$

$$(\sqrt{2})^2 + y^2 = 3 \qquad (-\sqrt{2})^2 + y^2 = 3$$

$$y^2 = 1 \qquad y^2 = 1$$

$$y = 1, y = -1 \qquad y = 1, y = -1$$

The solution set is $\{(\sqrt{2}, 1), (\sqrt{2}, -1), (-\sqrt{2}, 1), (-\sqrt{2}, -1)\}$.

Study Exercise Three (Frame 17)

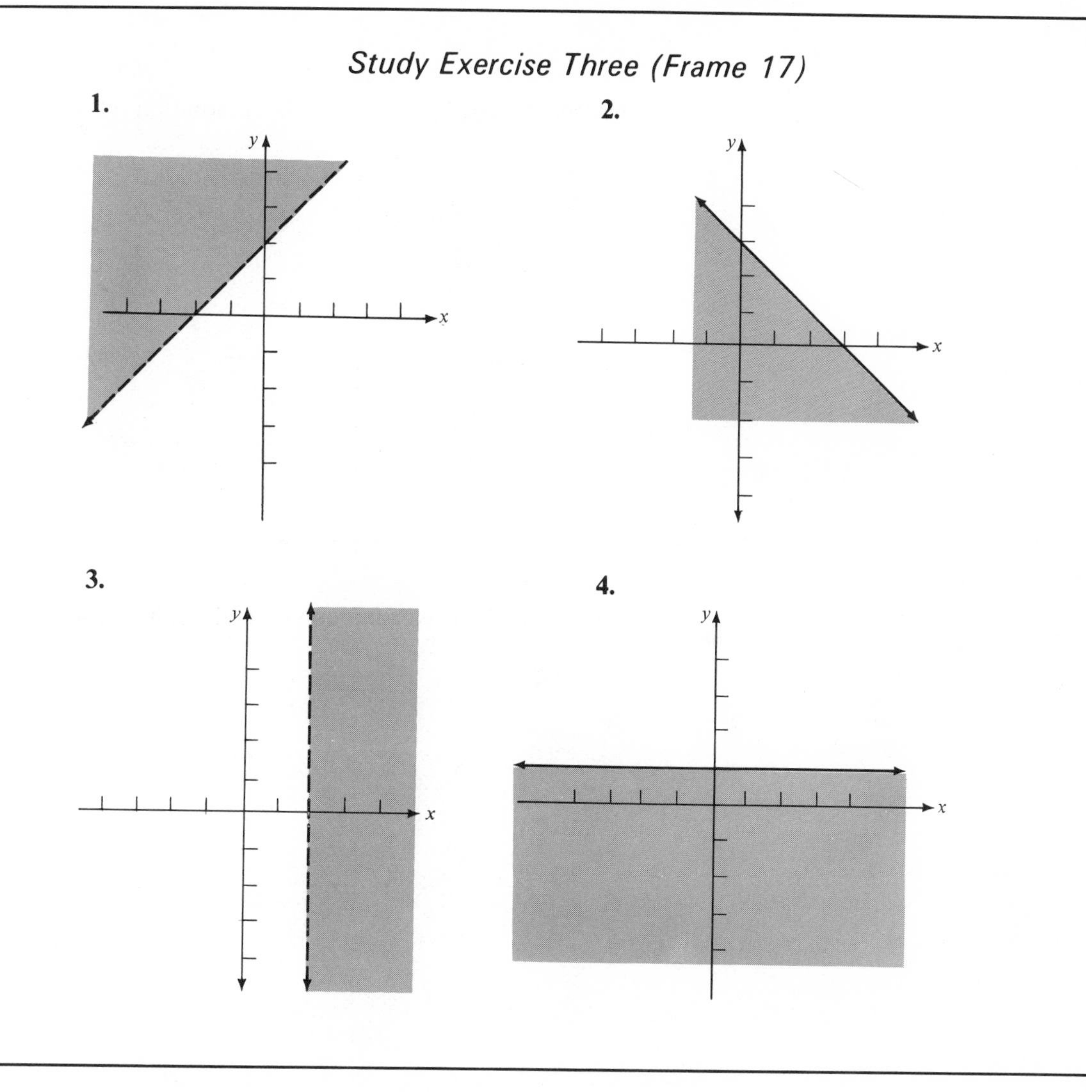

17A

Study Exercise Four (Frame 20)

1.

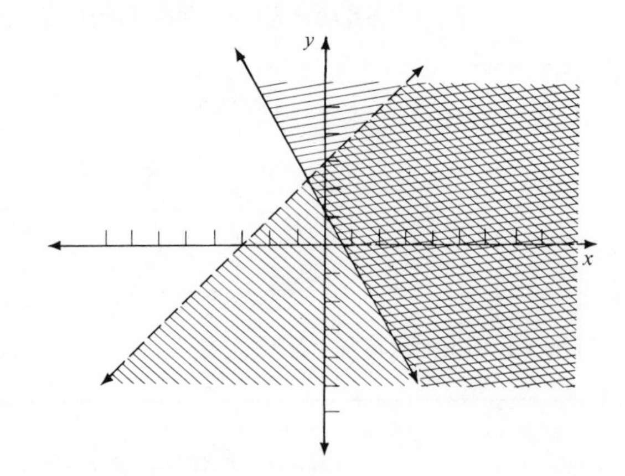

The cross hatched part represents the solution set.

2.

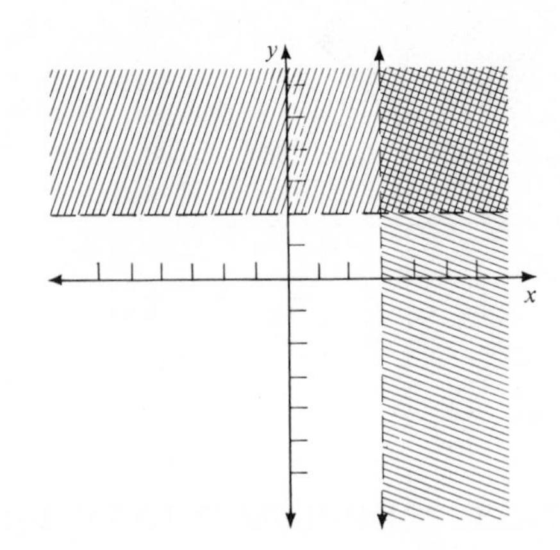

The cross hatched part represents the solution set.

3.

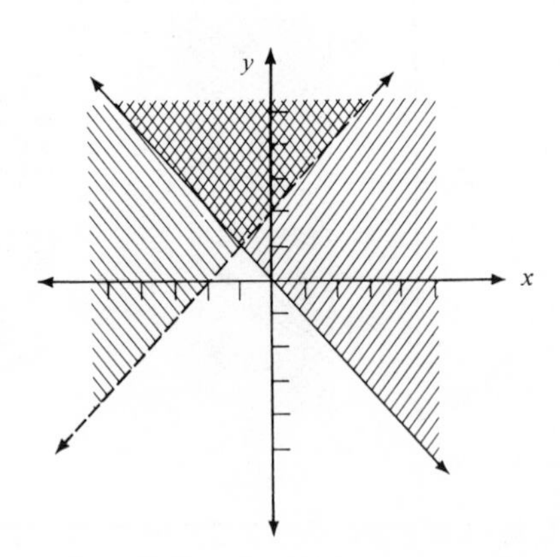

The cross hatched part represents the solution set.

20A

SOLUTIONS TO STUDY EXERCISES, CONTD.

Study Exercise Five (Frame 22)

1. **2.**

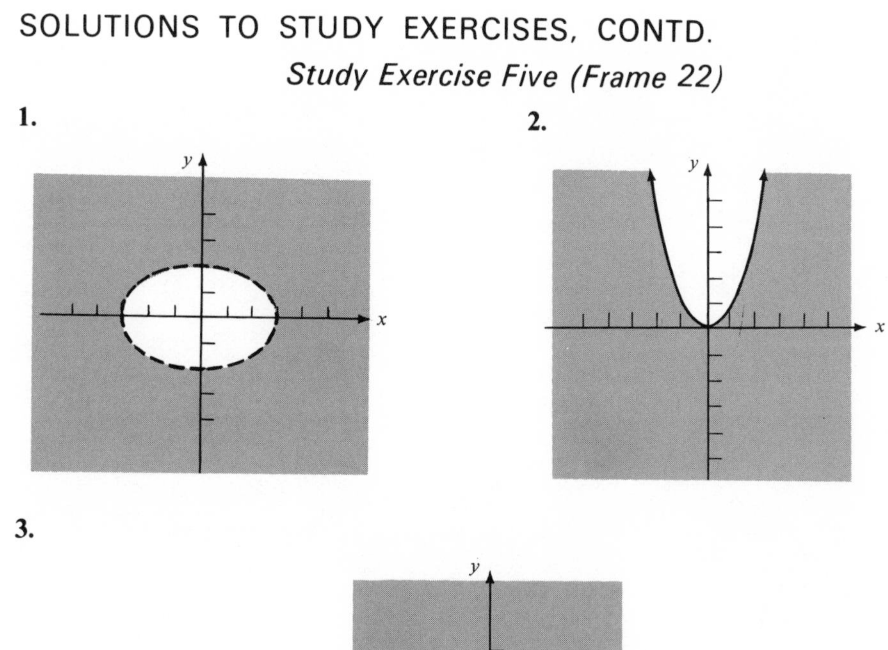

3.

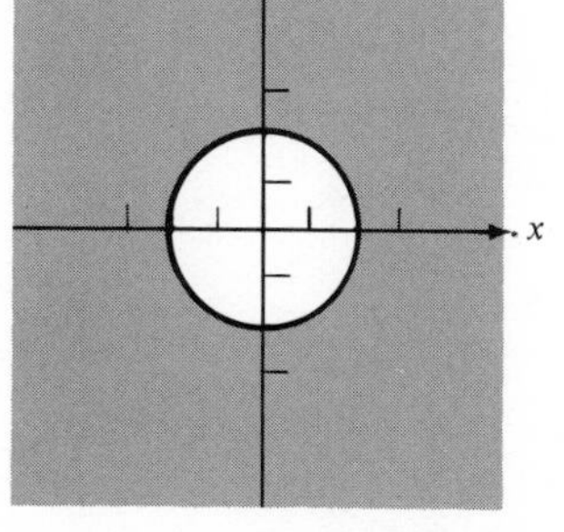

22A

UNIT 23—SUPPLEMENTARY PROBLEMS

A. Find the solution set of each system.

1. $\begin{cases} y = x^2 + 1 \\ y = 2x \end{cases}$ **2.** $\begin{cases} x^2 + y^2 = 25 \\ y - x = 7 \end{cases}$ **3.** $\begin{cases} x^2 + y^2 = 106 \\ x + 6 = 3y \end{cases}$

4. $\begin{cases} 4x^2 + 8xy - y^2 = 11 \\ 2x + 3y = 3 \end{cases}$ **5.** $\begin{cases} 4x^2 + y^2 = 25 \\ x^2 - y^2 = -5 \end{cases}$ **6.** $\begin{cases} xy = 2 \\ x - 3y = 5 \end{cases}$

B. Miscellaneous

7. Find $\{(x, y) | (x + 2)(y + 2) = 40\} \cap \{(x, y) \mid xy = 18\}$

8. Graph $\{(x, y) | 2x + y > 4\}$

9. Graph $\{(x, y) | x + 2y \leq 4\}$

10. Graph $\{(x, y) | x + y \leq 3\} \cap \{(x, y) | y \geq x\}$

11. Graph $\{(x, y) | x \leq 3\} \cap \{(x, y) | y \leq 2\}$

12. Graph $\begin{cases} x + y < 2 \\ \quad x \geq 0 \end{cases}$

13. Graph $\{(x, y) | x > 2 \text{ or } x < -2\}$

14. Graph $\{(x, y) | x^2 + y^2 \leq 16\}$

15. Graph $\{(x, y) | 9x^2 + 4y^2 \leq 36\}$

16. Graph $\{(x, y) | y > x^2 - 2\}$

17. Graph $\begin{cases} x^2 + y^2 > 9 \\ 4x^2 + 36y^2 < 144 \end{cases}$

unit

The Binomial Theorem

24

Objectives

Learn how to expand by the binomial theorem expressions such as $(a + b)^5$, $(a - b)^4$ and $(1 - 2x)^3$.

Terms

factorial

(1)

Study Exercise One

Using the distributive property (dpma) write as basic sums.

1. $(x + y)^0$ where $x + y \neq 0$ **2.** $(x + y)^1$ **3.** $(x + y)^2$

4. $(x + y)^3$ **5.** $(x + y)^4$

(2)

$(x + y)^5$ = forget it! There must be a better way.

Actually $(x + y)^5$ equals $x^5 + 5x^4y + 10x^3y^2 + 10x^2y^3 + 5xy^4 + y^5$.

(3)

Study Exercise Two

How many terms does the expansion of each of the following binomials contain? To answer, refer back to previous frames.

1. $(x + y)^0$ **2.** $(x + y)^1$ **3.** $(x + y)^2$

4. $(x + y)^3$ **5.** $(x + y)^4$ **6.** $(x + y)^5$

(4)

Did you notice a pattern in the previous frame? Can you tell how many terms in the expansion of $(x + y)^{10}$?

(5)

Study Exercise Three

Fill in the blank with the correct response.

1. $(x + y)^8$ consists of ___ terms. **2.** $(x + y)^{14}$ consists of ___ terms.

3. $(x + y)^{100}$ consists of ___ terms. **4.** $(x + y)^n$ where $n \in N$ consists of ___ terms.

(6)

---

consecutive powers of x decrease

$$(x + y)^4 = x^4 + 4x^3y^1 + 6x^2y^2 + 4x^1y^3 + y^4$$

consective powers of y increase

(7)

---

sum of the exponents always four

$$(x + y)^4 = x^4 + 4x^3y^1 + 6x^2y^2 + 4x^1y^3 + y^4$$

In each term, the sum of the powers of x and y is always the exponent n on the binomial.

(8)

Study Exercise Four

Fill in the blanks with the correct response.

1. $(x + y)^7$ has ___ terms in its expansion.

2. In the third term of the expansion of $(x + y)^7$, the sum of the exponents of x and y is ___.

3. In the fourth term of $(x + y)^7$, the exponent on y is ___.

(9)

In the expansion of $(a + b)^4$, we can write that $(a + b)^4 = (?)\,a^4 + (?)\,a^3b^1 + (?)\,a^2b^2 + (?)\,a^1b^3 + (?)\,b^4$.

Also $(x + y)^6 = (?)\,x^6 + (?)\,x^5y^1 + (?)\,x^4y^2 + (?)\,x^3y^3 + (?)\,x^2y^4 + (?)\,xy^5 + (?)\,y^6$.

How can we determine the coefficients?

(10)

Question: Is there any other way to determine the coefficient?

Answer: Yes, a particular coefficient may be determined provided the preceding term is known.

(11)

The first term of $(x + y)^n$ is x^n. The second term is $nx^{n-1}y$.

Example: $(x + y)^9$ has x^9 as its first term and $9x^8y$ as its second.

(12)

$$(x + y)^4 = x^4 + 4x^3y + 6x^2y^2 + 4xy^3 + y^4$$

To determine the 6 (in the third term) from the preceeding term, $4x^3y^1$, multiply 4 by 3 and divide by one more than the power of y.

$$6 = \frac{4 \cdot 3}{2}$$

(13)

$$(x + y)^5 = x^5 + 5x^4y + 10x^3y^2 + 10x^2y^3 + 5xy^4 + y^5$$

To determine the coefficient of the third term, look at the second term, multiply 4 by 5 and then divide by 2.

$$10 = \frac{4 \cdot 5}{2}$$

(14)

$$(x + y)^n,\ n \in N$$

1. The expansion begins with x^n
2. The second term is always $nx^{n-1}y^1$
3. From then on, each succeeding coefficient is found by multiplying the previous coefficient of x by the power of x and dividing by one more than the power of y.
4. The last term will be y^n.

(15)

$$(x + y)^5 = x^5 + 5x^4y + 10x^3y^2 + 10x^2y^3 + 5xy^4 + y^5$$

Observations.

1. There are six terms in the expansion.
2. The sum of the exponents on x and y in any term is 5.
3. The coefficients of terms equidistant from the ends are the same.

(16)

Summary

$$(x + y)^n, \quad n \in N$$

1. The number of terms is $n + 1$.
2. The first term is x^n and the second term is $nx^{n-1}y$.
3. The exponents of x decrease by one and those of y increase by one from term to term and the sum of the exponents of x and y in each term is equal to n.
4. If in any term the numerical coefficient is multiplied by the exponent of x and this product divided by one more than the exponent of y, the result is the numerical coefficient of the next term.

(17)

$$\begin{aligned}(2x + b)^3 &= (2x)^3 + 3(2x)^2(b) + 3(2x)^1(b)^2 + (b)^3\\ &= 8x^3 + 12x^2b + 6xb^2 + b^3\end{aligned}$$

$$\begin{aligned}(x - 1)^4 &= [x + (-1)]^4\\ &= x^4 + 4(x)^3(-1) + 6(x)^2(-1)^2 + 4(x)(-1)^3 + (-1)^4\\ &= x^4 - 4x^3 + 6x^2 - 4x + 1\end{aligned}$$

(18)

If $n \in N$, $\quad (x + y)^n =$

$$x^n + nx^{n-1}y + \frac{n(n-1)}{2}x^{n-2}y^2 + \frac{n(n-1)(n-2)}{6}x^{n-3}y^3 + \dots + y^n$$

The numerical coefficient of the fourth term is determined as follows:

$$\left[\frac{n(n-1)}{2} \cdot (n-2)\right] \div 3 = \frac{n(n-1)(n-2)}{2} \cdot 1/3$$

$$= \frac{n(n-1)(n-2)}{6}$$

(19)

Study Exercise Five

Expand by the Binomial Theorem.

1. $(x + y)^4$ **2.** $(x - 1)^3$ **3.** $(2x + 1)^4$

(20)

$(x + y)^0 = 1$
$(x + y)^1 = 1x + 1y$
$(x + y)^2 = 1x^2 + 2xy + 1y^2$
$(x + y)^3 = 1x^3 + 3x^2y + 3xy^2 + y^3$
$(x + y)^4 = 1x^4 + 4x^3y + 6x^2y^2 + 4xy^3 + 1y^4$

Power	*Coeffieients*
$n = 0$	1
$n = 1$	1 1
$n = 2$	1 2 1
$n = 3$	1 3 3 1
$n = 4$	1 4 6 4 1

(21)

Power	*Coefficients*
$n = 0$	1
$n = 1$	1 1
$n = 2$	1 2 1
$n = 3$	1 3 3 1
$n = 4$	1 4 6 4 1

Notice that the second number in each row is the same as the power n.

(22)

Power	*Coefficients*
$n = 0$	1
$n = 1$	1 1
$n = 2$	1 [2 + 1]
$n = 3$	1 3 3 1
$n = 4$	1 4 6 4 1

Any coefficient is the sum of the number above it and the one immediately to the left of the one above it. For example, 3 above equals 1 + 2.

(23)

Pascal's Triangle

This triangle, known as *Pascal's Triangle,* may be extended as far as one wishes.

1
1 1
1 2 1
1 3 3 1
1 4 6 4 1
1 5 10 10 5 1
1 6 15 20 15 6 1

Remember the second number in each row always gives the power n.

(24)

Study Exercise Six

Using Pascal's Triangle, write out the expansion of the following binomials.

1. $(a + b)^4$ **2.** $(x + y)^6$

(25)

By means of Pascal's Triangle, $(x + 2)^3$ may be expanded as follows:

$$(x + 2)^3 = (1)\,x^3 + (3)\,x^2\,(2)^1 + (3)\,x^1\,(2)^2 + (1)\,(2)^3$$
$$= x^3 + 6x^2 + 12x + 8$$

(26)

Notice the symmetry in the table.

$$
\begin{array}{llllll}
1 \\
1 & 1 \\
1 & 2 & 1 \\
1 & 3 & 3 & 1 \\
1 & 4 & 6 & 4 & 1 \\
1 & 5 & 10 & 10 & 5 & 1
\end{array}
$$

If the number of terms in the expansion is odd, there is one middle coefficient, if the number of terms is even, there are two middle coefficients. The numbers on either side of the middle term(s) are the same.

(27)

Factorial N

If $n \in N$, we define *factorial n* to be the product of all consecutive natural numbers from 1 to n. The symbol for factorial n is $n!$

$$
\begin{aligned}
1! &= 1 \\
2! &= 1 \cdot 2 \\
&= 2 \\
3! &= 1 \cdot 2 \cdot 3 \\
&= 6 \\
4! &= 1 \cdot 2 \cdot 3 \cdot 4 \\
&= 24 \\
5! &= 1 \cdot 2 \cdot 3 \cdot 4 \cdot 5 \\
&= 120 \\
6! &= 1 \cdot 2 \cdot 3 \cdot 4 \cdot 5 \cdot 6 \\
&= 720
\end{aligned}
$$

(28)

Since

$$6! = 6 \cdot 5 \cdot 4 \cdot 3 \cdot 2 \cdot 1 \text{ then, } \quad 6! = 6 \cdot [5 \cdot 4 \cdot 3 \cdot 2 \cdot 1] = 6 \cdot 5!$$

(29)

Similarly, $4! = 4 \cdot 3 \cdot 2 \cdot 1 = 4 \cdot 3!$; $8! = 8 \cdot 7!$; $10! = 10 \cdot 9!$

(30)

$$n! = n\,(n - 1)!$$

Let $n = 5$, then $5! = 5\,(5 - 1)! = 5 \cdot 4!$
Let $n = 3$, then $3! = 3\,(3 - 1)! = 3 \cdot 2!$
Let $n = 1$, then $1! = 1\,(1 - 1)! = 1 \cdot 0!$
$= 1 \cdot 0!$, for consistency $0!$

But what is $0!$ Since $1! = 1.0!$, for consistency $0!$ must be the identity element. Thus, $0! = 1$.

(31)

Example 1. Simplify: $8!/6!$

Solution: $\dfrac{8!}{6!} = \dfrac{8 \cdot 7!}{6!} = \dfrac{8 \cdot 7 \cdot \not{6}!}{\not{6}!} = 8 \cdot 7 = 56$

Example 2: Simplify $\dfrac{4!}{7!}$

Solution: $\dfrac{4!}{7!} = \dfrac{4!}{7 \cdot 6 \cdot 5 \cdot 4!} = \dfrac{1}{7 \cdot 6 \cdot 5} = \dfrac{1}{210}$

(32)

Study Exercise Seven

Simplify:

1. $\frac{3!}{0!}$ **2.** $\frac{9!}{6!}$ **3.** $\frac{(8!)(6!)}{10!}$ **4.** $\frac{(n+2)!}{n!}$

(33)

The binomial expansion of $(x + y)^n$, $n \in N$ may be written

$$x^n + \frac{n}{1!}x^{n-1}y + \frac{n(n-1)}{2!}x^{n-2}y^2 + \frac{n(n-1)(n-2)}{3!}x^{n-3}y^3 + \ldots + y^n$$

(34)

Finding a Particular Term

$$(x + y)^n = x^n + \frac{n}{1!}x^{n-1}y + \frac{n(n-1)}{2!}x^{n-2}y^2 + \frac{n(n-1)(n-2)}{3!}x^{n-3}y^3 + \ldots + y^n$$

2nd term is $\frac{n}{1!}x^{n-1}y^1$

3rd term is $\frac{n(n-1)}{2!}x^{n-2}y^2$

4th term is $\frac{n(n-1)(n-2)}{3!}x^{n-3}y^3$

5th term is $\frac{n(n-1)(n-2)(n-3)}{4!}x^{n-4}y^4$

What do you think the 6th term is?

(35)

6th term is $\frac{n(n-1)(n-2)(n-3)(n-4)}{5!}x^{n-5}y^5$

rth term is $\frac{n(n-1)(n-2)\cdot\ldots(n-[r-2])}{(r-1)!}x^{n-[r-1]}y^{r-1}$

In simplified form the rth term is;

$$\frac{n(n-1)(n-2)\ldots(n-r+2)}{(r-1)!}x^{n-r+1}y^{r-1}$$

(36)

The preceeding frame gives us a method for finding a single term in the binomial expansion without expanding the binomial.

In the expansion of $(x + y)^n$, $n \in \mathrm{N}$,

the rth term is $\frac{n(n-1)(n-2)\ldots(n-r+2)}{(r-1)!}x^{n-r+1}y^{r-1}$

(37)

Example 1: Find the 4th term of $(x + y)^6$

Solution: The rth term is $\frac{n(n-1)(n-2)\ldots(n-r+2)}{(r-1)!}x^{n-r+1}y^{r-1}$

$n = 6$, $r = 4$; hence $n - r + 2 = 6 - 4 + 2$
$= 4$

and $n - r + 1 = 6 - 4 + 1$
$= 3$

The 4th term is $\frac{6 \cdot 5 \cdot 4}{3!}x^3y^3$ and in simplified form $20x^3y^3$.

(38)

Example 2: Find the 5th term of $(2x - 1)^8$

Solution: $n = 8, r = 5$; hence $n - r + 2 = 5$ and $n - r + 1 = 4$.

The 5th term is $\frac{8 \cdot 7 \cdot 6 \cdot 5}{4!}(2x)^4(-1)^4$

In simplified form, the 5th term is $\frac{\not{8} \cdot 7 \cdot \overset{2}{\not{6}} \cdot 5}{\not{4} \cdot 3 \cdot \not{2} \cdot 1}(2x)^4(-1)^4 = 1120x^4$

(39)

Study Exercise Eight

Write and simplify the indicated term in the expansions of the following.

1. Fifth term of $(2x + y)^7$ **2.** Seventh term of $(1 - x)^9$ **3.** Middle term of $(x^2 - 1)^8$

(40)

Recall that $(x + y)^{10}$ has eleven terms in its expansion.
Let us write the first four terms using the binomial theorem:

$$(x + y)^{10} = x^{10} + 10x^9y + \frac{10 \cdot 9}{2!}x^8y^2 + \frac{10 \cdot 9 \cdot 8}{3!}x^7 \cdot y^3 + \ldots$$
$$= x^{10} + 10x^9y + 45x^8y^2 + 120x^7y^3 + \ldots$$

(41)

The binomial theorem can be used to find the value of $(1.01)^8$ by expanding a binomial whose terms have the sum 1.01.

$$(1.01)^8 = (1 + .01)^8$$
$$= 1^8 + 8(1)^7(.01) + \frac{8 \cdot 7}{2!}(1)^6(.01)^2 + \frac{8 \cdot 7 \cdot 6}{3!}(1)^5(.01)^3 + \ldots$$
$$= 1 + .08 + 28(.01)^2 + 56(.01)^3 + \ldots$$
$$= 1 + .08 + .0028 + .000056 + \ldots$$
$$= 1.082856 \ldots$$

(42)

Study Exercise Nine

A. Write the first four terms in the expansion and simplify.

1. $(x + 1)^9$ **2.** $(x - 2)^{11}$

B. Find the numerical value to 3 decimal places.

3. $(1.02)^6$ **4.** $(.99)^5$

(43)

REVIEW EXERCISES

A. Fill in the blanks with the correct response.

1. $(2x + 1)^6$ contains ___ terms in its expansion.

2. In the 5th term of $(5x + 2y)^8$ the sum of the exponents of x and y is___.

3. $0! =$ ___

B. Expand by the binomial theorem.

4. $(a + b)^5$ **5.** $(a - b)^4$ **6.** $(2x + 1)^3$

7. $(1 - 2x)^4$ **8.** $(2x/3 + 1/2)^3$

C. Miscellaneous:

9. Simplify $\dfrac{8!\,4!}{6!\,7!}$

10. Find the 4th term of $(2x + 2)^{11}$. (do not simplify)

11. Find the numerical value of $(.99)^6$ to three decimal places.

44

SOLUTIONS TO REVIEW EXERCISES

A. **1.** 7 **2.** 8 **3.** 1

B. **4.** $a^5 + 5a^4b + 10a^3b^2 + 10a^2b^3 + 5ab^4 + b^5$

5. $a^4 - 4a^3b + 6a^2b^2 - 4ab^3 + b^4$

6.
$$\begin{aligned}(2x + 1)^3 &= (2x)^3 + 3(2x)^2(1) + 3(2x)(1)^2 + (1)^3\\ &= 8x^3 + 12x^2 + 6x + 1\end{aligned}$$

7.
$$\begin{aligned}(1 - 2x)^4 &= [1 + (-2x)]^4\\ &= 1^4 + 4(1)^3(-2x) + 6(1)^2(-2x)^2 + 4(1)(-2x)^3 + (-2x)^4\\ &= 1 - 8x + 24x^2 - 32x^3 + 16x^4\end{aligned}$$

8.
$$\begin{aligned}\left(\frac{2x}{3} + \frac{1}{2}\right)^3 &= \left(\frac{2x}{3}\right)^3 + 3\left(\frac{2x}{3}\right)^2\left(\frac{1}{2}\right) + 3\left(\frac{2x}{3}\right)^1\left(\frac{1}{2}\right)^2 + \left(\frac{1}{2}\right)^3\\ &= \frac{8x^3}{27} + 3\left(\frac{4x^2}{9}\right)\left(\frac{1}{2}\right) + 3\left(\frac{2x}{3}\right)\left(\frac{1}{4}\right) + \frac{1}{8}\\ &= \frac{8}{27}x^3 + \frac{2}{3}x^2 + \frac{1}{2}x + \frac{1}{8}\end{aligned}$$

C. **9.**
$$\frac{8 \cdot \not{7}! \cdot \not{4}!}{6 \cdot 5 \cdot \not{4}! \cdot \not{7}!} = \frac{8}{30} = \frac{4}{15}$$

10. $n = 11, r = 4$ and $n - r + 2 = 9$, $n - r + 1 = 8$

the 4th term is $\dfrac{11 \cdot 10 \cdot 9}{3!}(2x)^8(2)^3$

11.
$$\begin{aligned}(.99)^6 &= (1 - .01)^6 = [1 + (-.01)]^6\\ &= 1^6 + 6(1)^5(-.01) + 15(1)^4(-.01)^2 + 20(1)^3(-.01)^3 + \ldots\\ &= 1 + 6(-.01) + 15(.0001) + 20(-.000001) + \ldots\\ &= 1 - .06 + .0015 - .000020 + \ldots\\ &\doteq .941\end{aligned}$$

45

SOLUTIONS TO STUDY EXERCISES

Study Exercise One (Frame 2)

1. $1, x + y \neq 0$ **2.** $x + y$ **3.** $x^2 + 2xy + y^2$

4.
$$\begin{aligned}(x + y)^3 &= (x + y)(x + y)^2\\ &= (x + y)(x^2 + 2xy + y^2)\\ &= (x + y)x^2 + (x + y)2xy + (x + y)y^2\\ &= x^3 + x^2y + 2x^2y + 2xy^2 + xy^2 + y^3\\ &= x^3 + 3x^2y + 3xy^2 + y^3\end{aligned}$$

5. $x^4 + 4x^3y + 6x^2y^2 + 4xy^3 + y^4$

(2A)

Study Exercise Two (Frame 4)

1. 1 **2.** 2 **3.** 3 **4.** 4 **5.** 5 **6.** 6

(4A)

Study Exercise Three (Frame 6)

1. 9 **2.** 15 **3.** 101 **4.** $n + 1$

(6A)

Study Exercise Four (Frame 9)

1. 8 (one more than the power).
2. 7 (the power on the binomial).
3. 3 (powers of y increase beginning with 1 in the second term).

(9A)

Study Exercise Five (Frame 20)

1. $(x + y)^4 = x^4 + 4x^3y + 6x^2y^2 + 4xy^3 + y^4$

2.
$$\begin{aligned}(x - 1)^3 &= [x + (-1)]^3\\ &= x^3 + 3(x)^2(-1) + 3(x)^1(-1)^2 + (-1)^3\\ &= x^3 - 3x^2 + 3x - 1\end{aligned}$$

3.
$$\begin{aligned}(2x + 1)^4 &= (2x)^4 + 4(2x)^3(1) + 6(2x)^2(1)^2 + 4(2x)(1)^3 + (1)^4\\ &= 16x^4 + 32x^3 + 24x^2 + 8x + 1\end{aligned}$$

(20A)

Study Exercise Six (Frame 25)

1. $(a + b)^4 = a^4 + 4a^3b + 6a^2b^2 + 4ab^3 + b^4$
2. $(x + y)^6 = x^6 + 6x^5y + 15x^4y^2 + 20x^3y^3 + 15x^2y^4 + 6xy^5 + y^6$

(25A)

Study Exercise Seven (Frame 33)

1. $\dfrac{3!}{0!} = \dfrac{3 \cdot 2 \cdot 1}{1} = 6$

2. $\dfrac{9!}{6!} = \dfrac{9 \cdot 8 \cdot 7 \cdot \cancel{6!}}{\cancel{6!}} = 504$

3. $\dfrac{8! \cdot 6!}{10!} = \dfrac{\cancel{8!} \cdot \cancel{6} \cdot \cancel{5} \cdot 4 \cdot \cancel{3} \cdot \cancel{2} \cdot 1}{\cancel{10} \cdot \cancel{9} \cdot \cancel{8!}} = 8$ (with 2 written above)

4. $\dfrac{(n + 2)!}{n!} = \dfrac{(n + 2)(n + 1)(\cancel{n!})}{\cancel{n!}} = (n + 2)(n + 1)$

(33A)

SOLUTIONS TO STUDY EXERCISES, CONTD.

Study Exercise Eight (Frame 40)

1. $\dfrac{7 \cdot 6 \cdot 5 \cdot 4 \cdot}{4!}(2x)^3 (y)^4 = 280x^3y^4$ **2.** $\dfrac{9 \cdot 8 \cdot 7 \cdot 6 \cdot 5 \cdot 4}{6!}(1)^3 (-x)^6 = 84x^6$

3. The 5th term. $\dfrac{8 \cdot 7 \cdot 6 \cdot 5}{4!}(x^2)^4 (-1)^4 = 70x^8$

40A

Study Exercise Nine (Frame 43)

A. 1. $(x + 1)^9 = x^9 + 9x^8(1) + 36x^7(1)^2 + 84x^6(1)^3 + \ldots$
$= x^9 + 9x^8 + 36x^7 + 84x^6 + \ldots$

2. $(x - 2)^{11} = x^{11} + 11(x)^{10}(-2) + 55(x)^9(-2)^2 + 165(x)^8(-2)^3 + \ldots$
$= x^{11} - 22x^{10} + 220x^9 - 1320x^8 + \ldots$

B. 3. $(1 + .02)^6 = 1^6 + 6(1)^5(.02) + 15(1)^4(.02)^2 + 20(1)^3(.02)^3 + \ldots$
$= 1 + .12 + .0060 + .00016 + \ldots = 1.126 \ldots$

4. $(1 - .01)^5 = 1^5 + 5(1)^4(-.01) + 10(1)^3(-.01)^2 + 10(1)^2(-.01)^3 + \ldots$
$= 1 - .05 + .001 - .00001 + \ldots$
$= .951 \ldots$

43A

UNIT 24—SUPPLEMENTARY PROBLEMS

A. Expand by the binomial theorem.

1. $(x + r)^4$ **2.** $(x - 1)^6$ **3.** $(2x + 1)^5$ **4.** $(\frac{1}{x} - x)^3$

5. $(\frac{x}{2} + y)^4$ **6.** $(x^2 + 2)^5$ **7.** $(3a + b)^4$

B. Write and simplify the first four terms.

8. $(x + 1)^{13}$ **9.** $(2x - 1)^{10}$ **10.** $(x^2 + 1)^8$

C. Simplify:

11. $\dfrac{7!}{5!}$ **12.** $\dfrac{2! \cdot 3!}{4!}$ **13.** $\dfrac{6! - 4!}{0!}$

14. $\dfrac{(n - 1)!}{(n + 1)!}$ **15.** $\dfrac{(n + r)!}{(n + r - 2)!}$

D. Find the numerical value to 3 decimal places by the binomial theorem.

16. $(1.02)^5$ **17.** $(1.002)^8$ **18.** $(.998)^4$

E. Miscellaneous:

19. What numbered term in the expansion $(a - x)^6$ involves x^3?
20. What is the power of x for the middle term of $(x + y)^6$?
21. Find the 5th term of $(x + 2)^7$
22. Find the 4th term of $(x^2 + y)^9$
23. Find the 8th term of $(x - 1)^{10}$
24. Find the middle term of $(y^2 - \frac{1}{2})^8$

unit

25

Sequences and Series

Objectives

1. Recognize a sequence and a series.
2. Know the definitions of an A. P. and a G. P.
3. Memorize the formulas for the nth term and the sum of n terms of both an A. P. and G. P.
4. Memorize the formula for the sum of an infinite G. P.
5. Simplify problems involving summation notation.

Terms

sequence function | arithmetic sequence | geometric sequence

①

Sequence Function

A *sequence function* is a function whose domain is the set of natural numbers.

A *finite sequence* of n terms is a function s whose domain is the set $\{1, 2, 3, \dots n\}$. The *range* is the set $\{s(1), s(2), s(3), \dots, s(n)\}$ which is usually written $s_1, s_2, s_3 \dots, s_n$.

The elements of the range are called *terms* of the *sequence.*

②

Consider $s(n) = 2n + 1$ with domain $\{1, 2, 3, 4, 5, 6\}$;

$s(1) = 2 \cdot 1 + 1 = 3$ $\qquad s(4) = 2 \cdot 4 + 1 = 9$

$s(2) = 2 \cdot 2 + 1 = 5$ $\qquad s(5) = 2 \cdot 5 + 1 = 11$

$s(3) = 2 \cdot 3 + 1 = 7$ $\qquad s(6) = 2 \cdot 6 + 1 = 13$

The range is $\{3, 5, 7, 9, 11, 13\}$.

Thus, $s_1 = 3$ $\qquad s_4 = 9$

$s_2 = 5$ $\qquad s_5 = 11$

$s_3 = 7$ $\qquad s_6 = 13$

③

Infinite Sequence

An *infinite sequence* is a function s whose domain is the set $\{1, 2, 3, \dots, n, \dots\}$ of natural numbers. The range is the set $\{s(1), s(2), s(3), \dots, s(n), \dots\}$, usually written $s_1, s_2, s_3, \dots, s_n, \dots$

The element s_n of the range is called the nth term of the sequence.

④

Series

Let $\{s_1, s_2, s_3, \dots, s_n\}$ be a finite sequence of real numbers. The indicated sum $s_1 + s_2 + s_3 + \dots + s_n$ is called a *finite series.*

Let $\{s_1, s_2, s_3, \dots s_n, \dots\}$ be an infinite sequence of real numbers. The indicated sum $s_1 + s_2 + s_3 + \dots s_n + \dots$ is called an *infinite series.*

⑤

Example: A sequence function is given by $s_n = 3n - 1$. Write the first four terms of the sequence.

Solution: $s_1 = 3 \cdot 1 - 1 = 2$
$s_2 = 3 \cdot 2 - 1 = 5$
$s_3 = 3 \cdot 3 - 1 = 8$
$s_4 = 3 \cdot 4 - 1 = 11$
the first four terms are 2, 5, 8, 11

(6)

Summation Notation

When writing a series, the Greek letter Σ (sigma) may be used to denote a sum.

$\sum_{x=1}^{3} x^2$, where x may be replaced only with integers beginning with 1 and ending with 3 means $(1)^2 + (2)^2 + (3)^2$.

$\sum_{n=1}^{4} (2n)$ means $2(1) + 2(2) + 2(3) + 2(4)$.

(7)

$\sum_{j=0}^{3} (3j + 1) = (3 \cdot 0 + 1) + (3 \cdot 1 + 1) + (3 \cdot 2 + 1) + (3 \cdot 3 + 1).$

In using the summation notation $\sum_{n=a}^{b}$, $a \le b$, $a, b \in I$.

(8)

To show that a series has an infinite number of terms, the following notation is used: $\sum_{n=a}^{\infty}$

Example 1: $\sum_{i=1}^{\infty} (2i) = (2 \cdot 1) + (2 \cdot 2) + (2 \cdot 3) + \ldots$

Example 2: $\sum_{K=1}^{\infty} K = 1 + 2 + 3 + 4 + \ldots$

(9)

Study Exercise One

A. Write the first four terms in the sequence with general term as given.

1. $s_n = 2n - 1$ **2.** $s_n = \dfrac{3}{n + 1}$ **3.** $s_n = 3^{n+1}$

B. Write each series in expanded form.

4. $\sum_{n=1}^{5} n^2$ **5.** $\sum_{n=1}^{3} (3n - 2)$

C. Find the exact value.

6. $\sum_{k=3}^{5} k^2$ **7.** $\sum_{k=1}^{2} \dfrac{k}{1 + k}$ **8.** $\sum_{n=2}^{4} (-1)^n(n - 1)$

(10)

Arithmetic Progression

A sequence in which each term after the first is obtained by adding the same constant to the preceeding term is called an arithmetic sequence or *arithmetic progression* (abbreviated A. P.).

The constant is called the *common difference* and is denoted by d.

The first term in the sequence (s_1) is sometimes denoted by a.

(11)

Consider the sequence 3, 7, 11, 15, ...
Since each term after the first is obtained by adding 4 to the preceeding term, the sequence is an arithmetic sequence.

The difference is $4(7 - 3 = 4, 11 - 7 = 4, 15 - 11 = 4)$ and the first term is 3. That is, $d = 4$ and $a = 3$.

(12)

Study Exercise Two

For the arithmetic sequences below, find the common difference and continue the sequence for two more terms.

1. 5, 8, 11, ... **2.** $-6, -1, \ldots$ **3.** $a, a + 5, \ldots$

(13)

Let a be the first term in an arithmetic sequence and d the common difference.

$$\begin{array}{cccccc} a, & a + d, & a + 2d, & a + 3d, & a + 4d, & \ldots \\ \uparrow & \uparrow & \uparrow & \uparrow & \uparrow & \\ s_1 & s_2 & s_3 & s_4 & s_5 & \end{array}$$

What would s_{10} be?

What would be s_n where $n \in N$?

(14)

$$s_n = a + (n - 1)d$$

The formula for the nth term of an arithmetic sequence is given above. The first term is a and the common difference is d.

(15)

Example: Find the 11th term in the sequence 5, 7, 9, ...

Solution: 5, 7, 9, ... is an A. P.

$d = 2, a = 5, n = 11$

Line (1) $s_n = a + (n - 1)d$

Line (2) $s_{11} = 5 + (11 - 1) \cdot 2$

Line (3) $s_{11} = 5 + 20$

Line (4) $s_{11} = 25$

(16)

Example: Find the number of multiples of 7 between 30 and 283.

Solution: The first multiple of 7 is 35 and the last multiple is 280. Now suppose there are n multiples of 7. They form an A. P. with $a = 35$, $s_n = 280$, and $d = 7$.

Line (*1*) $s_n = a + (n - 1)d$

Line (*2*) $280 = 35 + (n - 1) \cdot 7$

Line (*3*) $280 = 35 + 7n - 7$

Line (*4*) $252 = 7n$

Line (*5*) $36 = n$

There are 36 multiples of 7 between 30 and 283.

(17)

Study Exercise Three

1. Find s_{20} for 7, 9, 11, ...
2. Find s_{10} for 12, 9, 6, ...
3. Find the number of multiples of 3 between 31 and 182.
4. If $s_6 = 18$ and $s_{12} = 30$, find s_{25}.

(18)

Problem: Add all the natural numbers from 1 to 100 inclusive.

That is, find $\sum_{n=1}^{100} n$ or find $1 + 2 + 3 + 4 + \ldots + 100$.

Solution: Let S_n represent the sum of n terms.

Line (1) $S_{100} = 1 + 2 + 3 + \ldots + 100$

Line (2) $S_{100} = 100 + 99 + 98 + \ldots + 1$

Line (3) $2S_{100} = (101) + (101) + (101) + \ldots + (101)$

Line (4) $2S_{100} = 100(101)$

Line (5) $S_n = 50(101)$

Line (6) $S_n = 5{,}050$

(19)

Problem: Find the sum of n terms of an arithmetic sequence.

Solution:

Line (1) $S_n = a + (a + d) + (a + 2d) + \ldots + s_n$

Line (2) $S_n = s_n + (s_n - d) + (s_n - 2d) + \ldots + a$

Line (3) $2S_n = (a + s_n) + (a + s_n) + (a + s_n) + \ldots + (a + s_n)$

Line (4) $2S_n = n(a + s_n)$

Line (5) $S_n = \frac{n}{2}(a + s_n)$

(20)

Thus, we have developed two formulas for an A. P.

$s_n = a + (n - 1)d$ — nth term of an A. P.

$S_n = \frac{n}{2}(a + s_n)$ — sum of n terms of an A. P.

(21)

Example: Find the sum of 30 terms of 3, 6, 9, ...

Solution:

$$s_n = a + (n - 1)d$$
$$s_{30} = 3 + (29)(3)$$
$$s_{30} = 90$$
$$S_n = \frac{n}{2}(a + s_n)$$
$$S_{30} = \frac{30}{2}(3 + 90)$$
$$= 15(93)$$
$$= 1395$$

(22)

Study Exercise Four

1. Find S_{10} for 5, 8, 11, ...
2. Find $13 + 8 + 3 + \ldots$ for 20 terms
3. Find the sum of $3 + 7 + 11 + \ldots + 59 + 63$
4. How many terms of the series $25 + 19 + 13 + \ldots$ are needed to make the sum -20?

(23)

Geometric Progression

A sequence in which each term after the first is obtained by multiplying the preceding term by the same constant is called a geometric sequence or *geometric progression* (abbreviated G. P.). The constant is called the *common ratio* and is denoted by r.

(24)

Example 1: 1, 2, 4, 8, ... is a geometric sequence. The ratio is 2 $(2/1 = 2, 4/2 = 2, 8/4 = 2)$

Example 2: 4, -2, 1, $-1/2$, ... is a geometric sequence. The ratio is $-1/2$.

(25)

Study Exercise Five

For the geometric sequences below, find the ratio r and continue the sequence for two more terms.

1. 4, 8, 16, ...

2. 4, -2, 1, ...

3. $\frac{1}{2}, -\frac{3}{2}, \frac{9}{2}, \ldots$

4. $\frac{3}{10}, \frac{3}{100}, \frac{3}{1000}, \ldots$

(26)

Let a be the first term of a geometric sequence and r the common ratio.

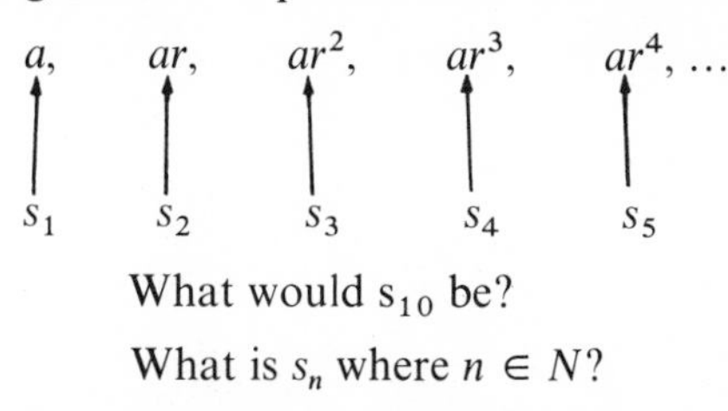

What would s_{10} be?

What is s_n where $n \in N$?

(27)

The formula for the *nth* term of a geometric sequence with first term a and ratio r is:

$$s_n = ar^{n-1}$$

(28)

Example: Find the seventh term in the sequence -24, 12, -6, ...

Solution: Since $12/-24 = 6/12 = -1/2$, the sequence is geometric with $r = -1/2$.

Line (1) $s_n = ar^{n-1}$

Line (2) $n = 7, a = -24, r = -1/2.$

Line (3) $s_7 = (-24)(-1/2)^6$

Line (4) $s_7 = (-24)(1/64)$

Line (5) $s_7 = -3/8$

(29)

Study Exercise Six

1. Find the 5th term in the sequence 48, 96, 192, ...

2. Find the 8th term in the sequence -81, -27, -9, ...

3. Find x so that 4, x, 9 will be in a geometric sequence. (Hint: use the definition)

(30)

Problem: To find the sum of n terms of a geometric sequence.

Solution:

Line (1) $S_n = a + ar + ar^2 + \dots + ar^{n-2} + ar^{n-1}$

Line (2) $r \cdot S_n = ar + ar^2 + \dots + ar^{n-2} + ar^{n-1} + ar^n$

Line (3) $S_n - S_n \cdot r = a + 0 + 0 + \dots + 0 - ar^n$

Line (4) $(1 - r) \cdot S_n = a(1 - r^n)$

Line (5) $S_n = \dfrac{a(1 - r^n)}{1 - r}, \quad r \neq 1$

(31)

The formula for the sum of n terms of a geometric sequence is

$$S_n = \frac{a(1 - r^n)}{1 - r}, \quad r \neq 1$$

It is important to note that the exponent n has r as its base. The minus sign does not go with the exponent.

(32)

Example: Find the sum of six terms of 4, 8, 16, ...

Solution: 4, 8, 16, ... is a G. P. with $r = 2$

Line (1) $S_n = \dfrac{a(1 - r^n)}{1 - r}$

Line (2) $a = 4, r = 2, n = 6$

Line (3) $S_6 = \dfrac{4(1 - 2^6)}{1 - 2}$

Line (4) $S_6 = \dfrac{4(1 - 64)}{-1}$

Line (5) $S_6 = \dfrac{4(-63)}{-1}$

Line (6) $S_6 = 252$

(33)

Study Exercise Seven

1. Find the sum of 5 terms of $3, 3^2, 3^3, \dots$
2. Find S_8 for $1 + (-2) + (4) + \dots$
3. If you were to be given one cent on the first day, two cents on the second day, four cents on the third day. etc., write the expression that would give the total amount of money you would have at the end of the 30th day. Do not attempt to evaluate.

(34)

Let us consider $1 + 1/2 + 1/4 + 1/8 + \dots$
This is a G. P. with $a = 1$ and $r = 1/2$.

Line (1) $S_n = \dfrac{a(1 - r^n)}{1 - r}$

Line (2) $S_n = \dfrac{1[1 - (1/2)^n]}{1 - 1/2}$

Line (3) $S_n = \dfrac{1 - (1/2)^n}{1/2}$

Line (4) $S_n = 2 - 2(1/2)^n$

(35)

For the series $1 + 1/2 + 1/4 + 1/8 + \ldots$, $S_n = 2 - 2(1/2)^n$;

$$S_2 = 2 - 2(1/2)^2 \qquad S_4 = 2 - 2(1/2)^4$$
$$= 2 - 1/2 \qquad = 2 - 2/16$$
$$= 1.5 \qquad = 1.875$$
$$S_6 = 2 - 2(1/2)^6 \qquad S_8 = 2 - 2(1/2)^8$$
$$= 2 - 2/64 \qquad = 2 - 2/256$$
$$\doteq 1.97 \qquad \doteq 1.99$$

(36)

We notice that for $1 + 1/2 + 1/4 + 1/8 + \ldots$

$S_2 = 1.5$, $S_4 = 1.875$, $S_6 \doteq 1.97$, $S_8 \doteq 1.99$

As n becomes larger, $(1/2)^n$ in the formula $S_n = 2 - 2(1/2)^n$ gets smaller and approaches zero.

Thus, as n becomes larger, S_n approaches 2.

We therefore define the sum of the infinite G. P. $1 + 1/2 + 1/4 + 1/8 + \ldots$ to be 2.

(37)

$$\text{Consider: } S_n = \frac{a(1 - r^n)}{1 - r}$$

If $-1 < r < 1$, r^n becomes smaller and smaller for larger and larger n. r^n approaches zero as n becomes larger.

We define the sum of the infinite geometric progression with $-1 < r < 1$ to be $\frac{a}{1 - r}$.

(38)

The symbol S_∞ will be used to indicate the sum of an infinite geometric progression.

$$S_\infty = \frac{a}{1 - r}, \quad -1 < r < 1$$

A sum exists only for r between -1 and 1.

(39)

Example: Find the sum of the infinite G. P. $2 + 4/3 + 8/9 + \ldots$

Solution:

Line (1) $r = 2/3$

Line (2) $S = \frac{a}{1 - r}$

Line (3) $S = \frac{2}{1 - 2/3}$

Line (4) $= \frac{2}{1/3}$

Line (5) $= 6$

(40)

We now look at repeating decimals such as $.33\overline{3}$, $.54\overline{54}$, $1.12\overline{12}$.

Since repeating decimals are rational numbers, how are such decimals expressed as a fraction?

(41)

Example: Find the fraction equivalent to $0.55\overline{5}$

Solution: 0.555 may also be written as .5 + .05 + .005 + .0005 + ...

$$\text{or} \quad \frac{5}{10} + \frac{5}{100} + \frac{5}{1000} + \frac{5}{10{,}000} + \ldots$$

This is an infinite G. P. with $r = 1/10$.

$$S_\infty = \frac{a}{1 - r}$$
$$= \frac{5/10}{1 - 1/10}$$
$$= \frac{5/10}{9/10}$$
$$= 5/9$$

thus $0.55\overline{5} = 5/9$

42

Example: Find the fraction equivalent to $1.122\overline{2}$

Solution: $1.122\overline{2}$ may be written 1 + .1 + .02 + .002 + .0002 + ...

$$\text{or} \quad 1 + \frac{1}{10} + \frac{2}{100} + \frac{2}{1000} + \frac{2}{10{,}000} + \ldots$$

$$\frac{11}{10} + \underbrace{\frac{2}{100} + \frac{2}{1000} + \frac{2}{10{,}000} + \ldots}_{\text{infinite G. P. with } r = \frac{1}{10}}$$

$$S_\infty = \frac{\frac{2}{100}}{1 - \frac{1}{10}}$$
$$= \frac{\frac{2}{100}}{\frac{9}{10}}$$
$$= \frac{2}{90}$$

$$\frac{11}{10} + \frac{2}{90} = \frac{101}{90}$$

Thus $1.122\overline{2} = \dfrac{101}{90}$

43

Study Exercise Eight

1. Find 3 + 1 + 1/3 + ...
2. Find the sum of an infinite number of terms of:
   1 − 1/2 + 1/4 − 1/8 + ...
3. Find the rational fraction in lowest terms equivalent to

   (a) $0.22\overline{2}$ **(b)** $1.12\overline{12}$ **(c)** $0.136\overline{36}$

44

Review Formulas

Arithmetic Progression

$$s_n = a + (n - 1)d$$

$$S_n = \frac{n}{2}(a + s_n)$$

Geometric Progression

$$s_n = ar^{n-1}$$

$$S_n = \frac{a(1 - r^n)}{1 - r}, r \neq 1$$

Infinite Geometric Progression

$$S_\infty = \frac{a}{1 - r}, \qquad -1 < r < 1$$

45

REVIEW EXERCISES

A. Which of the following sequences are arithmetic, which are geometric, and which are neither?

1. 3, −6, 9, −12, ... **2.** 25, 19, 13, 7, ... **3.** 4, 8, 12, ...

4. 4, 8, 16, ... **5.** 4/3, 2/3, 1/3, ...

B. Miscellaneous:

6. Find s_{12} and S_{12} for 2, 6, 10, 14, ...
7. Find s_6 and S_6 for 4, 12, 36, ...
8. Find the sum of 25 − 20 + 16 − ...
9. Find the fraction in lowest terms equivalent to $0.42\overline{42}$
10. Express in sigma notation $(7 \cdot 1 + 2) + (7 \cdot 2 + 2) + (7 \cdot 3 + 2) + (7 \cdot 4 + 2)$
11. Find x so that 3, x, 9 are in an A. P.
12. Find x so that 3, x, 9 are in a G. P.

46

SOLUTIONS TO REVIEW EXERCISES

A. **1.** neither A. P. nor G. P. **2.** A. P. $(d = -6)$ **3.** A. P. $(d = 4)$

4. G. P. $(r = 2)$ **5.** G. P. $(r = 1/2)$

B. **6.**
$$s_{12} = 2 + 11(4) = 2 + 44 = 46$$
$$S_{12} = \frac{12}{2}(2 + 46) = 6(48) = 288$$

7.
$$s_6 = 4(3)^5 = 4(243) = 972$$
$$S_6 = \frac{4(1 - 3^6)}{1 - 3} = \frac{4(1 - 729)}{-2} = -2(-728) = 1456$$

8.
$$r = -4/5$$
$$S_\infty = \frac{25}{1 - (-4/5)} = \frac{25}{1 + 4/5} = \frac{125}{9}$$

9.
$$.42 + .0042 + .000042 + \dots$$
or
$$\frac{42}{100} + \frac{42}{10{,}000} + \frac{42}{1{,}000{,}000} + \dots$$
$$S_\infty = \frac{\frac{42}{100}}{1 - \frac{1}{100}} = \frac{42}{100 - 1} = 42/99 = 14/33$$

10.
$$\sum_{n=1}^{4} (7n + 2)$$

11.
$$x - 3 = 9 - x$$
$$2x = 12$$
$$x = 6$$

12.
$$x/3 = 9/x$$
$$x^2 = 27$$
$$x = \pm\sqrt{27}$$
$$x = \pm 3\sqrt{3}$$

47

SOLUTIONS TO STUDY EXERCISES

Study Exercise One (Frame 10)

A. 1. 1, 3, 5, 7 **2.** 3/2, 1, 3/4, 3/5

3. $3^2, 3^3, 3^4, 3^5$ or 9, 27, 81, 243

B. 4. $1^2 + 2^2 + 3^2 + 4^2 + 5^2$ **5.** $1 + 4 + 7$

C. 6. $3^2 + 4^2 + 5^2 = 9 + 16 + 25 = 50$ **7.** $1/2 + 2/3 = 7/6$

8. $(-1)^2(1) + (-1)^3(2) + (-1)^4(3) = 1 - 2 + 3 = 2$

(10A)

Study Exercise Two (Frame 13)

1. $d = 3$ 5, 8, 11, 14, 17 **2.** $d = 5$ −6, −1, 4, 9

3. $d = 5$ $a, a + 5, a + 10, a + 15$

(13A)

Study Exercise Three (Frame 18)

1.
$$\begin{aligned} s_{20} &= 7 + 19(2) \\ &= 7 + 38 \\ &= 45 \end{aligned}$$

2.
$$\begin{aligned} s_{10} &= 12 + 9(-3) \\ &= 12 - 27 \\ &= -15 \end{aligned}$$

3. 33, 36, ..., 180
$$\begin{aligned} 180 &= 33 + (n - 1)\,3 \\ 180 &= 33 + 3n - 3 \\ 150 &= 3n \\ 50 &= n \end{aligned}$$
There are 50 multiples of 3 between 31 and 182.

4.
$$\begin{aligned} 18 &= a + 5d \\ 30 &= a + 11d \\ \hline -12 &= -6d \\ 2 &= d \\ 18 &= a + 5(2) \\ 8 &= a \end{aligned}$$

$$\begin{aligned} s_{25} &= 8 + 24(2) \\ &= 8 + 48 \\ &= 56 \end{aligned}$$

(18A)

Study Exercise Four (Frame 23)

1.
$$\begin{aligned} S_{10} &= \frac{n}{2}(a + s_{10}) \\ s_{10} &= 5 + 9(3) \\ &= 32 \\ S_{10} &= \frac{10}{2}(5 + 32) \\ &= 185 \end{aligned}$$

2.
$$\begin{aligned} s_{20} &= 13 + 19(-5) \\ &= -82 \\ S_{20} &= \frac{20}{2}(13 + [-82]) \\ &= -690 \end{aligned}$$

3.
$$\begin{aligned} 63 &= 3 + (n - 1)4 \\ 63 &= 3 + 4n - 4 \\ 64 &= 4n \\ 16 &= n \\ S_{16} &= \frac{16}{2}(3 + 63) \\ &= 8(66) \\ &= 528 \end{aligned}$$

Study Exercise Four (Frame 23, contd.)

4. $S_n = n/2(a + s_n)$

$-20 = n/2(25 + a + [n - 1]d)$

$-20 = n/2[25 + 25 + [n - 1](-6)]$

$-40 = n(50 - 6n + 6)$

$-40 = 56n - 6n^2$

$6n^2 - 56n - 40 = 0$

$3n^2 - 28n - 20 = 0$

$(3n + 2)(n - 10) = 0$

~~$n = -\frac{2}{3}$~~, $n = 10$

$n = 10$

10 terms are needed.

23A

Study Exercise Five (Frame 26)

1. $r = 2;\quad 4, 8, 16, 32, 64, \ldots$

2. $r = 1/2;\ 4, -2, 1, -1/2, 1/4, \ldots$

3. $r = -3;\quad 1/2, -3/2, 9/2, -27/2, 81/2, \ldots$

4. $r = 1/10;\quad \frac{3}{10}, \frac{3}{100}, \frac{3}{1{,}000}, \frac{3}{10{,}000}, \frac{3}{100{,}000}, \ldots$

26A

Study Exercise Six (Frame 30)

1. a G. P. with $r = 2$

$s_5 = 48(2)^4$

$= 768$

2. a G. P. with $r = 1/3$

$s_8 = (-81)(1/3)^7$

$= -\dfrac{1}{27}$

3. The ratio must be the same for any two terms:

$x/4 = 9/x$

$x^2 = 36$

$x = \pm 6$

4, 6, 9 *or* 4, −6, 9

30A

Study Exercise Seven (Frame 34)

1. $S_5 = \dfrac{3(1 - 3^5)}{1 - 3}$

$= \dfrac{3(1 - 243)}{-2}$

$= 3(121)$

$= 363$

2. $S_8 = \dfrac{1(1 - [-2]^8)}{1 - (-2)}$

$= \dfrac{1 - 256}{3}$

$= -85$

3. $S_{30} = \dfrac{1(1 - 2^{30})}{1 - 2}$

34A

Study Exercise Eight (Frame 44)

1. $S_\infty = \dfrac{3}{1 - 1/3}$

$= \dfrac{3}{2/3}$

$= 9/2$

2. $S_\infty = \dfrac{1}{1 - (-1/2)}$

$= \dfrac{1}{3/2}$

$= 2/3$

Study Exercise Eight (Frame 44, contd.)

3. (a) $0.22\overline{2} = .2 + .02 + .002 + \ldots$

$= 2/10 + 2/100 + \dfrac{2}{1{,}000} + \ldots$

$S_\infty = \dfrac{\frac{2}{10}}{1 - \frac{1}{10}}$

$= \dfrac{2/10}{9/10}$

$= 2/9$

$0.22\overline{2} = 2/9$

(b) $1.12\overline{12} = 1 + .12 + .0012 + \ldots$

$= 1 + \dfrac{12}{100} + \dfrac{12}{10{,}000}$

$\dfrac{12}{100} + \dfrac{12}{10{,}000} + \ldots$ is a G. P.

$S_\infty = \dfrac{\frac{12}{100}}{1 - \frac{1}{100}}$

$= \dfrac{12}{99}$

$= \dfrac{4}{33}$

$1.12\overline{12} = 1 + \dfrac{4}{33}$ or $\dfrac{37}{33}$

(c) $.136\overline{36} = .1 + .036 + .00036 + \ldots$

$= 1/10 + \dfrac{36}{1{,}000} + \dfrac{36}{100{,}000} + \ldots$

$\dfrac{36}{1{,}000} + \dfrac{36}{100{,}000} + \ldots$ is an infinite G. P. with $r = \dfrac{1}{100}$

$S_\infty = \dfrac{\frac{36}{1{,}000}}{1 - \frac{1}{100}}$

$= \dfrac{36}{990}$

$= \dfrac{4}{110}$

$.136\overline{36} = \dfrac{1}{10} + \dfrac{4}{110}$

$= \dfrac{15}{110}$

$= \dfrac{3}{22}$

44A

UNIT 25—SUPPLEMENTARY PROBLEMS

A. Continue for two more terms and state if an A. P or G. P.

1. $4 + 3 + \frac{9}{4} + \ldots$

2. $-1 + 4 + 9 + \ldots$

3. $1 + 1.1 + 1.21 + \ldots$

4. $-10, -4, 2, 8, \ldots$

B. Miscellaneous:

5. Find s_{10} for the sequence 5, 9, 13, ...

6. Find s_8 for 64, -32, 16, ...

7. Find S_8 for the series $1 + 1.5 + 2 + \ldots$

8. In an A. P., $a = -12$, $n = 15$, and $S_{15} = 135$. Find d.

9. How many terms of the series $1 + 1^1/_2 + 2 + \ldots$ are needed to make the sum 45?

10. Find the sum of the first 150 positive integers.

11. In an A. P., $s_6 = -2$ and $s_{40} = 100$. Find s_{25}.

12. Find the exact value of $\sum_{x=1}^{5} x^2$

13. Find the exact value of $\sum_{x=1}^{20} 9x$

14. Write the series $1 \cdot 3 + 1 \cdot 5 + 1 \cdot 7 + 1 \cdot 9$ in summation notation.

15. Write the infinite series $1/2 + 2/3 + 3/4 + 4/5 + \ldots$ in summation notation.

16. How many numbers between 50 and 150 are exactly divisible by 7?

17. Find m so that $8m + 4$, $6m - 2$, and $2m - 6$ will form an A. P.

18. In a G. P., $n = 8$, $r = -2$, and $s_n = -640$. Find a.

19. If one cent is saved the first day, two cents the second day, three cents the third day, etc., find the sum that will accumulate at the end of 365 days.

20. Find S_∞ for $3 + 2/3 + 4/27 + \ldots$

21. Find the fraction equivalent to $.\overline{5252}$

22. Find the fraction equivalent to $1.0\overline{202}$

23. A rubber ball that is dropped from a height of 9 feet always rebounds one-third of the distance of the previous fall. Approximately how far does it travel before it comes to rest?

24. The number of bacteria in milk doubles every 3 hours. If there are n bacteria at a given time, how many will there be at the end of 24 hours from then? (careful, the number of terms is not 8)

25. Show that if $x, y, z > 0$ and x, y, z is a G. P., then $\log x, \log y, \log z$ is an A. P.

unit

The Complex Number System

26

Objectives

1. Understand the development of the complex number system.
2. Be able to add, subtract, multiply, and divide complex numbers, and express the result in standard form.
3. Be able to find the solution set of any quadratic equation.

Terms

real and imaginary part — conjugate

(1)

In previous units we learned that the quadratic equation $x^2 = -1$ has no solution since no real number when squared gives one.

(2)

We desire to extend the number system to include numbers of such a character that every quadratic equation has a non empty solution set regardless of the value of its discriminant.

$$N \subset I \subset F \subset R \subset ?$$

(3)

The Complex Number System

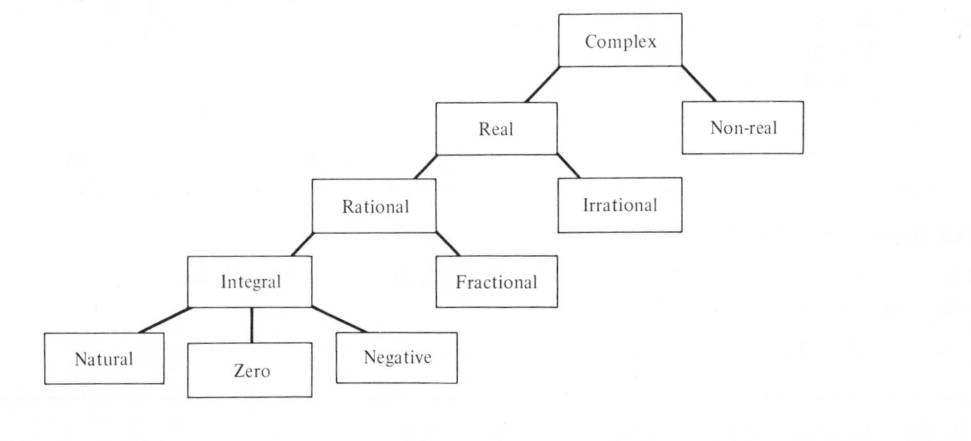

(4)

We desire to construct a system which:

1. Contains new members of the kind we need.
2. Includes all the numbers we have already.
3. Has the algebraic laws of the real number system.
4. Has the operations (addition, multiplication, etc.) agreeing with the operations of the old system.

(5)

We will call the new system the *complex number system*.

We desire definitions for addition and multiplication such that the laws for closure, commutativity, associativity, identity, inverse, and distributivity hold.

(6)

We designate the complex number system by C.
We will assume these properties for C:

1. Addition and multiplication are defined in C and the field properties of the reals hold.
2. Each real number is a member of C.
3. The additive identity of C is zero.
4. The multiplicative identity of C is one.
5. C contains a distinguished member i, with the property $i \cdot i = i^2 = -1$.
6. Zero times any number in C is zero.

(7)

A property of C is that it contains a member i with the property $i \cdot i = i^2 = -1$.

The definition that $i^2 = -1$ suggests that you write $i = \sqrt{-1}$ and call i "a square root of -1".

(8)

If a and b are real numbers, then a and b are members of C.
Since $i \in C$ and closure for multiplication holds, then $b \cdot i \in C$.

Since closure for addition holds, $(a + b \cdot i) \in C$. Numbers of the form $a + b \cdot i$ will be sufficient.

(9)

Complex Numbers

The standard form of a complex number is $a + b \cdot i$; $a, b \in R$.

For the number $a + bi$, a is called the *real part* and b is called the *imaginary part*.

(10)

Examples:

Number	*Real Part*	*Imaginary Part*
$4 + 2i$	4	2
$-3 + i$	-3	1
$6 - 4i$	6	-4
$7 + 0 \cdot i$	7	0

(11)

Equality of Complex Numbers

Two complex numbers, $a + bi$ and $c + di$, $a, b, c, d \in R$ are equal if and only if their real parts are equal and their imaginary parts are equal.

That is, $a + bi = c + di$ if and only if $a = c$ and $b = d$.

(12)

Examples: Write the following complex numbers in standard form.

1. 2 Answer. $2 + 0 \cdot i$

2. i^2 Answer. $-1 + 0 \cdot i$

3. $2i$ Answer. $0 + 2i$

4. i^3 Answer. $0 + (-1) \cdot i$ since

$$\begin{aligned} i^3 &= i \cdot i^2 \\ &= i(-1) \\ &= -i \end{aligned}$$

(13)

Addition of Complex Numbers

Definition of addition.

If $a, b, c, d \in R$ and $a + bi$ and $c + di \in C$,
then $(a + bi) + (c + di) = (a + c) + (b + d)i$.

(14)

Example 1: $(4 + 2i) + (3 + 3i)$
Solution: $(4 + 3) + (2 + 3)i = 7 + 5i$

Example 2: $(6 + 3i) + (-2 + 5i)$
Solution: $(6 + [-2]) + (3 + 5)i = 4 + 8i$

Example 3: $(-1 + [-2]i) + (3 + i)$
$(-1 + 3) + ([-2] + 1)i = 2 + (-1)i$

(15)

Remember that the imaginary unit i is such that:

$$i^2 = -1 \quad \text{and}$$
$$-i^2 = -(-1) = 1$$

(16)

Multiplication of Complex Numbers

Definition of multiplication.

If $a, b, c, d \in R$ and $(a + bi)$ and $(c + di) \in C$,
then $(a + bi) \cdot (c + di) = (ac - bd) + (ad + bc)i$.

(17)

Recall that $(a + bi) \cdot (c + di) = (ac - bd) + (ad + bc)i$.

If $(a + bi)$ and $(c + di)$ are treated as binomials, then

$$\begin{aligned}(a + bi) \cdot (c + di) &= \\ &= (a + bi) \cdot c + (a + bi) \cdot di \\ &= ac + bci + adi + bdi^2 \text{ and since } i^2 = -1 \\ &= ac + bci + adi - bd \\ &= (ac - bd) + (bc + ad)\, i\end{aligned}$$

(18)

Example 1: $(3 + 2i) \cdot (2 + 4i)$
Solution:
$$\begin{aligned}&6 + 4i + 12i + 8i^2 \\ &= 6 + 16i + 8i^2 \\ &= 6 + 16i + 8(-1) \\ &= -2 + 16i\end{aligned}$$

Example 2: $(-1 + i) \cdot (-1 - i)$
Solution:
$$\begin{aligned}&1 + (-i) + (i) - i^2 \\ &= 1 + 0 - i^2 \\ &= 1 + 0 + 1 \\ &= 2 \quad \text{or} \quad 2 + 0 \cdot i\end{aligned}$$

(19)

Study Exercise One

Perform the indicated operations and express the result in standard form.

1. $(1 + 4i) + (3 + 5i)$ **2.** $(2 + 6i) + (2 - 6i)$ **3.** $(2 + 3i) \cdot (6 + 4i)$

4. $i(3 + 5i)$ **5.** $(3 + 4i)(3 - 4i)$ **6.** $(2 - 3i)^2$

7. $3 + 2i + 5 + i$ **8.** $6 + 5i - (2 - 3i)$ **9.** $(4 + i) - 7$

10. $-2i^2$

(20)

Reciprocal of i

Let us multiply i by $-i$:

Line (1) $\quad i(-i) = -i^2$
Line (2) $\quad = -(-1)$
Line (3) $\quad = 1$

Since $i \cdot (-1)$ is 1, the reciprocal of i is $-i$.
That is, $1/i = -i$.

(21)

Let us multiply $(4 + \mathrm{i})$ by $\left[\frac{4}{17} + \left(-\frac{1}{17}\right)\mathrm{i}\right]$:

Line (1) $\quad (4 + i)\left(\frac{4}{17} - \frac{1}{17}i\right) = \frac{16}{17} + \frac{4i}{17} - \frac{4i}{17} - \frac{1}{17}i^2$

Line (2) $\quad = \frac{16}{17} - \frac{1}{17}(-1)$

Line (3) $\quad = 1$

The reciprocal of $(4 + i)$ is $\left(\frac{4}{17} - \frac{1}{17}i\right)$ since

$(4 + i)\left(\frac{4}{17} - \frac{1}{17}i\right) = 1$. That is, $\frac{1}{4 + i}$ equals $\frac{4}{17} - \frac{1}{17}i$.

But how is the reciprocal actually found?

(22)

The reciprocal of $5 + i$ would be written $\frac{1}{5 + i}$. But what is the standard form of $\frac{1}{5 + i}$?

A procedure for finding the reciprocal of $5 + i$ will be developed in succeeding frames.

(23)

Conjugate

If $a + bi$ is a complex number, then $a - bi$ is called the *conjugate* of the complex number $a + bi$.

Examples:

Number	*Conjugate*
$2 + 3i$	$2 - 3i$
$3 - 2i$	$3 + 2i$
$4 + 7i$	$4 - 7i$
$-1 - 2i$	$-1 + 2i$
5	5

(24)

We will now multiply a complex number by its conjugate.

$$\begin{aligned}(a + bi)(a - bi) &= a^2 + abi - abi - b^2i^2 \\ &= a^2 + 0 - b^2i^2 \\ &= a^2 - b^2(-1) \\ &= a^2 + b^2 \\ &= (a^2 + b^2) + 0 \cdot i\end{aligned}$$

(25)

The product of a complex number by its conjugate is a real number, since $(a + bi)(a - bi) = (a^2 + b^2) + 0 \cdot i$.

For example; $(2 + 3i)(2 - 3i) = 4 - 9i^2$
$= 4 - 9(-1)$
$= 4 + 9$
$= 13$

(26)

To find the reciprocal of $5 + i$ in standard form, we multiply numerator and denominator by the conjugate of the denominator.

$$\frac{1}{5+i} = \frac{(1)\cdot(5-i)}{(5+i)\cdot(5-i)}$$
$$= \frac{5-i}{25-i^2}$$
$$= \frac{5-i}{25+1}$$
$$= \frac{5-i}{26}$$
$$= \frac{5}{26} - \frac{1}{26}i$$

(27)

Example: Find $\frac{2}{3-2i}$

Solution: $\frac{2}{3-2i} = \frac{2(3+2i)}{(3-2i)(3+2i)}$
$$= \frac{2(3+2i)}{9-4i^2}$$
$$= \frac{2(3+2i)}{9-4(-1)}$$
$$= \frac{6+4i}{13}$$

(28)

Study Exercise Two

Evaluate each quotient and write the result in standard form.

1. $\frac{1}{3+i}$ **2.** $\frac{3}{2i}$ **3.** $\frac{5+i}{2+3i}$

4. $\frac{2i}{1-2i}$ **5.** $\frac{\sqrt{2}+\sqrt{3}i}{\sqrt{2}-\sqrt{3}i}$

(29)

Let x be a real number.

If $x > 0$, then $\sqrt{x}$ is a unique non negative real number.

If $x < 0$, we define $\sqrt{x} = i\sqrt{-x}$.
Remember that if $x < 0$, $-x$ is a positive number.

(30)

In evaluating radicals, it is important to remember if the radicand is negative you must use the definition.

$$\sqrt{x} = i\sqrt{-x}, \quad x < 0.$$
$$\sqrt{-4} = i\sqrt{-(-4)}$$
$$= i\sqrt{4}$$
$$= 2i$$

(31)

Do not assume that laws studied earlier for radicals are valid when the radicand is negative.

$$\sqrt{2} \cdot \sqrt{3} = \sqrt{6}$$

but

$$\sqrt{-2} \cdot \sqrt{-3} \neq \sqrt{6}$$

(32)

$$\sqrt{-2} \cdot \sqrt{-3} = i\sqrt{2} \cdot i\sqrt{3}$$
$$= i^2 \cdot \sqrt{2}\sqrt{3}$$
$$= i^2 \cdot \sqrt{6}$$
$$= -1\sqrt{6}$$
$$= -\sqrt{6}$$

(33)

Example 1: Simplify $\sqrt{-25}$

Solution: $\sqrt{-25} = i\sqrt{25}$
$= 5i$

Example 2: Simplify $\sqrt{-20}$

Solution: $\sqrt{-20} = i\sqrt{20}$
$= i\sqrt{4} \cdot \sqrt{5}$
$= 2i\sqrt{5}$

Example 3: Simplify $\sqrt{-50}$

Solution: $\sqrt{-50} = i\sqrt{50}$
$= i\sqrt{25}\sqrt{2}$
$= 5i\sqrt{2}$

(34)

Example: Reduce $\frac{20 + \sqrt{-20}}{4}$

Solution: $\frac{20 + \sqrt{-20}}{4} = \frac{20 + i\sqrt{20}}{4}$

$$= \frac{20 + 2i\sqrt{5}}{4}$$

$$\frac{\not{2}(10 + i\sqrt{5})}{\not{4}_2}$$

$$= \frac{10 + i\sqrt{5}}{2}$$

(35)

Study Exercise Three

Write the results in standard form. All answers should be reduced.

1. $5 + \sqrt{-36}$
2. $-\sqrt{-8}$
3. $\frac{2 + \sqrt{-4}}{2}$
4. $\frac{\sqrt{-50} + \sqrt{-2}}{6}$
5. $\sqrt{-4}\sqrt{-9}$

(36)

We are now in a position to solve all quadratic equations of the form $ax^2 + bx + c = 0$, $a, b, c \in R, a \neq 0$.

Example: Find the solution set of $x^2 + x + 1 = 0$.

Solution: $a = 1, b = 1, c = 1$ in the quadratric formula; $x = \frac{-b \pm \sqrt{b^2 - 4ac}}{2a}$

$$x = \frac{-(1) \pm \sqrt{(1)^2 - 4(1)(1)}}{2(1)}$$

$$x = \frac{-1 \pm \sqrt{1 - 4}}{2}$$

$$x = \frac{-1 \pm \sqrt{-3}}{2} = \frac{-1 \pm i\sqrt{3}}{2}$$

The solution set is $\left\{\frac{-1 + i\sqrt{3}}{2}, \frac{-1 - i\sqrt{3}}{2}\right\}$

(37)

Study Exercise Four

Find the solution set of each of the following over C.

1. $x^2 + 2x + 2 = 0$
2. $3x^2 + 2x + 4 = 0$
3. $x^2 - 4x + 8 = 0$
4. $x^2 + 4 = 0$
5. $2x^2 + x - 1 = 0$
6. $x^4 - 1 = 0$

(38)

Recall that the discriminant of the quadratic equation $ax^2 + bx + c = 0, a, b, c \in R, a \neq 0$ is given by $D = b^2 - 4ac$.

We now can make a more complete statement by saying;

if $D > 0$, there are two unequal real roots;

$D = 0$, there is one real root;

$D < 0$, there are two complex number roots.

(39)

Example: Classify the roots of $x^2 + 3x + 8 = 0$.

Solution: $D = b^2 - 4ac$

$= (3)^2 - 4(1)(8)$

$= 9 - 32$

$= -23$

Since $D < 0$, there are two complex number roots.

(40)

REVIEW EXERCISES

A. True or False:

1. Since 4 is a real number, 4 is a complex number.

2. The real numbers are a subset of the complex numbers.

3. $i^3 = -i$ where $i = \sqrt{-1}$

4. $(i\sqrt{2})^2 = (\sqrt{-2})^2$ where $i = \sqrt{-1}$

5. $a + bi = b + ai$ where $a, b \in R$

6. $(2 + i)(2 - i) = 5$ where $i = \sqrt{-1}$

7. $-i^2 = -1$ where $i = \sqrt{-1}$

B. Perform the indicated operations and express the result in standard form:

8. $(5 + 3i) + (-6 + i)$ **9.** $(-1 - 2i) - (3 - i)$

10. $(5 + 3i)(-6 + i)$ **11.** $(1 - i)^2$

12. $\dfrac{5 + 3i}{-6 + i}$ **13.** $\sqrt{-20} + 2\sqrt{5} - 3i - (4)^2$

C. Find the solution sets over C.

14. $2x^2 - 2x + 3 = 0$ **15.** $x^2 + 18 = 0$

16. $x^3 + 2x^2 + x + 2 = 0$ (hint: factor)

(41)

SOLUTIONS TO REVIEW EXERCISES

A. **1.** True **2.** True **3.** True **4.** True

5. False **6.** True **7.** False

B. **8.** $(5 + [-6]) + (3 + 1)i = -1 + 4i$

9. $(-1 + [-3]) + (-2 + 1)i = -4 + (-1)i$

10. $-30 - 18i + 5i + 3i^2 = -30 - 13i - 3$
$= (-33) + (-13)i$

11. $1 - 2i + i^2 = 1 - 2i - 1$
$= 0 + (-2)i$

12. $$\frac{(5 + 3i)(-6 - i)}{(-6 + i)(-6 - i)} = \frac{-30 - 23i - 3i^2}{36 - i^2} = \frac{-27 - 23i}{37}$$
$$= \frac{-27}{37} + \left(\frac{-23}{37}\right)i$$

13. $-2i\sqrt{5} + 2\sqrt{5} - 3i - 16 = 2\sqrt{5} - 16 + (-2\sqrt{5} - 3)i$

C. **14.** $$x = \frac{-(-2) \pm \sqrt{(-2)^2 - 4(2)(3)}}{2(2)}$$

$$x = \frac{2 \pm \sqrt{-20}}{4} = \frac{2 \pm 2i\sqrt{5}}{4} = \frac{\cancel{2}(1 \pm i\sqrt{5})}{\cancel{4}_2}$$

The solution set is $\left\{\dfrac{1 \pm i\sqrt{5}}{2}\right\}$.

15. $x = \pm\sqrt{-18}$

$x = \pm 3i\sqrt{2}$

The solution set is $\{\pm 3i\sqrt{2}\}$.

SOLUTIONS TO REVIEW EXERCISES, CONTD.

(Frame 42, contd.)

16. $x^2(x+2) + (x+2) = 0$

$(x^2+1)(x+2) = 0$

$x^2 + 1 = 0$ *or* $x + 2 = 0$

$x^2 = 1$ or $x = -2$

The solution set is $\{i, -i, -2\}$.

SOLUTIONS TO STUDY EXERCISES

Study Exercise One (Frame 20)

1. $4 + 9i$

2. $4 + 0 \cdot i$

3. $12 + 18i + 8i + 12i^2 = 0 + 26i$

4. $3i + 5i^2 = (-5) + 3i$

5. $9 - 16i^2 = 25 + 0 \cdot i$

6. $4 - 12i + 9i^2 = -5 + (-12)i$

7. $8 + 3i$

8. $4 + 8i$

9. $-3 + i$

10. $2 + 0 \cdot i$

20A

Study Exercise Two (Frame 29)

1. $\dfrac{1(3-i)}{(3+i)(3-i)} = \dfrac{3-i}{9-i^2} = \dfrac{3-i}{10} = \dfrac{3}{10} + \left(-\dfrac{1}{10}\right)i$

2. $\dfrac{3(i)}{2i(i)} = \dfrac{3i}{2i^2} = \dfrac{3i}{-2} = 0 + (-3/2)\,i$

3. $\dfrac{(5+i)(2-3i)}{(2+3i)(2-3i)} = \dfrac{10 - 13i - 3i^2}{4 - 9i^2} = \dfrac{13 - 13i}{13} = 1 + (-1)\,i$

4. $\dfrac{(2i)(1+2i)}{(1-2i)(1+2i)} = \dfrac{2i + 4i^2}{1 - 4i^2} = \dfrac{-4 + 2i}{5} = -\dfrac{4}{5} + \dfrac{2}{5}i$

5. $\dfrac{(\sqrt{2} + \sqrt{3}i)(\sqrt{2} + \sqrt{3}i)}{(\sqrt{2} - \sqrt{3}i)(\sqrt{2} + \sqrt{3}i)} = \dfrac{2 + 2\sqrt{6}i}{\sqrt{4} - \sqrt{9}i^2} + 3i^2 = \dfrac{-1 + 2\sqrt{6}i}{5}$

29A

Study Exercise Three (Frame 36)

1. $5 + 6i$

2. $-i\sqrt{8} = -2i\sqrt{2}$

3. $\dfrac{2 + 2i}{2} = \dfrac{\cancel{2}(1+i)}{\cancel{2}} = 1 + i$

4. $\dfrac{5i\sqrt{2} + i\sqrt{2}}{6} = \dfrac{6i\sqrt{2}}{6} = 0 + \sqrt{2} \cdot i$

5. $i\sqrt{4} \cdot i\sqrt{9} = 6i^2 = -6 = -6 + 0i$

36A

Study Exercise Four (Frame 38)

1. $x = \dfrac{-2 \pm \sqrt{4-8}}{2} = \dfrac{-2 \pm 2i}{2} = -1 \pm i$

The solution set is $\{-1 + i, -1 - i\}$.

SOLUTIONS TO STUDY EXERCISES, CONTD.

Study Exercise Four (Frame 38, contd.)

2. $x = \dfrac{-2 \pm \sqrt{4 - 48}}{6} = \dfrac{-2 \pm 2i\sqrt{11}}{6} = \dfrac{-1 \pm i\sqrt{11}}{3}$

The solution set is $\left\{\dfrac{-1 + i\sqrt{11}}{3}, \dfrac{-1 - i\sqrt{11}}{3}\right\}$.

3. $x = \dfrac{4 \pm \sqrt{16 - 32}}{2} = \dfrac{4 \pm 4i}{2} = 2 \pm 2i$

The solution set is $\{2 + 2i, 2 - 2i\}$.

4. $x = \pm\sqrt{-4} = \pm 2i$

The solution set is $\{2i, -2i\}$.

5. $x = \dfrac{-1 \pm \sqrt{1 + 8}}{4} = \dfrac{-1 \pm 3}{4}$

The solution set is $\left\{\dfrac{1}{2}, -1\right\}$.

6. $(x^2 + 1)(x + 1)(x - 1) = 0$

$x^2 + 1 = 0$ or $x + 1 = 0$ or $x - 1 = 0$

$x = \pm i$ or $x = -1$ or $x = 1$

The solution set is $\{i, -i, 1, -1\}$.

38A

UNIT 26—SUPPLEMENTARY PROBLEMS

A. True or false.

1. $i + \sqrt{-1} = 0$

2. $\sqrt{-2}\sqrt{-2} = 2$

3. $-i^2 = 1$

4. $\sqrt[3]{-64} = 4i$

5. $1/i = -i$

B. Write in standard form as a single complex number.

6. $2 + \sqrt{-3}$

7. $-\sqrt{-25}$

8. $(\sqrt{2} + \sqrt{-1})^2$

9. $\sqrt{-50} + 2\sqrt{-32}$

10. $\sqrt{9} - \sqrt{-16} + \sqrt{-25}$

11. $\sqrt{-3}\,(2 - \sqrt{-12})$

12. $(3 - 2i)(-1 + 5i)$

13. $(2i\sqrt{5})^2$

14. $\dfrac{-1 + 2i}{4 + i}$

15. $\dfrac{-3 - i}{2 - 3i}$

C. Find the solution sets over C.

16. $x^2 = -25$

17. $(x - 3)^2 = -4$

18. $2x^2 + x + 3 = 0$

19. $x^2 + 8 = 0$

20. $3x^2 + 5x - 2 = 0$

21. $x^2 - 3x + 1 = 0$

22. $3x^2 + 2x + 4 = 0$

23. $x^2 + x + 3 = 0$

D. Miscellaneous:

24. Find the complex number which is the reciprocal of $-2 + 4i$.

25. If $3x - 4yi = 9 + 8i$, solve for x and y by using the definition of the equality of complex numbers.

26. Find the value of $3x^2 - 4x - 5$ if $x = 4 + 3i$.

UNIT 26—SUPPLEMENTARY PROBLEMS, CONTD.

27. Find the three cube roots of -1. (Hint: solve $x^3 = -1$ by factoring).
28. Find the three cube roots of 8.
29. Find the three cube roots of 27.
30. a) Evaluate $\left(\frac{1}{2} + \frac{i\sqrt{3}}{2}\right)^3$

b) What does $\left(\frac{1}{2} + \frac{i\sqrt{3}}{2}\right)$ represent in relation to your answer?

31. Write a quadratic equation with solution set $\{2i, -2i\}$.
32. Prove that the sum of two conjugate complex numbers is a real number.

Permutations

unit

27

Objectives

1. Know the multiplication principle.
2. Know the formula for permutations and be able to work problems involving permutations.

Terms

permutation

①

Students are often intrigued by problems that require them to find the number of ways of arranging a set of objects.

How many license plates can be made using 3 letters followed by a 3-digit number? In how many ways can 5 books be arranged on a bookshelf? How many 4-digit odd numbers can be made from the digits 1, 2, 3, 4, 5?

②

We will attempt to discover a general principle that will enable us to determine the number of possible arrangements of objects in a set by considering an example.

③

Example: In how many ways can three floats be arranged for a parade?

Solution: Let us call the three floats X, Y, and Z. An organized way of listing all the arrangements is called a *tree diagram.*

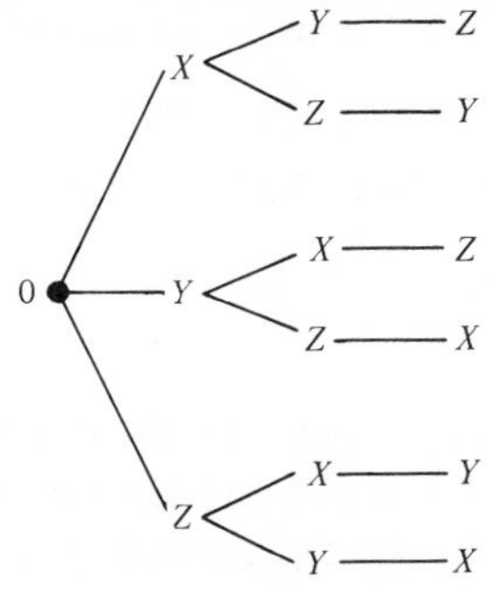

Possible Arrangement

$X\ Y\ Z$

$X\ Z\ Y$

$Y\ X\ Z$

$Y\ Z\ X$

$Z\ X\ Y$

$Z\ Y\ X$

The initial point is denoted by 0. If we follow all possible branches, we get the 6 possible arrangements. Note that order is taken into account. *A change in order is a different arrangement.*

④

Another way to solve the preceeding problem is to consider that 3 spaces are to be filled
___, ___, ___.

In the first space, we can put X, Y, or Z. Hence, the first space can be filled in 3 ways:
3, ___, ___.

After the first space is filled, we have 2 other possibilities for filling the second space:
3, 2, ___.

After the first two spaces are filled, there remains only one possibility for filling the third space: 3, 2, 1.

We obtain the total number of arrangements by multiplication: $3 \cdot 2 \cdot 1 = 6$.

(5)

The Multiplication Principle

If an operation can be performed in m ways, and after it is performed, another operation can be performed in n ways, and after it is performed, another can be performed in p ways, etc., then the number of ways all operations can be performed is $m \cdot n \cdot p \cdot \ldots$

(6)

Permutations

An arrangement of the elements in a set in a specified order is called a *permutation* of the elements.

To permute a set of objects means to arrange them in a definite order.

(7)

Example: Find all possible permutations of the elements of $\{r, s, t\}$.

Solution: There are three spaces to fill, ___, ___, ___. The first can be filled in any one of 3 ways, the second in two ways, and the third in one way. Hence, there are 3 · 2 · 1 or 6 permutations. They are:

rst, *rts*, *srt*, *str*, *tsr*, *trs*

(8)

Example: A football stadium has 10 gates. In how many ways can a spectator enter by one gate and leave by another?

Solution: A spectator can enter in any one of 10 ways. After he has entered, he may leave in any one of 9 ways. Hence, by the multiplication principle, there are $10 \cdot 9$ or 90 ways he may enter by one gate and leave by another.

(9)

Example: Given the digits 1, 2, 3, 4. How many 3-digit numbers can be formed from them
(a) if no digit may be repeated
(b) if repetition of a digit is allowed?

Solution:

(a) *No repetitions*. There are 3 places to fill. The first may be filled with any one of 4 digits. Then, since no digit may be used more than once, the second place may be filled in 3 ways, and the third place in any one of 2 ways. By the multiplication principle, there are $4 \cdot 3 \cdot 2$ *or* 24 three-digit numbers.

(b) *Repetitions allowed.* Since repetitions are allowed each of the three places may be filled with any one of the 4 digits. Hence, by the multiplication principle, there are $4 \cdot 4 \cdot 4$ *or* 64 possible three-digit numbers.

(10)

Study Exercise One

1. A nickel and a dime are tossed onto a sidewalk. In how many ways can they fall? Actually write out these ways.
2. In how many ways can 5 people line up at a theater box office?
3. The Greek alphabet consists of 24 letters. How many names of fraternities can be formed by using 3 letters at a time:
   (a) if no repetitions are permitted
   (b) if repetitions are permitted?
4. A passenger train has 9 coaches. In how many ways can 4 people be assigned to coaches if they must ride in different coaches?
5. In how many ways can the letters of the word "math" be arranged?

(11)

Example: Given the digits 1, 2, 3, 4, 5. How many three-digit numbers may be formed if the numbers must be odd and no repetitions are allowed?

Solution: If the number must be odd, the final digit has to be 1, 3, or 5. There are 3 places to fill and the third place can be filled in 3 ways. After this has been done in any one of these ways, the first place can be filled in 4 ways and the second place in 3 ways. Thus, the number of odd, 3-digit numbers is $4 \cdot 3 \cdot 3$ or 36.

(12)

In the preceding example, we filled the third place first. *If some operation must be performed in a special way, it is advisable to do it first.*

(13)

Example: If two dice are thrown, in how many ways will they fall?

Solution: A die has six faces labeled 1 to 6. The first die can fall in any one of 6 ways and the second die can fall in any one of 6 ways. Hence, two dice may fall in any one of $6 \cdot 6$ or 36 ways.

(14)

Study Exercise Two

1. In how many ways can the letters of the word "poker" be arranged if the last place must be occupied by a vowel?
2. There are 6 seats available in a sedan. In how many ways can 6 persons be seated for a journey if only 3 are able to drive?
3. In how many ways can 3 dice fall?

(15)

Formulas for Permutations

The multiplication principle provides a general method for finding the number of permutations of sets of objects.

Certain formulas can be developed based on the multiplication principle.

(16)

By use of the multiplication principle,

(a) Five people can be arranged in a line in
$\underline{5} \cdot \underline{4} \cdot \underline{3} \cdot \underline{2} \cdot \underline{1}$ ways.

(b) Eleven books can be arranged on a shelf in
$\underline{11} \cdot \underline{10} \cdot \underline{9} \cdot \underline{8} \cdot \underline{7} \cdot \underline{6} \cdot \underline{5} \cdot \underline{4} \cdot \underline{3} \cdot \underline{2} \cdot \underline{1}$ ways.

(c) n objects can be arranged in a line in
$\underline{n} \cdot \underline{(n-1)} \cdot \underline{(n-2)} \cdot \underline{(n-3)} \ldots \underline{(1)}$ ways.
The short way of writing $n(n-1)(n-2)\ldots(1)$ is $n!$

Theorem. The number of permutations of a set of n different objects, taken all together is $n!$

The notation for the permutation of n objects taken all together is $P(n, n)$. Thus, $P(n, n) = n!$

(17)

Example: In how many ways can 5 books be arranged?

Solution:
$$P(n, n) = 5! = 5 \cdot 4 \cdot 3 \cdot 2 \cdot 1 = 120$$

(18)

In working with permutations, it will be necessary to simplify factorials.

Example 1: Simplify $4! \cdot 5!$

Solution:
$$4! \cdot 5! = (4 \cdot 3 \cdot 2 \cdot 1)(5 \cdot 4 \cdot 3 \cdot 2 \cdot 1) = (24)(120) = 2880$$

Example 2: Simplify $\frac{8!}{4!}$

Solution:
$$\frac{8!}{4!} = \frac{8 \cdot 7 \cdot 6 \cdot 5 \cdot \cancel{4}!}{\cancel{4}!} = 8 \cdot 7 \cdot 6 \cdot 5 = 1680$$

(19)

Study Exercise Three

1. Evaluate $P(7, 7)$.
2. In how many ways can 6 books be arranged on a shelf?
3. Simplify $\frac{8! \cdot 3!}{6!}$
4. Simplify $\frac{6!}{2! \cdot 4!}$

(20)

Problem: In how many ways can 3 books be chosen from 5 books and arranged in 3 spaces on a bookshelf?

Solution: The first space can be filled in 5 ways, the second in 4 ways and the third space in 3 ways. By the multiplication principle, there are $5 \cdot 4 \cdot 3$ or 60 ways. We can also use factorial symbols to denote the product $5 \cdot 4 \cdot 3$.

$$5 \cdot 4 \cdot 3 = \frac{5 \cdot 4 \cdot 3 \cdot [2 \cdot 1]}{[2 \cdot 1]} = \frac{5!}{2!}$$

(Frame 21, contd.)

The number of permutations of 5 books taken 3 at a time is denoted by $P(5, 3)$.

$$P(5, 3) = 5 \cdot 4 \cdot 3 = \frac{5!}{2!}$$

(21)

Theorem. The number of permutations of a set of n different objects taken r at a time, without repetition is;

$$P(n, r) = \frac{n!}{(n - r)!}$$

Proof. There are r spaces to fill. The first space can be filled in any one of n ways, the second space in $(n - 1)$ ways, the third space in $(n - 2)$ ways, etc. to the rth space which can be filled in $[n - (r - 1)]$ ways.

Line (1) $\quad P(n, r) = n(n - 1)(n - 2) \ldots [n - (r - 1)]$

Line (2) $\quad = \dfrac{n(n - 1)(n - 2) \ldots (n - r + 1)[(n - r)(n - r - 1)(n - r - 2) \ldots (1)]}{[(n - r)(n - r - 1)(n - r - 2) \ldots (1)]}$

Line (3) $\quad = \dfrac{n!}{(n - r)!}$

(22)

The formula for $P(n, r)$ is valid for $r \leq n$ where n and r are non-negative integers.

$$\text{If } r = n,\ P(n, n) = \frac{n!}{(n - n)!} = \frac{n!}{0!} = \frac{n!}{1} = n!$$

(23)

Example: Find the number of permutations of 5 things taken 3 at a time.

Solution: $P(n, r) = \dfrac{n!}{(n - r)!}$

$n = 5, \quad r = 3$

$$P(5, 3) = \frac{5!}{(5 - 3)!} = \frac{5!}{2!} = \frac{5 \cdot 4 \cdot 3 \cdot \not{2}!}{\not{2}!} = 60$$

(24)

Example: How many 4-letter words can be formed from the word "study"? (A "word" in this sense means any arrangement of letters.)

Solution: The problem is that of finding the permutation of 5 things taken 4 at a time.

$$P(5, 4) = \frac{5!}{(5-4)!} = \frac{5!}{1!} = 120$$

(25)

Study Exercise Four

1. Evaluate $P(8, 5)$.
2. Evaluate $P(4, 1)$.
3. How many words can be formed from the letters of the word "grades"
   (a) taken all at a time
   (b) taken 3 at a time?
4. Three people get on a bus with 10 vacant seats. In how many ways can they take their places?

(26)

Permutations of Things Not All Different

Up to now we have considered permutations of sets of objects that were different from each other. What would be the permutations of a set of n objects if some of the objects are alike?

(27)

Example: In how many ways can the letters of the word "error" be arranged?

Solution: If the 3 r's were different, there would be 5! permutations of 5 different objects.

Let x be the total number of permutations of the letters of "error". Let's consider any one of these arrangements; for example, rrreo.

If we replace the 3 r's by r_1, r_2, r_3, the original arrangement gives rise to 3! arrangements by permuting the 3 r's with subscripts. In the same way, each of the original x permutations give rise to 3! permutations. The total number of permutations is $x \cdot (3!)$.

$$\text{Thus, } x \cdot (3!) = 5!$$

$$x = \frac{5!}{3!}$$

$$x = \frac{5 \cdot 4 \cdot 3!}{3!}$$

$$x = 20$$

(28)

Theorem. Given a set of n objects having n_1 elements alike of one kind, and n_2 elements alike of another kind, and n_3 elements alike of a third kind, and so on for k kinds of objects; then the number of permutations of the n objects, taken all together, is;

$$\frac{n!}{n_1!\, n_2!\, n_3! \ldots n_k!}$$

(29)

Example: How many arrangements can be made of the letters of the word "Toronto"?

Solution: The total number of letters is 7. The letter *o* occurs 3 times and the letter *t* two times. Hence, the number of arrangements is;

$$\frac{7!}{3!\,2!} = \frac{7 \cdot 6 \cdot 5 \cdot \overset{2}{\cancel{4}} \cdot \cancel{3}!}{\cancel{3}!\,\cancel{2} \cdot 1}$$
$$= 420$$

(30)

Study Exercise Five

Find the number of permutations of the letters of each of the following words:

1. teeth **2.** mamma **3.** Tennessee

(31)

REVIEW EXERCISES

1. What is a permutation?
2. Evaluate $\frac{10!}{6!\,2!}$
3. Evaluate:
   (a) $P(5, 1)$ (b) $P(4, 4)$ (c) $P(8, 4)$
4. In how many ways can the letters of the following words be arranged?
   (a) tables (b) hippie
5. Given the digits, 2, 3, 4, 5, 6. How many two digit numbers can be made?
   (a) if no repetition is allowed
   (b) if repetition is allowed
   (c) if the number must be odd and no repetition is allowed?
6. Six persons are applicants for 3 different positions in a store, each person being qualified for each position. In how many ways is it possible to fill the positions?
7. If 7 horses run in the Kentucky Derby, in how many different orders can they finish? (exclude ties).

(32)

SOLUTIONS TO REVIEW EXERCISES

1. An arrangement of the elements in a set in a definite order.
2. $$\frac{10!}{6!\,2!} = \frac{10 \cdot 9 \cdot \overset{4}{\cancel{8}} \cdot 7 \cdot \cancel{6}!}{\cancel{6}!\,\cancel{2} \cdot 1} = 2520$$
3. (a) $P(5, 1) = \frac{5!}{(5-1)!} = \frac{5!}{4!} = 5$

   (b) $P(4, 4) = \frac{4!}{(4-4)!} = \frac{4!}{0!} = \frac{4 \cdot 3 \cdot 2 \cdot 1}{1} = 24$

   (c) $P(8, 4) = \frac{8!}{4!} = \frac{8 \cdot 7 \cdot 6 \cdot 5 \cdot \cancel{4!}}{\cancel{4!}} = 1680$
4. (a) $6!$ *or* 720 (b) $\frac{6!}{2!\,2!}$ *or* 180
5. (a) $5 \cdot 4$ *or* 20 (b) $5 \cdot 5$ *or* 25 (c) $4 \cdot 2$ *or* 8
6. $P(6, 3)$ *or* $6 \cdot 5 \cdot 4 = 120$
7. $7!$ *or* 5040

(33)

SOLUTIONS TO STUDY EXERCISES

Study Exercise One (Frame 11)

1. Each coin may fall in one of 2 ways. The two coins may fall in $2 \cdot 2$ or 4 ways
(heads (nickel), heads (dime)) (heads (nickel), tails (dime))
(tails (nickel), tails (dime)) (tails (nickel), heads (dime))

2. $5 \cdot 4 \cdot 3 \cdot 2 \cdot 1$ or 120

3. **(a)** $(24)(23)(22) = 12{,}144$
(b) $(24)(24)(24) = 13{,}824$

4. $9 \cdot 8 \cdot 7 \cdot 6 = 3{,}024$

5. $4 \cdot 3 \cdot 2 \cdot 1 = 24$

(11A)

Study Exercise Two (Frame 15)

1. The last place may be filled in any one of 2 ways.
$\underline{4} \cdot \underline{3} \cdot \underline{2} \cdot \underline{1} \cdot \underline{2} = 48$ ways.

2. The drivers seat may be filled in any one of three ways.
$3 \cdot 5 \cdot 4 \cdot 3 \cdot 2 \cdot 1 = 360$ ways.

3. Each die may fall in any one of 6 ways. $6 \cdot 6 \cdot 6 = 216$ ways.

(15A)

Study Exercise Three (Frame 20)

1. $P(7, 7) = 7!$
$= 5040$

2. $6!$ or 720

3. $\dfrac{8 \cdot 7 \cdot \cancel{6}! \cdot 3 \cdot 2 \cdot 1}{\cancel{6}!} = 336$

4. $\dfrac{\cancel{6} \cdot 5 \cdot \cancel{4}!}{\cancel{2} \cdot 1 \cdot \cancel{4}!} = 15$

(20A)

Study Exercise Four (Frame 26)

1.
$$P(8, 5) = \frac{8!}{(8-5)!} = \frac{8 \cdot 7 \cdot 6 \cdot 5 \cdot 4 \cdot \cancel{3}!}{\cancel{3}!} = 6{,}720$$

2.
$$P(4, 1) = \frac{4!}{3!} = 4$$

3. **(a)**
$$P(6, 6) = \frac{6!}{(6-6)!} = \frac{6!}{0!} = \frac{6 \cdot 5 \cdot 4 \cdot 3 \cdot 2 \cdot 1}{1} = 720$$

(b)
$$P(6, 3) = \frac{6!}{(6-3)!} = \frac{6 \cdot 5 \cdot 4 \cdot \cancel{3}!}{\cancel{3}!} = 120$$

SOLUTIONS TO STUDY EXERCISES, CONTD.

Study Exercise Four (Frame 26, contd.)

4. $P(10, 3) = \frac{10!}{(10-3)!} = \frac{10 \cdot 9 \cdot 8 \cdot \cancel{7}!}{\cancel{7}!} = 720$

26A

Study Exercise Five (Frame 31)

1. $\frac{5!}{2!\,2!} = \frac{5 \cdot \overset{2}{\cancel{4}} \cdot 3 \cdot \cancel{2}!}{\cancel{2}!\cancel{2}} = 30$

2. $\frac{5!}{3!\,2!} = \frac{5 \cdot \overset{2}{\cancel{4}} \cdot \cancel{3}!}{\cancel{3}!\,\cancel{2}\ 1} = 10$

3. $\frac{9!}{4!\,2!\,2!} = \frac{9 \cdot \overset{4}{\cancel{8}} \cdot 7 \cdot \overset{3}{\cancel{6}} \cdot 5 \cdot \cancel{4}!}{\cancel{4}!\,\cancel{2} \cdot 1 \cdot \cancel{2} \cdot 1} = 3{,}780$

31A

UNIT 27—SUPPLEMENTARY PROBLEMS

A. Evaluate the following.

1. $P(12, 3)$ **2.** $P(8, 8)$ **3.** $P(k, 2)$ where $k \geq 2$

B. **4.** In how many ways may 5 books be arranged on a shelf?

5. Find the number of permutations of the letters in the word "gargling".

6. How many different committees consisting of one Democrat and one Republican can be formed from seven Democrats and five Republicans?

7. If there are 11 milers entered in a race, in how many ways can first, second, and third place be awarded?

8. How many numbers of three digits each can be formed from the digits 1, 2, 3, 4, 5, 6, 7;

(a) if repetition is not allowed **(b)** if repetition is allowed?

9. A person is to mail a letter, pay a bill, and buy an article. In how many orders may these things be done?

10. How many words can be formed with all the letters of "Cincinnati"?

11. There are 5 different colored flags. How many different signals are possible using 3 flags at a time, one below the other?

12. How many telephone exchange symbols, each consisting of two letters, can be formed from the letters of the English alphabet?

13. In how many ways can six men stand in a row if two of them always stand together?

14. Find n if $P(n, 4) = 12P(n, 2)$.

15. How many different four-digit numbers can be formed with the digits 0, 1, 2, 3 if the numbers must be divisible by 5?

16. A chemist has 8 test tubes to examine. In how many orders can he do this?

17. How many integers greater than 300,000 may be formed from the digits 1, 2, 3, 4, 5, 9, no digit being repeated?

18. A man tries to choose the winner of each of 12 football games. Excluding ties, how many different predictions are possible?

unit

28

Combinations

Objectives

1. Know the formula for combinations and be able to work problems involving combinations.

2. Be able to expand $(a + b)^n$ by the binomial theorem using combinatorial notation.

Terms

combination

①

In order to see the difference between permutations and combinations, we will consider an example.

Example: In how many ways can 3 books be selected from 4 books without regard to order.

Solution: We know that $P(4, 3) = \frac{4!}{(4 - 3)!}$ *or* 24. But, in these permutations, order counts. Let us assume the 4 books are *A*, *B*, *C*, and *D*. To make a selection of 3 of the 4 books, we may choose *ABC*, *ABD*, *ACD*, *BCD*. We do not list *CBA* because the selection *ABC* is the same as *CBA* since order does not count.

②

Combinations

Definition. A *combination* is a selection of objects considered without regard to their order.

The difference between a combination and a permutation is that in a permutation, order counts and in a combination, order does not count.

③

Ordinarily, we must decide from the nature of the problem whether permutations or combinations are involved.

The whole thing hinges on whether order does or does not count.

④

The number of combinations of n objects taken all together will be denoted by the symbol $\binom{n}{n}$. This symbol has no fraction bar in the middle; it is not a fraction.

The number of combinations of n objects taken r at a time will be denoted by $\binom{n}{r}$.

⑤

Let us consider the set $\{a, b, c\}$. There are six permutations of the six letters, namely *abc*, *acb*, *bac*, *bca*, *cab*, *cba*.

There is only one combination of the three letters *abc* since a different arrangement of the letters is still the same combination. Therefore, $\binom{3}{3} = 1$

(6)

Let us consider the set $\{a, b, c\}$ and write the permutations and combinations taken two at a time.

combinations	*permutations*
ab	*ab*, *ba*
ac	*ac*, *ca*
bc	*bc*, *cb*

Each combination will yield 2! permutations.
(Number of combinations) · 2! = number of permutations.

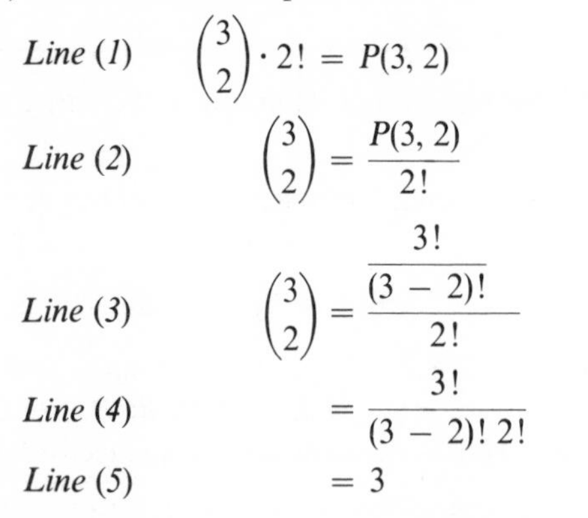

Line (1) $\binom{3}{2} \cdot 2! = P(3, 2)$

Line (2) $\binom{3}{2} = \frac{P(3, 2)}{2!}$

Line (3) $\binom{3}{2} = \frac{\frac{3!}{(3-2)!}}{2!}$

Line (4) $= \frac{3!}{(3-2)!\,2!}$

Line (5) $= 3$

(7)

Theorem. The number of combinations of a set of n different objects taken r at a time is

$$\binom{n}{r} = \frac{n!}{r!(n-r)!};\ r \leq n,\ r \text{ and } n \text{ are non-negative integers}$$

Proof. Each combination of r objects can be arranged in $r!$ permutations.

Line (1) $\binom{n}{r} \cdot r! = P(n, r)$

Line (2) $\binom{n}{r} \cdot r! = \frac{n!}{(n-r)!}$

Line (3) $\binom{n}{r} = \frac{n!}{r!(n-r)!}$

(8)

Example: Evaluate $\binom{8}{3}$.

Solution:
$$\binom{8}{3} = \frac{8!}{3!(8-3)!}$$
$$= \frac{8!}{3!5!}$$
$$= \frac{8 \cdot 7 \cdot \not{6} \cdot \not{5}!}{\not{3} \cdot \not{2} \cdot 1 \cdot \not{5}!}$$
$$= 56$$

(9)

Study Exercise One

Evaluate:

1. $\binom{4}{3}$ 2. $\binom{6}{6}$ 3. $\binom{9}{3}$ 4. $\binom{9}{6}$

(10)

Example: How many different committees of 4 could be appointed in a club containing 10 members?

Solution: Order is unimportant. We need to find $\binom{10}{4}$.

$$\binom{10}{4} = \frac{10!}{4!6!}$$
$$= \frac{10 \cdot \overset{3}{\not{9}} \cdot \not{8} \cdot 7 \cdot \not{6}!}{\not{4} \cdot \not{3} \cdot \not{2} \cdot 1 \cdot \not{6}!}$$
$$= 210$$

(11)

Example: In how many ways can a hand of 13 cards be selected from a standard deck of 52 cards?

Solution: We need to find the number of combinations of 52 things taken 13 at a time.

$$\binom{52}{13} = \frac{52!}{13!39!}$$
$$= \frac{52 \cdot 51 \cdot 50 \cdot 49 \cdot 48 \cdot 47 \cdot 46 \cdot 45 \cdot 44 \cdot 43 \cdot 42 \cdot 41 \cdot 40 \cdot 39!}{13 \cdot 12 \cdot 11 \cdot 10 \cdot 9 \cdot 8 \cdot 7 \cdot 6 \cdot 5 \cdot 4 \cdot 3 \cdot 2 \cdot 1 \cdot 39!}$$
$$= 635,013,559,600$$

(12)

Study Exercise Two

1. In how many ways may a committee of 5 be chosen from 8 people?
2. In how many ways can 2 white and 5 red marbles be drawn from a jar containing 5 white and 6 red marbles?
3. A track coach has a group of 7 men from which to pick a squad of 3 men to enter an event. How many different squads can he select?
4. In how many ways can a pile of 48 cards be dealt from a deck of 52 cards?

(13)

Example: How many committees of 3 boys and 2 girls can be formed from 8 boys and 10 girls?

Solution: The 3 boys can be selected from 8 boys in $\binom{8}{3}$ ways. The 2 girls can be selected from 10 girls in $\binom{10}{2}$ ways. Hence the number of combinations is;

$$\binom{8}{3}\cdot\binom{10}{2} = \frac{8!}{3!5!}\cdot\frac{10!}{2!8!}$$

$$= \frac{8\cdot 7\cdot \not{6}\cdot \not{5}!}{\not{3}\cdot\not{2}\cdot 1\cdot\not{5}!}\;\frac{\overset{5}{\not{10}}\cdot 9\cdot\not{8}!}{\not{2}\cdot 1\cdot\not{8}!}$$

$$= 2{,}520$$

(14)

Example: From a set of 12 different books, how many selections of 5 books can be made if book A is to be included in every selection?

Solution: Book A can be selected in 1 way (by selecting it; $\binom{1}{1} = 1$). The other 4 books can be selected from the remaining 11 books in $\binom{11}{4}$. The total number of selections is;

$$\binom{1}{1}\binom{11}{4} = 1\cdot\frac{11\cdot 10\cdot \overset{3}{\not{9}}\cdot\not{8}\cdot\not{7}!}{\not{4}\cdot\not{3}\cdot\not{2}\cdot 1\cdot\not{7}!}$$

$$= 330$$

(15)

Study Exercise Three

1. How many committees of 2 boys and 1 girl can be selected from 4 boys and 3 girls?
2. In how many ways can a committee of 4 be formed from 8 students if student A (one of the students) must be a member?
3. In a deck of 52 cards, how many hands are there of 3 aces and 2 cards not aces?
4. How many hands of 5 cards are possible with a set of 52 different cards?

(16)

Let us consider $\binom{n+1}{r}$, $\binom{n}{r-1}$, and $\binom{n}{r}$.

Line (1) $\binom{n+1}{r} = \frac{(n+1)!}{r!(n+1-r)!}$

Line (2) $\binom{n}{r-1} = \frac{n!}{(r-1)!\,(n-r+1)!}$

Line (3) $\binom{n}{r} = \frac{n!}{r!(n-r)!}$

Line (4) $\frac{n!}{(r-1)!\,(n-r+1)!} + \frac{n!}{r!(n-r)!} = \frac{n!}{(r-1)!\,(n-r+1)\,(n-r)!} + \frac{n!}{r(r-1)!\,(n-r)!}$

(Frame 17, contd.)

Line (5) $$= \frac{n!\cdot r}{(r-1)!\,(n-r+1)\,(n-r)!\cdot r} + \frac{n!(n-r+1)}{r(r-1)!\,(n-r)!\,(n-r+1)}$$

Line (6) $$= \frac{n!\cdot r + n!\,(n-r+1)}{(r-1)!\,(n-r+1)\,(n-r)!\cdot r}$$

Line (7) $$= \frac{n!(r+n-r+1)}{(r-1)!\,(n-r+1)!r}$$

Line (8) $$= \frac{(n+1)!}{r!(n-r+1)!}$$

Line (9) Thus $\binom{n+1}{r} = \binom{n}{r-1} + \binom{n}{r}$.

(17)

$$\binom{n}{r} = \frac{n!}{r!(n-r)!}$$

Let us build a table for various values of n and r.

n \ r	**0**	**1**	**2**	**3**	**4**
0	1				
1	1	1			
2	1	2	1		
3	1	3	3	1	
4	1	4	6	4	1

The theorem in the previous frame stated that; $\binom{n+1}{r} = \binom{n}{r-1} + \binom{n}{r}$

$$\text{Thus } \binom{4}{2} = \binom{3}{1} + \binom{3}{2}.$$

Any entry in the table equals the sum of the entry above it and the number to the left of the entry above. Do you recognize the numbers in this table?

(18)

The entries in the previous table are from *Pascal's triangle.* This means that the binomial coefficients can be formed by combinations.

$$(x+y)^3 = x^3 + 3x^2y + 3xy^2 + y^3$$
$$= \binom{3}{0}x^3 + \binom{3}{1}x^2y + \binom{3}{2}xy^3 + \binom{3}{3}y^3$$

(19)

$$(x+y)^4 = \binom{4}{0}x^4 + \binom{4}{1}x^3y + \binom{4}{2}x^2y^2 + \binom{4}{3}xy^3 + \binom{4}{4}y^4$$

$$(x+y)^n = \binom{n}{0}x^n + \binom{n}{1}x^{n-1}y + \binom{n}{2}x^{n-2}y^2 + \ldots + \binom{n}{n}y^n$$

(20)

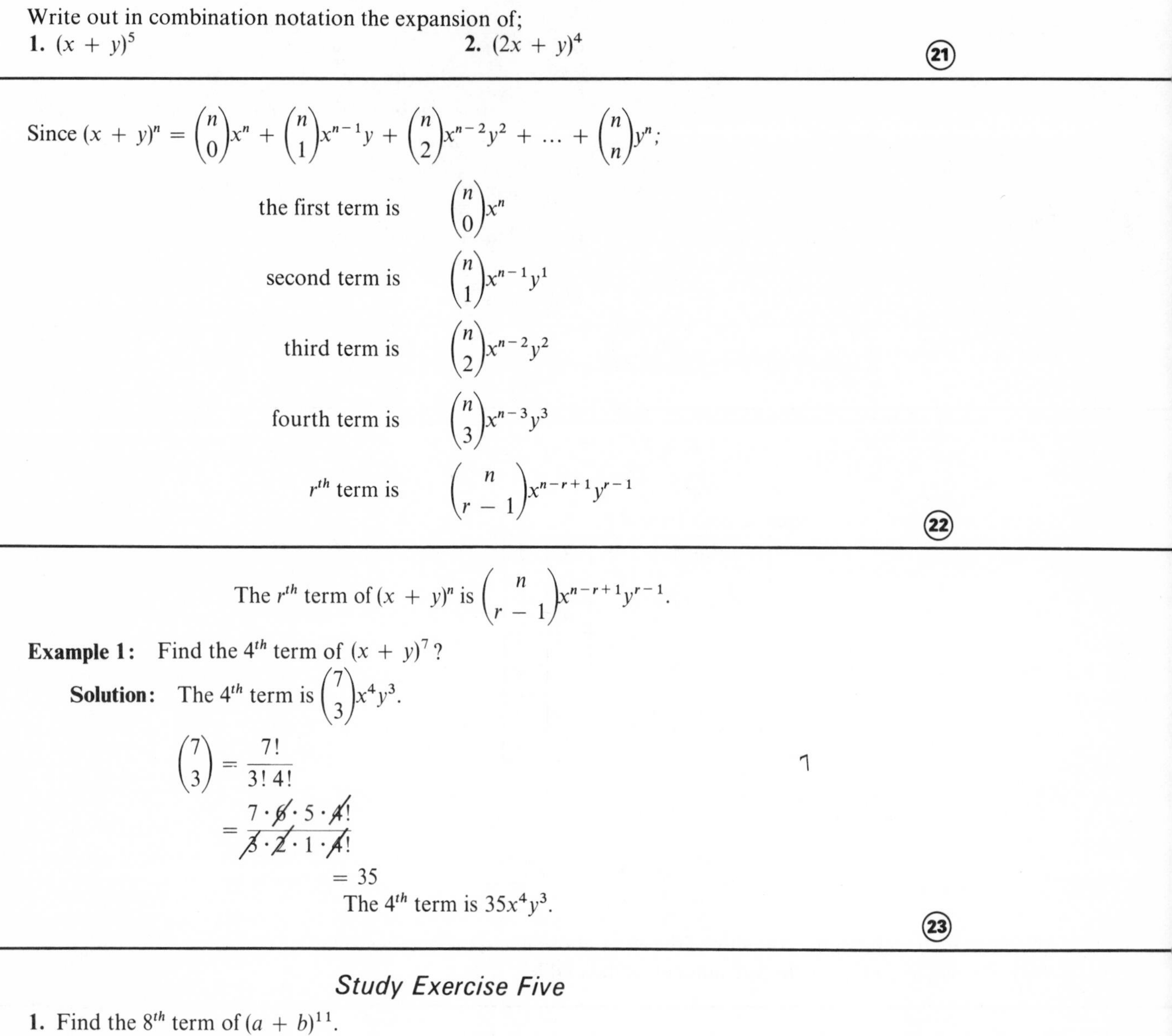

Study Exercise Four

Write out in combination notation the expansion of;

1. $(x + y)^5$ **2.** $(2x + y)^4$

(21)

Since $(x + y)^n = \binom{n}{0}x^n + \binom{n}{1}x^{n-1}y + \binom{n}{2}x^{n-2}y^2 + \ldots + \binom{n}{n}y^n$;

the first term is $\binom{n}{0}x^n$

second term is $\binom{n}{1}x^{n-1}y^1$

third term is $\binom{n}{2}x^{n-2}y^2$

fourth term is $\binom{n}{3}x^{n-3}y^3$

r^{th} term is $\binom{n}{r-1}x^{n-r+1}y^{r-1}$

(22)

The rth term of $(x + y)^n$ is $\binom{n}{r-1}x^{n-r+1}y^{r-1}$.

Example 1: Find the 4th term of $(x + y)^7$?

Solution: The 4th term is $\binom{7}{3}x^4y^3$.

$$\binom{7}{3} = \frac{7!}{3!\,4!}$$

$$= \frac{7 \cdot \not{6} \cdot 5 \cdot \not{4}!}{\not{3} \cdot \not{2} \cdot 1 \cdot \not{4}!}$$

$$= 35$$

The 4th term is $35x^4y^3$.

(23)

Study Exercise Five

1. Find the 8th term of $(a + b)^{11}$.
2. Find the 6th term of $(2x + 1)^8$.
3. Write the first three terms of $(a + b)^{12}$.

(24)

REVIEW EXERCISES

1. What is a combination?
2. Find $\binom{6}{6}$
3. Find $\binom{7}{2}$
4. In how many ways can a battery (pitcher and catcher) be formed if a team has 4 pitchers and 2 catchers?
5. Twenty politicians meet at a party. How many handshakes are exchanged if each person shakes hands with each other person once and only once?
6. Find the 4th term of $(2x + 1)^6$.

(25)

SOLUTIONS TO REVIEW EXERCISES

1. An arrangement of objects without regard to order.

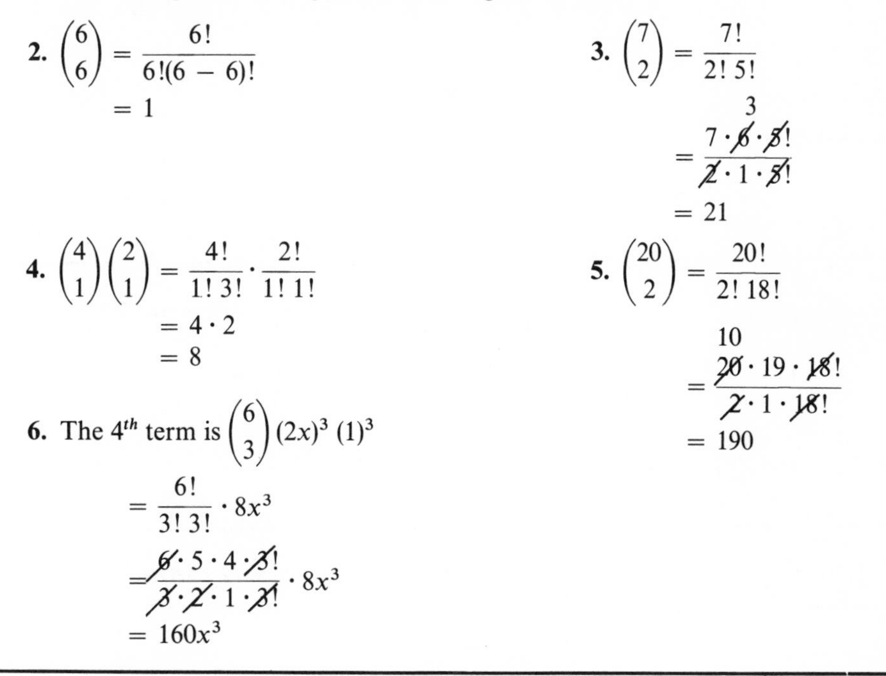

2. $\binom{6}{6} = \dfrac{6!}{6!(6-6)!}$
$= 1$

3. $\binom{7}{2} = \dfrac{7!}{2!\,5!}$
$= \dfrac{7 \cdot \overset{3}{\cancel{6}} \cdot \cancel{5}!}{\cancel{2} \cdot 1 \cdot \cancel{5}!}$
$= 21$

4. $\binom{4}{1}\binom{2}{1} = \dfrac{4!}{1!\,3!} \cdot \dfrac{2!}{1!\,1!}$
$= 4 \cdot 2$
$= 8$

5. $\binom{20}{2} = \dfrac{20!}{2!\,18!}$
$= \dfrac{\overset{10}{\cancel{20}} \cdot 19 \cdot \cancel{18}!}{\cancel{2} \cdot 1 \cdot \cancel{18}!}$
$= 190$

6. The 4th term is $\binom{6}{3}(2x)^3\,(1)^3$
$= \dfrac{6!}{3!\,3!} \cdot 8x^3$
$= \dfrac{\cancel{6} \cdot 5 \cdot 4 \cdot \cancel{3}!}{\cancel{3} \cdot \cancel{2} \cdot 1 \cdot \cancel{3}!} \cdot 8x^3$
$= 160x^3$

(26)

SOLUTIONS TO STUDY EXERCISES

Study Exercise One (Frame 10)

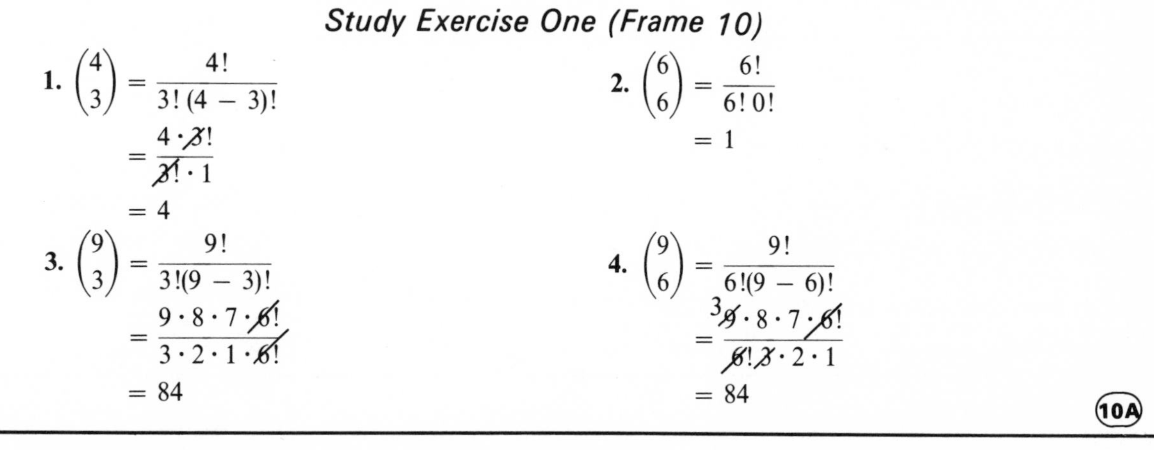

1. $\binom{4}{3} = \dfrac{4!}{3!\,(4-3)!}$
$= \dfrac{4 \cdot \cancel{3}!}{\cancel{3}! \cdot 1}$
$= 4$

2. $\binom{6}{6} = \dfrac{6!}{6!\,0!}$
$= 1$

3. $\binom{9}{3} = \dfrac{9!}{3!(9-3)!}$
$= \dfrac{9 \cdot 8 \cdot 7 \cdot \cancel{6}!}{3 \cdot 2 \cdot 1 \cdot \cancel{6}!}$
$= 84$

4. $\binom{9}{6} = \dfrac{9!}{6!(9-6)!}$
$= \dfrac{\overset{3}{\cancel{9}} \cdot 8 \cdot 7 \cdot \cancel{6}!}{\cancel{6}!\,\cancel{3} \cdot 2 \cdot 1}$
$= 84$

(10A)

Study Exercise Two (Frame 13)

1. $\binom{8}{5} = \frac{8!}{5!\,3!}$
$= \frac{8 \cdot 7 \cdot 6 \cdot \cancel{5!}}{\cancel{5!} \cdot \cancel{3} \cdot \cancel{2} \cdot 1}$
$= 56$

2. $\binom{5}{2}\binom{6}{5} = \frac{5!}{2!\,3!}\ \frac{6!}{5!\,1!}$
$= 10 \cdot 6$
$= 60$

3. $\binom{7}{3} = \frac{7!}{3!\,4!}$
$= 35$

4. $\binom{52}{48} = \frac{52!}{48!\,4!}$
$= 270{,}725$

13A

Study Exercise Three (Frame 16)

1. $\binom{4}{2}\binom{3}{1} = \frac{4!}{2!\,2!} \cdot \frac{3!}{1!\,2!}$
$= 6 \cdot 3$
$= 18$

2. $\binom{1}{1}\binom{7}{3} = \frac{1!}{1!\,0!} \cdot \frac{7!}{3!\,4!}$
$= 1 \cdot 35$
$= 35$

3. $\binom{4}{3}\binom{48}{2} = \frac{4!}{3!\,1!} \cdot \frac{48!}{2!\,46!}$
$= 4 \cdot 1128$
$= 4{,}512$

4. $\binom{52}{5} = \frac{52!}{5!\,47!}$
$= 2{,}598{,}960$

16A

Study Exercise Four (Frame 21)

1. $\binom{5}{0}x^5 + \binom{5}{1}x^4y + \binom{5}{2}x^3y^2 + \binom{5}{3}x^2y^3 + \binom{5}{4}xy^4 + \binom{5}{5}y^5$

2. $\binom{4}{0}(2x)^4 + \binom{4}{1}(2x)^3y + \binom{4}{2}(2x)^2y^2 + \binom{4}{3}(2x)^1y^3 + \binom{4}{4}y^4$

21A

Study Exercise Five (Frame 24)

1. The 8th term is $\binom{11}{7}(a)^4(b)^7$

$\frac{11!}{7!\,4!}a^4b^7 = 330a^4b^7$

2. The 6th term is $\binom{8}{5}(2x)^3(1)^5$

$\frac{8!}{5!\,3!} \cdot 8x^3 \cdot 1 = 56 \cdot 8x^3$
$= 448x^3$

3. $\binom{12}{0}a^{12} + \binom{12}{1}a^{11}b + \binom{12}{2}a^{10}b^2 + \ldots$
$= \frac{12!}{0!\,12!}a^{12} + \frac{12!}{1!\,11!}a^{11}b + \frac{12!}{2!\,10!}a^{10}b^2 + \ldots$
$= a^{12} + 12a^{11}b + 66a^{10}b^2 + \ldots$

24A

UNIT 28—SUPPLEMENTARY PROBLEMS

1. Evaluate:

(a) $\binom{8}{2}$ **(b)** $\binom{8}{6}$ **(c)** $\binom{12}{2}$

2. From a group of 6 phonograph records, a boy has enough money to buy 3 of them. How many different ways can he make a choice?

3. From a group of 8 students, how many different committees of 4 can be formed?

4. How many 2 card hands are possible from a deck of 52 cards?

5. A student is instructed to answer any 8 out of 10 questions. How many different ways are there for him to choose the questions he answers?

6. How many committees of 2 Democrats and 2 Republicans may be formed from a set of 7 Democrats and 6 Republicans?

7. From a deck of 52 cards, how many possible hands of 5 cards are there which would contain exactly 1 ace?

8. Find the number of poker hands (5 cards) that consist of 2 hearts and 3 black cards.

9. **(a)** How many 3 letter combinations can be made from the word "integral"?
(b) How many of these include the letter "*g*"?

10. In how many ways may a man, woman, and child be selected from 7 men, 5 women and 9 children?

11. From 3 Republicans and 3 Democrats, find the number of committees of 3 which can be formed
(a) with 3 Republicans and no Democrats
(b) with 2 Republicans and 1 Democrat
(c) with 1 Republican and 2 Democrats?

12. There are 10 points on a plane, no three of them on the same straight line. How many lines are determined by the points?

13. Find n if $\binom{n}{7} = \binom{n}{5}$

14. Find n if $\binom{n}{3} = 84$

15. Find the 7th term of $(a + b)^{15}$

16. Find the 5th term of $(1/2 + x)^{10}$

17. Find the middle term of $(1/x - x^2)^{12}$

unit

Introduction to Probability

29

Objectives

1. To know the meaning of probability and be able to work problems dealing with probability.

Terms

sample space	sample point	event
probability	odds	independent events

①

Definition. A *sample space* of an experiment is the set of all possible outcomes of the experiment. An element in a sample space is called a *sample point*. Any subset of the sample space is called an *event*.

②

Example 1: Consider the experiment of tossing two coins, a penny and dime. List the possible outcomes or sample space of this experiment.

Solution: We will let H stand for the coin coming up heads and T for the coin coming up tails. Then the sample space is $\{HH, HT, TH, TT\}$ where the first letter in the pair designates the outcome for the penny and the second the outcome for the dime.

③

Study Exercise One

1. Two letters are chosen randomly, one after another, from the word cat. List a sample space.
2. Three coins are tossed, a penny, dime and quarter. List the possible outcomes.

④

Probability

Definition. If an experiment can result in any one of n different, equally likely outcomes and if exactly k of these outcomes correspond to an event A, then the *probability* of event A is given by:

$$P(A) = \frac{k}{n}$$

⑤

Consider a well shuffled deck of cards. When we draw a card from the deck, the probability of getting the ace of diamonds is the number of favorable outcomes divided by the number of possible outcomes. In this case, the probability is 1/52.

⑥

If there are no favorable outcomes in the set of possible outcomes, then the probability of a favorable outcome is zero. If all possible outcomes are favorable, then the probability of a favorable outcome is one. It then follows;

$$0 \leq \text{probability of favorable outcome} \leq 1$$

(7)

Example: Four faces of an ordinary die are painted green and the other two faces are painted blue. If the die is rolled once, what is the probability the top face is **(a)** green; **(b)** blue?

Solution: **(a)** Since the six faces are equally likely, the probability that the top face is green is given by;

$$\frac{\text{number of favorable outcomes}}{\text{number of possible outcomes}} = 4/6$$
$$= 2/3$$

(b) Similarly, the probability of blue $= 2/6$
$= 1/3$

(8)

Study Exercise Two

1. If a die is thrown, what is the probability the upper face shows 3?
2. When 2 coins are tossed, what is the probability that both show heads?
3. One card is drawn at random from a well-shuffled deck. Find the probability that it is; **(a)** an ace; **(b)** a black card.

(9)

If there are n different equally likely outcomes, and if exactly k of these correspond to event A, then $P(A) = \frac{k}{n}$.

If we desire the probability of not A, the number of favorable cases is $n - k$ and $P(\text{not } A) = \frac{n-k}{n}$

$$= 1 - \frac{k}{n}$$

$$\text{Thus, } P(\text{not } A) = 1 - P(A)$$

(10)

Example: A bag contains 6 white and 3 black marbles. If 2 marbles are drawn at random, find the probability that they are both white.

Solution: The sample space consists of all possible pairs of marbles. This number is $\binom{9}{2}$. The event "both are white" contains all possible pairs of marbles given 6 white marbles. This number is $\binom{6}{2}$

(Frame 11, contd.)

$$P\text{ (both are white)} = \frac{\binom{6}{2}}{\binom{9}{2}}$$

$$= \frac{\frac{6!}{2!\,4!}}{\frac{9!}{2!\,7!}}$$

$$= \frac{\frac{6\cdot 5\cdot \cancel{4!}}{2\cdot 1\cdot \cancel{4!}}}{\frac{9\cdot 8\cdot \cancel{7!}}{2\cdot 1\cdot \cancel{7!}}}$$

$$= 15/36 \quad \text{or} \quad 5/12$$

(11)

Consider the sample space E with sample points $e_1, e_2, e_3, \ldots e_n$.

$$E = \{e_1, e_2, e_3, \ldots e_n\}$$

Then $P(E) = P(e_1) + P(e_2) + P(e_3) + \ldots + P(e_n)$
$= 1$

(12)

Study Exercise Three

A jar contains 6 white, 4 black, and 2 red marbles. If a marble is drawn at random, find the probability of the following events.

1. It is white **2.** It is black **3.** It is red
4. It is not red **5.** It is not white

(13)

Sometimes the relative chances of an event are expressed in terms of the odds in favor of the event.

$$\text{Odds in favor of event } A = \frac{P\text{ (event } A)}{P\text{ (not } A)}$$

(14)

Example: A card is drawn at random from a deck. Find the probability and odds that the card is an ace.

Solution: $P\text{ (ace)} = \dfrac{\text{number of favorable outcomes}}{\text{total possible outcomes}}$

$= 4/52$ or $1/13$

$P\text{ (not an ace)} = 1 - \dfrac{1}{13}$ or $\dfrac{12}{13}$

Odds in favor of an ace $= \dfrac{\frac{1}{13}}{\frac{12}{13}}$ or $1/12$.

Odds in favor of an ace are 1 to 12.

(15)

Example: A committee of 4 is to be chosen by lot from a group of 5 men and 7 women. Find the probability that the committee will consist of 2 men and 2 women.

Solution: A committee of 4 can be selected from 12 persons in $\binom{12}{4}$ ways. The 2 men can be selected in $\binom{5}{2}$ ways and the 2 women in $\binom{7}{2}$ ways all equally likely. Thus,

$$P\,(2 \text{ men and } 2 \text{ women}) = \frac{\binom{5}{2}\binom{7}{2}}{\binom{12}{4}} = \frac{\frac{5!}{2!\,3!}\cdot\frac{7!}{2!\,5!}}{\frac{12!}{4!\,8!}} = \frac{10\cdot 21}{495} = \frac{210}{495} = \frac{14}{33}$$

(16)

Study Exercise Four

1. Out of a group of 10 girls, three have blue eyes. If two girls are selected at random, what is the probability and odds both have blue eyes?
2. What is the probability that 2 cards drawn at random from a deck will both be aces?

(17)

Independent Events

When we say in everyday language that two events have nothing to do with each other, we are describing independent events. If the occurence of one event has no effect on the occurence of another, the events are called independent.

For example, successive coin flips produce independent results. Whether a coin comes up heads or tails does not depend on what came up on previous throws.

(18)

For independent events, the probability of the occurence of two or more events is the product of their individual probabilities.

Thus, if A and B represent independent events;

$$P(A \cap B) = P(A) \cdot P(B)$$

(19)

On the other hand, consider the experiment of drawing two marbles from a jar containing 3 white and 5 red marbles. If the first marble is not replaced in the jar prior to drawing the second marble, the sample space has been reduced by one and thus the probability the second marble being a particular color has been altered from that of the first.

(20)

Example: A jar contains 4 white and 6 red marbles. If the first marble is not replaced prior to drawing the second, find the probability the second is white given that the first was white.

Solution: P (1st marble is white) $= \dfrac{\text{number of favorable cases}}{\text{total number of cases}}$

$= 4/10$ or $2/5$

P (2nd marble is white) $= 3/9$ or $1/3$

P (1st marble and 2nd marble is white) $= 2/5 \cdot 1/3$

$= 2/15$

(21)

We consider the previous example of the jar with 4 white and 6 red marbles.

Let: A_1 be the event the first marble is white.
B_1 be the event the first marble is red.
A_2 be the event the second marble is white.
B_2 be the event the second marble is red.

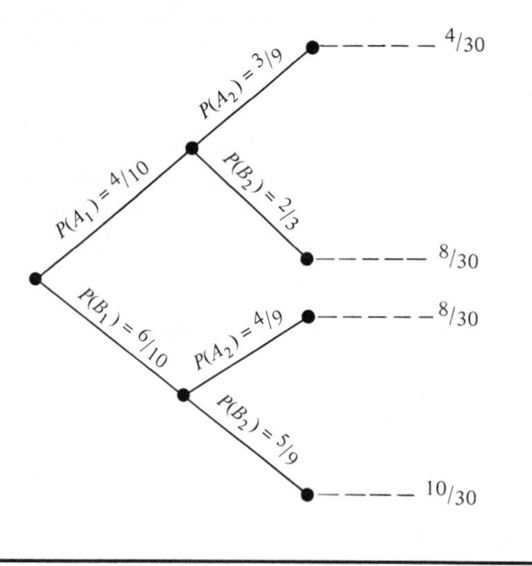

(22)

REVIEW EXERCISES

1. What is the probability that a card drawn from a set of 50 cards numbered 1 to 50 will display a number divisible by 5?
2. What is the probability that a playing card drawn from a deck of cards will be a king?
3. If 2 coins are tossed, find the probability that one shows heads and the other tails.
4. If a die is thrown, find the probability the upper face shows 6.
5. A bag contains 4 red balls and 6 green balls.
   (a) if 4 balls are drawn, find the probability that all will be red.
   (b) if 4 balls are drawn, find the probability that 2 will be red and 2 will be green.
6. What are the odds that a coin will fall heads?
7. Find the probability of obtaining an ace on both the first and second draws from a deck of cards when the first card is not replaced before the second is drawn.

(23)

SOLUTIONS TO REVIEW EXERCISES

1. There are 10 numbers between 1 and 50 inclusive that are divisible by 5; hence, the required probability is 10/50 or 1/5.
2. 4/52 or 1/13
3. There are 4 possible ways two coins may fall; the required probability is 2/4 or 1/2.
4. There are 6 possible ways a die may fall; the required probability is 1/6.
5. **(a)** $$\frac{\binom{4}{4}}{\binom{10}{4}} = \frac{\frac{4!}{4!\,0!}}{\frac{10!}{4!\,6!}} = \frac{1}{210}$$
   (b) $$\frac{\binom{4}{2}\binom{6}{2}}{\binom{10}{4}} = \frac{\frac{4!}{2!\,2!}\cdot\frac{6!}{2!\,4!}}{\frac{10!}{4!\,6!}} = \frac{6 \cdot 15}{210} = \frac{90}{210} \text{ or } \frac{3}{7}$$
6. P (heads) $= 1/2$, P (not heads) $= 1 - 1/2$ or $1/2$

   Odds in favor of heads $= \dfrac{\frac{1}{2}}{\frac{1}{2}}$ or 1 to 1.
7. $$\frac{4}{52} \cdot \frac{3}{51} = \frac{1}{221}$$

(24)

SOLUTIONS TO STUDY EXERCISES

Study Exercise One (Frame 4)

1. $\{ca, ct, at, ac, ta, tc\}$
2. $\{HHH, HHT, HTH, THH, HTT, THT, TTH, TTT\}$ where the first letter in each triple designates the outcome for the penny, the second letter the outcome for the dime, and the third letter the outcome for the quarter.

(4A)

Study Exercise Two (Frame 9)

1. 1/6
2. 1/4
3. **(a)** 4/52 or 1/13
   (b) 26/52 or 1/2

(9A)

SOLUTIONS TO STUDY EXERCISES, CONTD.

Study Exercise Three (Frame 13)

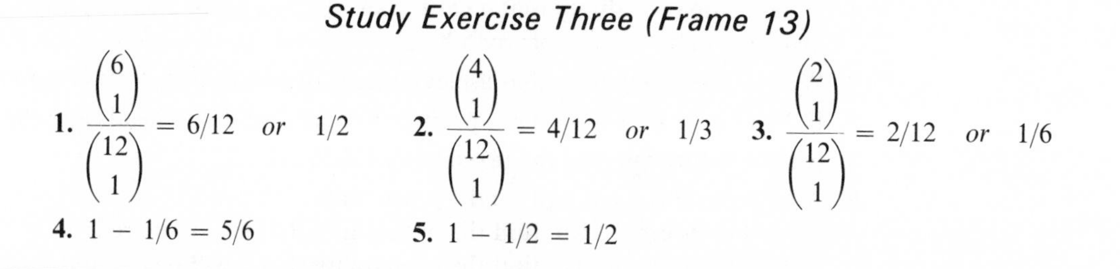

1. $\dfrac{\binom{6}{1}}{\binom{12}{1}} = 6/12$ *or* $1/2$ **2.** $\dfrac{\binom{4}{1}}{\binom{12}{1}} = 4/12$ *or* $1/3$ **3.** $\dfrac{\binom{2}{1}}{\binom{12}{1}} = 2/12$ *or* $1/6$

4. $1 - 1/6 = 5/6$ **5.** $1 - 1/2 = 1/2$

13A

Study Exercise Four (Frame 17)

1. $\dfrac{\binom{3}{2}}{\binom{10}{2}} = \dfrac{\frac{3!}{2!\,1!}}{\frac{10!}{2!\,8!}}$

$= 3/45$ or $1/15$

P (not selecting two with blue eyes) $= 1 - 1/15$ or $14/15$

The odds in favor are $\dfrac{1/15}{14/15}$ or 1 to 14.

2. $\dfrac{\binom{4}{2}}{\binom{52}{2}} = \dfrac{6}{1326}$ or $\dfrac{1}{221}$

17A

UNIT 29—SUPPLEMENTARY PROBLEMS.

1. What is the probability of drawing a face card (king, queen, or jack) from a deck of cards?
2. A box contains 6 white balls and 4 black balls. Three balls are taken at random. Find the probability that they are all white.
3. A number between 1 and 30 inclusive is selected at random. Find the probability it is prime.
4. A box contains 10 white balls and 5 red balls. If 4 balls are drawn at random, what is the probability 2 are white and 2 are red.
5. A jar contains 6 white and 4 black marbles. If 2 marbles are drawn in succession what is the probability that both are black;
   (a) if the first one is replaced before the second one is drawn
   (b) if the first one is not replaced before the second one is drawn?
6. What is the probability of throwing 3 sixes with 3 throws of a single die?
7. Find the probability that both coins will not come up heads when two coins are tossed.
8. A jar contains 6 white, 5 black, and 4 red marbles. What is the probability of first drawing a white marble and then drawing a red one if the first marble is not replaced?
9. What is the probability that 2 dice on one throw
   (a) will show a sum of 2
   (b) will show a sum of 11
   (c) will show a sum of 7?
10. What is the probability of randomly selecting a red apple and a green apple from a box containing 6 red apples and 4 green apples?

VERBAL PROBLEMS

Directions: Use either one or two unknowns.

1. The sum of 3 consecutive integers is 105. Find the smallest of these integers.
2. There are 1260 students at a certain junior college. The ratio of boys to girls is 6 to 4. How many girls are there?
3. A jar of pennies and nickels contains 110 coins worth \$2.50. How many nickels are in the jar?
4. Find, if possible, two positive integers such that one is 5 more than the other and their product is 84.
5. A rectangular garden is surrounded by a walk 5 feet wide. The area of the garden is 800 square feet and the area of the walk is 700 square feet. What are the dimensions of the garden?
6. Six years ago, a man was four times as old as his son; in ten more years he will be twice as old. Find their ages now.
7. An open box with a square base and a volume of 675 cubic inches is to be constructed from a square sheet of cardboard by cutting 3-inch squares from the corners and bending up the sides. What size of cardboard should be used?
8. Separate 24 into two parts whose product will be a maximum. (Hint: Set up a quadratic function and find the maximum.)
9. The sum of the reciprocals of two numbers is 7/12. The product of the numbers is 24. Find the numbers.
10. Find three positive integers such that their sum is 44, the second is twice the first, and the third is twice the square of the first.
11. Find the dimensions of a rectangle if its perimeter is 64 feet and its area is 252 square feet.
12. Joe has twice as much cash as Bill. If he lent Bill a dollar, they would have the same amount. How much does Bill have?
13. If either 25 pounds flour and 10 pounds sugar or 16 pounds flour and 16 pounds sugar can be purchased for \$3.20, find the price per pound of each.
14. The sum of two positive numbers is 24. Their squares differ by 48. What are the numbers?
15. A grocer sold oranges at \$1 a bag and raised the price per dozen by 10¢ by reducing the number of oranges in the bag by 4. Find the original number of oranges in the bag.

PART II—Short Answer

12. $\log_{10} 1$

13. $2/i$ where $i = \sqrt{-1}$

14. $5!$

15. $(\sqrt{64})^{1/3}$

16. -5^2

17. $4 + 6 \div 2$

18. $\sqrt{0}$

19. $|4| + |-2|$

20. $(2^{-1} + 2^{-2})^{-1}$

21. The y intercept of $3y = x^2 + 4x + 16$

22. List all x so that $x < 4$ and $x \in N$.

23. $\dfrac{1^{-1}}{2^{-1} + 3^{-1}}$

24. $\log_3 9 + \log_9 3$

25. $\dfrac{1}{x} + \dfrac{2}{x^2}$

26. Factor $8x^3 + 16$

27. Factor $xy + 3y - 2x - 6$

PART III—Find the solution sets.

28. $\dfrac{1}{x-3} - \dfrac{1}{x+1} = \dfrac{3x-2}{(x-3)(x+1)}$

29. $a(x + 3) + b(x - 2) = c(x - 1)$
Solve for x.

30. $\sqrt{5x - 1} - \sqrt{x} = 1$

31. $|x + 2| = 6$

32. $5x - 3 < 8x - 12$

33. $x = \text{Log}_8 2$

34. $(x - 2)(x + 3) = 6$

35. $x^{2/3} + 2x^{1/3} - 8 = 0$

36. $4(1 - x^2) + 2x = 3$

37. $\begin{vmatrix} x & 1 \\ 0 & x \end{vmatrix} > 25$

38. $|x - 3| < 5$

PART IV—Miscellaneous

39. Find the domain of $f(x) = \dfrac{1}{x + 5}$

40. Find the minimum value of $y = x^2 + 2x - 5$.

41. Find the 4th term of $(2x + 1)^8$.

42. Change $3.1\overline{1}$ to a reduced fraction.

43. Divide $(2x^3 - x + 1)$ by $(x - 2)$.

44. Is the following system consistent or inconsistent?

$$\begin{cases} 3x + 4 = 4y \\ 9x + 2y = 9 \end{cases}$$

45. If $f(x) = x - 10$, find $\dfrac{f(4) - f(2)}{4 - 2}$

46. Simplify:

$$7^0 - \frac{1}{\sqrt[3]{-27}} + 64^{2/3} + 25^{1/2}$$

47. Identify the graphs of:

a) $2x + y = 4$ **b)** $2x^2 + y^2 = 4$ **c)** $x^2 - 2y^2 = 4$
d) $2x^2 - y = 4$ **e)** $x^2 + y^2 = 4$

48. $\dfrac{1}{\frac{1}{x} + \frac{1}{y}}$

49. Find the distance from the x intercept of $y = 2x + 4$ to the point $(3, 4)$.

50. If y varies as the cube of x and inversely as the square of t and $x = 6$ when $y = 2$ and $t = 4$, find x when $y = 4$ and $t = 10$.

COMPREHENSIVE REVIEW EXERCISES

PART I—Evaluate or simplify.

1. $\frac{1}{x} + \frac{2}{y} + \frac{3}{z}$

2. $\begin{vmatrix} -1 & 3 & -4 \\ 0 & 2 & 0 \\ 2 & -3 & 5 \end{vmatrix}$

3. $0! + 1^0$

4. $(x + y^{-1})^2$

5. $(2^{-6})^{2/3}$

6. $\frac{3^0 x + 4x^{-1}}{x^{-2/3}}$ if $x = 8$

7. $\frac{\frac{1}{x} + \frac{1}{y}}{\frac{x+y}{y} + \frac{x+y}{x}}$

8. $\frac{3(-3)^2 + 4(-2)^3}{2^3 - 3^3}$

9. $\frac{-2}{2 - \sqrt{3}}$

10. $\sqrt{27} + \sqrt{48} - \sqrt{12}$

11. $\frac{3\sqrt{2} - 4\sqrt{3}}{4\sqrt{2} - 3\sqrt{3}}$

PART II—Short Answer

12. $\log_{10} 1$

13. $2/i$ where $i = \sqrt{-1}$

14. $5!$

15. $(\sqrt{64})^{1/3}$

16. -5^2

17. $4 + 6 \div 2$

18. $\sqrt{0}$

19. $|4| + |-2|$

20. $(2^{-1} + 2^{-2})^{-1}$

21. The y intercept of $3y = x^2 + 4x + 16$

22. List all x so that $x < 4$ and $x \in N$.

23. $\frac{1^{-1}}{2^{-1} + 3^{-1}}$

24. $\log_3 9 + \log_9 3$

25. $\frac{1}{x} + \frac{2}{x^2}$

26. Factor $8x^3 + 16$

27. Factor $xy + 3y - 2x - 6$

PART III—Find the solution sets.

28. $\frac{1}{x - 3} - \frac{1}{x + 1} = \frac{3x - 2}{(x - 3)(x + 1)}$

29. $a(x + 3) + b(x - 2) = c(x - 1)$ Solve for x.

30. $\sqrt{5x - 1} - \sqrt{x} = 1$

31. $|x + 2| = 6$

32. $5x - 3 < 8x - 12$

33. $x = \text{Log}_8 2$

34. $(x - 2)(x + 3) = 6$

35. $x^{2/3} + 2x^{1/3} - 8 = 0$

36. $4(1 - x^2) + 2x = 3$

37. $\begin{vmatrix} x & 1 \\ 0 & x \end{vmatrix} > 25$

38. $|x - 3| < 5$

PART IV—Miscellaneous

39. Find the domain of $f(x) = \frac{1}{x + 5}$

40. Find the minimum value of $y = x^2 + 2x - 5$.

41. Find the 4th term of $(2x + 1)^8$.

42. Change 3.111 to a reduced fraction.

43. Divide $(2x^3 - x + 1)$ by $(x - 2)$.

44. Is the following system consistent or inconsistent?

$$\begin{cases} 3x + 4 = 4y \\ 9x + 2y = 9 \end{cases}$$

45. If $f(x) = x - 10$, find $\dfrac{f(4) - f(2)}{4 - 2}$

46. Simplify:

$$7^0 - \frac{1}{\sqrt[3]{-27}} + 64^{2/3} + 25^{1/2}$$

47. Identify the graphs of:

a) $2x + y = 4$ **b)** $2x^2 + y^2 = 4$ **c)** $x^2 - 2y^2 = 4$
d) $2x^2 - y = 4$ **e)** $x^2 + y^2 = 4$

48. Simplify $\dfrac{1}{\frac{1}{x} + \frac{1}{y}}$

49. Find the distance from the x intercept of $y = 2x + 4$ to the point (3, 4).

50. If y varies as the cube of x and inversely as the square of t and $x = 6$ when $y = 2$ and $t = 4$, find x when $y = 4$ and $t = 10$.

PART V—True or False.

51. The graph of $3x - y = 6$ passes through $(4, -6)$.
52. $(a + b)^2 = a^2 + b^2$
53. $2^m \cdot 2^n = 4^{m+n}$
54. If $n \geq 5$ and $n \leq 5$, then $n = 5$.
55. If $-m = 8$, then m is a negative number.
56. If the quotient of two numbers is -1, their sum is zero.
57. The distance between $(-2, 0)$ and $(4, 8)$ is 14.
58. If the lengths of two adjacent sides of a rectangle are $(2\sqrt{5} - 3)$ and $(2\sqrt{5} + 3)$, the area is 11.
59. The equation of the line with slope 5 and passing through (0, 5) is $y = 5x + 5$.
60. $(x^{-1} + y^{-1})^{-1} = x + y, x \neq 0, y \neq 0$
61. If $K = 5 - 3 \cdot 2$, then $K = 4$.
62. $-6^2 = -36$
63. Zero is an integer.
64. $\sqrt{9} = \pm 3$
65. $\dfrac{x}{x} = 1$ for all x.
66. The parabola $y = x^2 + 2x + 5$ has a maximum value.
67. $\{x \mid |x| < 3\}$ is equivalent to $\{x \mid -3 < x < 3\}$
68. Besides $(x - 2)$, another factor of $x^3 - 8$ is $(x^2 + 2x + 4)$.
69. If $b \neq 0$, $b - 1 \div b = (b - 1) \div b$
70. $\sqrt{2} + \sqrt{3} = \sqrt{5}$
71. $\sqrt{4} = -2$
72. $-1 < -100$
73. $-x$ is a negative number.
74. $4^{2/3} \div 4^{1/6} = 2$

COMPREHENSIVE REVIEW EXERCISES, CONTD.

75. The limit of the sum of $2 + 1 + 1/2 + 1/4 + \ldots$ is 4.
76. $\text{Log}_8 2 = 1/4$
77. $(-2)^{-3} = 1/8$
78. A function is a relation.
79. Every real number has a reciprocal.
80. $(-1)^{-1} = -1$
81. $\dfrac{x^3 + 1}{x^3 - 1}$, $(x^3 \neq 1)$ is in lowest terms.
82. If $3^x = 7$, then $x = \log 7 - \log 3$.
83. The line through (3, 2) with zero slope is the line $x = 2$.
84. The radius of the circle $x^2 + y^2 = 10$ is 10.
85. There are no elements in the set $\{0\}$.

PART VI
Sketch the graphs of the following. Be sure to indicate your scale and use equal scales on the x and y axes.

86. $y = x^2 - 2$ **87.** $2x + y = 4$ **88.** $y > x + 3$ **89.** $x^2 + y^2 = 16$
90. $y = |x| - 1$ **91.** $9x^2 + y^2 = 36$ **92.** $xy = 4$ **93.** $x = 2$

PART VII—Define the following:

94. Rational number **95.** Function **96.** Relation
97. Domain of a function **98.** Range of a function **99.** Logarithm
100. Mantissa **101.** Abscissa
102. Ordinate **103.** Absolute value of a number x

PART VIII—State the following:

104. Distributive Property of the Reals
105. Fundamental Principal of Fractions
106. Quadratic Formula
107. Slope-Intercept Form of a Straight Line
108. Zero Factor Law
109. Identity Element for Addition in the Reals

SOLUTIONS TO THE COMPREHENSIVE REVIEW EXERCISES

1. $$1/x + 2/y + 3/z = \frac{1 \cdot yz}{x \cdot yz} + \frac{2 \cdot xz}{y \cdot xz} + \frac{3 \cdot xy}{x \cdot xy}$$
$$= \frac{yz + 2xz + 3xy}{xyz}, \; x, y, z \neq 0$$

2. $$\begin{vmatrix} -1 & 3 & -4 \\ 0 & 2 & 0 \\ 2 & -3 & 5 \end{vmatrix} = \begin{array}{|ccc|cc} -1 & 3 & -4 & -1 & 3 \\ 0 & 2 & 0 & 0 & 2 \\ 2 & -3 & 5 & 2 & -3 \end{array}$$
$$= (-10) + (0) + (0) - (-16) - (0) - (0)$$
$$= (-10) + 16$$
$$= 6$$

SOLUTIONS TO THE COMPREHENSIVE REVIEW EXERCISES, CONTD.

3. $0! + 1^0 = 1 + 1 = 2$; (zero factorial is equal to one and $x^0 = 1$ if $x \neq 0$)

4.
$$(x + y^{-1})^2 = \left(x + \frac{1}{y}\right)^2$$
$$= x^2 + 2(x)(1/y) + 1/y^2$$
$$= x^2 + \frac{2x}{y} + \frac{1}{y^2}, \quad y \neq 0$$

5.
$$(2^{-6})^{2/3} = (2)^{(-6)(2/3)}$$
$$= (2)^{-4}$$
$$= \frac{1}{2^4}$$
$$= \frac{1}{16}$$

6. Evaluate $\dfrac{3^0x + 4x^{-1}}{x^{-2/3}}$ if $x = 8$

$$\frac{3^0(8) + 4(8)^{-1}}{(8)^{-2/3}} = \frac{1(8) + 4\left(\frac{1}{8}\right)}{\frac{1}{8^{2/3}}}$$
$$= \frac{8 + \frac{4}{8}}{\frac{1}{4}}$$
$$= \frac{8 + \frac{1}{2}}{\frac{1}{4}}$$
$$= \frac{4 \cdot (8 + \frac{1}{2})}{4 \cdot (\frac{1}{4})}$$
$$= \frac{32 + 2}{1}$$
$$= 34$$

7.
$$\frac{\frac{1}{x} + \frac{1}{y}}{\frac{x+y}{y} + \frac{x+y}{x}} = \frac{xy \cdot (\frac{1}{x} + \frac{1}{y})}{xy \cdot (\frac{x+y}{y} + \frac{x+y}{x})}$$
$$= \frac{y + x}{x(x + y) + y(x + y)}$$
$$= \frac{y + x}{x^2 + xy + xy + y^2}$$
$$= \frac{x + y}{x^2 + 2xy + y^2}$$
$$= \frac{(x + y)}{(x + y)^2}$$
$$= \frac{1}{x + y}, xy \neq 0, \quad x \neq -y$$

8.
$$\frac{3(-3)^2 + 4(-2)^3}{2^3 - 3^3} = \frac{3(9) + 4(-8)}{8 - 27}$$
$$= \frac{27 + (-32)}{-19}$$
$$= \frac{-5}{-19}$$
$$= \frac{5}{19}$$

9. $\dfrac{-2}{2-\sqrt{3}} = \dfrac{(-2)(2+\sqrt{3})}{(2-\sqrt{3})(2+\sqrt{3})}$

$= \dfrac{-2(2+\sqrt{3})}{4-\sqrt{9}}$

$= \dfrac{-4-(2\sqrt{3})}{4-3}$

$= \dfrac{-4-2\sqrt{3}}{1}$

$= -4-2\sqrt{3}$

10. $\sqrt{27}+\sqrt{48}-\sqrt{12} = \sqrt{9\cdot 3}+\sqrt{16\cdot 3}-\sqrt{4\cdot 3}$

$= 3\sqrt{3}+4\sqrt{3}-2\sqrt{3}$

$= 7\sqrt{3}-2\sqrt{3}$

$= 5\sqrt{3}$

11. $\dfrac{3\sqrt{2}-4\sqrt{3}}{4\sqrt{2}-3\sqrt{3}} = \dfrac{(3\sqrt{2}-4\sqrt{3})(4\sqrt{2}+3\sqrt{3})}{(4\sqrt{2}-3\sqrt{3})(4\sqrt{2}+3\sqrt{3})}$

$= \dfrac{12\sqrt{4}-16\sqrt{6}+9\sqrt{6}-12\sqrt{9}}{16\sqrt{4}-9\sqrt{9}}$

$= \dfrac{12(2)-7\sqrt{6}-12(3)}{16(2)-9(3)}$

$= \dfrac{24-7\sqrt{6}-36}{32-27}$

$= \dfrac{-12-7\sqrt{6}}{5}$

12. $\log_{10} 1 = 0$ since $10^0 = 1$

13. $\dfrac{2}{i} = \dfrac{2\cdot i}{i\cdot i}$

$= \dfrac{2i}{i^2}$

$= \dfrac{2i}{-i}$

$= -2i$

14. $5! = 5\cdot 4\cdot 3\cdot 2\cdot 1$

$= 120$

15. $(\sqrt{64})^{1/3} = 8^{1/3}$

$= \sqrt[3]{8}$

$= 2$

SOLUTIONS TO THE COMPREHENSIVE REVIEW EXERCISES, CONTD.

16. $-5^2 = -(5^2)$
$= -(25)$
$= -25$

17. $4 + 6 \div 2 = 4 + [6 \div 2]$
$= 4 + 3$
$= 7$

18. $\sqrt{0} = 0$

19. $|4| + |-2| = 4 + 2$
$= 6$

20. $(2^{-1} + 2^{-2})^{-1} = \left(\frac{1}{2^1} + \frac{1}{2^2}\right)^{-1}$
$= (1/2 + 1/4)^{-1}$
$= (2/4 + 1/4)^{-1}$
$= (3/4)^{-1}$
$= \frac{1}{(3/4)^1}$
$= \frac{4 \cdot 1}{4 \cdot (3/4)}$
$= 4/3$

21. The y intercept of $3y = x^2 + 4x + 16$.
Set $x = 0$ and solve for y.
$3y = 0^2 + 4 \cdot 0 + 16$
$3y = 16$
$y = \frac{16}{3}$, The y intercept is $(0, 16/3)$

22. List all x so that $x < 4$ and $x \in N$
$\{1, 2, 3\}$

23. $\frac{1^{-1}}{2^{-1} + 3^{-1}} = \frac{\frac{1}{1}}{\frac{1}{2} + \frac{1}{3}}$
$= \frac{1}{\frac{1}{2} + \frac{1}{3}}$
$= \frac{6(1)}{6(\frac{1}{2} + \frac{1}{3})}$
$= \frac{6}{3 + 2}$
$= 6/5$

24. $\log_3 9 + \log_9 3$
$= 2 + 1/2$ $\qquad \log_3 9 = 2$ since $3^2 = 9$
$= 2\frac{1}{2}$ or $5/2$ $\qquad \log_9 3 = \frac{1}{2}$ since $9^{1/2} = 3$

25. $\frac{1}{x} + \frac{2}{x^2} = \frac{1x}{xx} + \frac{2}{x^2}$
$= \frac{x}{x^2} + \frac{2}{x^2}$
$= \frac{x + 2}{x^2}, \quad x \neq 0$

26. Factor $8x^3 + 16$.
$8(x^3 + 2)$

27. Factor $xy + 3y - 2x - 6$
$(xy + 3y) - (2x + 6)$
$y(x + 3) - 2(x + 3)$
$(x + 3)(y - 2)$

SOLUTIONS TO THE COMPREHENSIVE REVIEW EXERCISES, CONTD.

28. $\frac{1}{x-3} - \frac{1}{x+1} = \frac{3x-2}{(x-3)(x+1)}, \quad x \neq 3, x \neq -1$

$$(x-3)(x+1)\cdot\frac{1}{(x-3)} - (x-3)(x+1)\frac{1}{(x+1)} = \frac{(3x-2)}{(x-3)(x+1)}(x-3)(x+1)$$

$$\begin{aligned}(x+1)-(x-3) &= 3x-2\\ x+1-x+3 &= 3x-2\\ 4 &= 3x-2\\ 6 &= 3x\\ 2 &= x\end{aligned}$$

Solution set is: $\{2\}$.

29. $a(x+3)+b(x-2)=c(x-1)$. Solve for x.

$$\begin{aligned}ax+3a+bx-2b &= cx-c\\ ax+bx-cx &= 2b-3a-c\\ (a+b-c)x &= 2b-3a-c\\ x &= \frac{2b-3a-c}{a+b-c}, \quad (a+b-c)\neq 0\end{aligned}$$

30. $\sqrt{5x-1}-\sqrt{x}=1$

$$[\sqrt{5x-1}]^2 = [\sqrt{x}+1]^2$$

$$5x-1 = x+2\sqrt{x}+1$$

$$4x-2 = 2\sqrt{x}$$

$$[2x-1]^2 = [\sqrt{x}]^2$$

$$\begin{aligned}4x^2-4x+1 &= x\\ 4x^2-5x+1 &= 0\\ (4x-1)(x-1) &= 0\end{aligned}$$

$x = 1/4, \quad x = 1$

Check: $x = 1/4$ (crossed out) $\quad \sqrt{5/4}-1 = \sqrt{1/4}+1$

$$\sqrt{1/4} \neq \sqrt{1/4}+1$$

Check: $x=1 \quad \sqrt{5-1} = \sqrt{1}+1$

$$\sqrt{4} = 1+1$$

$$2 = 2$$

Solution set is: $\{1\}$.

31. $|x+2| = 6$

$(x+2) = 6$ *or* $(x+2) = -6$

$x = 4$ *or* $x = -8$

Solution set is: $\{4, -8\}$.

32. $5x-3 < 8x-12$

$$\begin{aligned}-3x-3 &< -12\\ -3x &< -9\\ x &> 3\end{aligned}$$

$\{x|x>3\}$

33. $x = \log_8 2$

$$\begin{aligned}8x &= 2\\ (2^3)^x &= 2\\ 2^{3x} &= 2^1\\ 3x &= 1\\ x &= 1/3\end{aligned}$$

Solution set is: $\{1/3\}$

34. $(x - 2)(x + 3) = 6$
$x^2 + x - 6 = 6$
$x^2 + x - 12 = 0$
$(x + 4)(x - 3) = 0$
$x = -4 \qquad x = 3$
Solution set is: $\{-4, 3\}$.

35. $x^{2/3} + 2x^{1/3} - 8 = 0$
$(x^{1/3} + 4)(x^{1/3} - 2) = 0$
$x^{1/3} = -4 \qquad x^{1/3} = 2$
$\sqrt[3]{x} = -4 \qquad \sqrt[3]{x} = 2$
$x = -64 \qquad x = 8$
Solution set is: $\{-64, 8\}$.

36. $4(1 - x^2) + 2x = 3$
$4 - 4x^2 + 2x = 3$
$-4x^2 + 2x + 1 = 0$
$4x^2 - 2x - 1 = 0$

$$x = \frac{-b \pm \sqrt{b^2 - 4ac}}{2a} = \frac{-(-2) \pm \sqrt{(-2)^2 - 4(4)(-1)}}{2(4)}$$

$$x = \frac{2 \pm \sqrt{4 + 16}}{8}$$

$$= \frac{2 \pm \sqrt{20}}{8}$$

$$= \frac{2 \pm 2\sqrt{5}}{8}$$

$$x = \frac{2(1 \pm \sqrt{5})}{8} = \frac{1 \pm \sqrt{5}}{4}$$

Solution set is: $\left\{\frac{1 + \sqrt{5}}{4}, \frac{1 - \sqrt{5}}{4}\right\}$.

37.
$$\begin{vmatrix} x & 1 \\ 0 & x \end{vmatrix} > 25$$
$x^2 - 0 > 25$
$x^2 > 25$
$x > 5$ *or* $x < -5$
$\{x|x > 5\} \cup \{x|x < -5\}$

38. $|x - 3| < 5$
$-5 < (x - 3) < 5$
$-2 < x < 8$
$\{x|-2 < x < 8\}$

39. Find the domain of $f(x) = \frac{1}{x + 5}$. The fraction $\frac{1}{x + 5}$ is undefined when the denominator is zero (that is, when $x = -5$). Therefore, x cannot equal -5.

$$D_f = \{x|x \neq -5\}$$

40. Find the minimum value of $y = x^2 + 2x - 5$.
$y = x^2 + 2x + 1 - 5 - 1$
$y = (x + 1)^2 - 6$
The minimum value is -6 (occurs when $x = -1$).

SOLUTIONS TO THE COMPREHENSIVE REVIEW EXERCISES, CONTD.

41. Find the 4th term of $(2x + 1)^8$

$$(2x + 1)^8 = (2x)^8 + 8(2x)^7(1) + 28(2x)^6(1)^2 + \frac{(28)(6)}{3}(2x)^5(1)^3 + \ldots$$

(4th term: $\frac{(28)(6)}{3}(2x)^5(1)^3$)

$$\frac{(28)(6)}{3}(2x)^5(1)^3 = (28)(2)(32x^5) = 1792x^5$$

42. Change $3.11\overline{1}$ to a reduced fraction.

$$3.11\overline{1} = 3 + .11\overline{1}$$

$$.11\overline{1} = .1 + .01 + .001 + \ldots = \frac{1}{10} + \frac{1}{100} + \frac{1}{1000} + \ldots$$

$$S_\infty = \frac{a}{1 - r} = \frac{\frac{1}{10}}{1 - \frac{1}{10}} = \frac{\frac{10}{1}}{\frac{9}{10}} = 1/9$$

$$3.11\overline{1} = 3 + .11\overline{1} = 3 + 1/9 = 3\tfrac{1}{9} = 28/9$$

43. Divide $(2x^3 - x + 1)$ by $(x - 2)$.

$$2x^2 + 4x + 7 + \frac{15}{x - 2}, \quad x \neq 2$$

$$\begin{array}{r l}
x - 2 \,\big)\, 2x^3 - x + 1 & \\
\underline{2x^3 - 4x^2} & \\
4x^2 - x + 1 & \\
\underline{4x^2 - 8x} & \\
7x + 1 & \\
\underline{7x - 14} & \\
15 &
\end{array}$$

44. Is $\begin{cases} 3x + 4 = 4y \\ 9x + 2y = 9 \end{cases}$ consistent or inconsistent?

$$\begin{cases} 3x - 4y = -4 \\ 9x + 2y = 9 \end{cases} \Rightarrow \begin{cases} 3x - 4y = -4 \\ 18x + 4y = 18 \end{cases}$$

$$21x = 14$$

$$x = 2/3$$

$$y = 3/2$$

The system is consistent.

SOLUTIONS TO THE COMPREHENSIVE REVIEW EXERCISES, CONTD.

45. If $f(x) = x - 10$, find $\dfrac{f(4) - f(2)}{4 - 2}$

$$f(4) = 4 - 10 \qquad f(2) = 2 - 10$$
$$= 6 \qquad = -8$$

$$\frac{f(4) - f(2)}{4 - 2} = \frac{(-6) - (-8)}{2}$$
$$= \frac{-6 + 8}{2}$$
$$= 2/2$$
$$= 1$$

46. Simplify: $7 - \dfrac{1}{\sqrt[3]{-27}} + 64^{2/3} + 25^{1/2}$

$$1 - \frac{1}{-3} + 16 + 5$$
$$1 + 1/3 + 16 + 5 = 22\tfrac{1}{3}$$

47. **(a)** Straight line **(b)** ellipse **(c)** hyperbola **(d)** parabola **(e)** circle

48. $$\frac{1}{\frac{1}{x} + \frac{1}{y}} = \frac{xy(1)}{xy(\frac{1}{x} + \frac{1}{y})}$$
$$= \frac{xy}{y + x}$$
$$x \neq 0, \quad y \neq 0, \quad x \neq -y$$

49. Find the distance from the x intercept of $y = 2x + 4$ to the point (3, 4). To find the x intercept, let $y = 0$.

$$0 = 2x + 4$$
$$-4 = 2x$$
$$-2 = x$$

Now find the distance from $(-2, 0)$ to (3, 4).

$$d = \sqrt{(x_2 - x_1)^2 + (y_2 - y_1)^2}$$
$$d = \sqrt{[3 - (-2)]^2 + (4 - 0)^2}$$
$$d = \sqrt{5^2 + 4^2}$$
$$d = \sqrt{25 + 16}$$
$$d = \sqrt{41}$$

SOLUTIONS TO THE COMPREHENSIVE REVIEW EXERCISES, CONTD.

50. $y = \dfrac{Kx^3}{t^2}$

$x = 6, \quad y = 2, \quad t = 4$

$2 = \dfrac{K(6)^3}{4^2}$

$2 = \dfrac{216K}{16}$

$216K = 32$

$K = \dfrac{32}{216} = \dfrac{4}{27}$

$y = \dfrac{\frac{4}{27}x^3}{t^2}$

Find x when $y = 4$, $t = 10$.

$y = \dfrac{\frac{4}{27}x^3}{t^2}$

$4 = \dfrac{\frac{4}{27}x^3}{10^2}$

$400 = \dfrac{4}{27}x^3$

$(400)(27) = 4x^3$

$(100)(27) = x^3$

$\sqrt[3]{(27)(100)} = x$

$3\sqrt[3]{100} = x$

51. $3(4) - (-6) \neq 6$

$12 + 6 \neq 6$

False

52. $(a + b)^2 = a^2 + 2ab + b^2$

53. $2^m \cdot 2^n = 2^{m+n}$

54. True

55. $-m = 8$

$m = -8$

True

56. Let $x =$ one number

$y =$ the other number

$x/y = -1$

$x = -y$

$x + y = 0$ Their sum is zero.

57. $d = \sqrt{[4 - (-2)]^2 + (8 - 0)^2} = \sqrt{6^2 + 8^2}$

$= \sqrt{36 + 64}$

$= \sqrt{100}$

$= 10$

58. Area $= (2\sqrt{5} - 3)(2\sqrt{5} + 3)$

$= 4\sqrt{25} - 9$

$= 20 - 9$

$= 11$

59. $y = mx + b$

$m = 5$

$(0, b) = (0, 5)$

$y = 5x + 5$

60. $(x^{-1} + y^{-1})^{-1} = \dfrac{1}{(x^{-1} + y^{-1})}$

$= \dfrac{1}{\frac{1}{x} + \frac{1}{y}}$

$= \dfrac{xy(1)}{xy(\frac{1}{x} + \frac{1}{y})}$

$= \dfrac{xy}{y + x}, \quad x \neq 0, \quad y \neq 0, \quad x \neq -y$

61. $K = 5 - 3 \cdot 2$

$= 5 - [3 \cdot 2]$

$= 5 - 6$

$= -1$

62. $-6^2 = -[6^2]$
$= -36$

63. $I = \{\dots -2, -1, 0, 1, 2 \dots\}$
True

64. $\sqrt{9} = 3$

65. $x/x = 1$ for all $x \neq 0$

66. $y = x^2 + 2x + 5$
Since $a > 0$, the parabola has a minimum.
False

67. True

68. $x^3 - 8 = (x - 2)(x^2 + 2x + 4)$

69. If $b \neq 0$, $b - 1 \div b = b - 1/b$

$$= \frac{b^2}{b} - 1/b$$

$$= \frac{b^2 - 1}{b} \neq \frac{b - 1}{b}$$

70. False; $\sqrt{2} + \sqrt{3} \neq \sqrt{5}$

71. $\sqrt{4} = 2$

72. $-1 \nless -100$
Since -1 is not further to the left than -100 on the number line.

73. If x is positive, $-x$ is negative
If x is negative, $-x$ is positive
If x is zero, $-x$ is zero

74.
$$4^{2/3} \div 4^{1/6} = \frac{4^{2/3}}{4^{1/6}}$$
$$= 4^{2/3 - 1/6}$$
$$= 4^{4/6 - 1/6}$$
$$= 4^{3/6}$$
$$= 4^{1/2}$$
$$= \sqrt{4}$$
$$= 2$$

75. $a = 2, \quad r = 1/2$
$$S_\infty = \frac{a}{1 - r}$$
$$= \frac{2}{1 - \frac{1}{2}}$$
$$= \frac{2}{\frac{1}{2}}$$
$$= 4$$

76. $\text{Log}_8 2 = 1/3$ since $8^{1/3} = 2$

77.
$$(-2)^{-3} = \frac{1}{(-2)^3}$$
$$= 1/-8$$
$$= -1/8$$

78. True

79. Every real number except zero has a reciprocal.

80.
$$(-1)^{-1} = \frac{1}{(-1)^1}$$
$$= 1/-1$$
$$= -1$$

81.
$$\frac{x^3 + 1}{x^3 - 1} = \frac{(x + 1)(x^2 - x + 1)}{(x - 1)(x^2 + x + 1)}, \quad x^3 - 1 \neq 0$$

82.
$$3^x = 7$$
$$\log 3^x = \log 7$$
$$x \log 3 = \log 7$$
$$x = \frac{\log 7}{\log 3} \neq \log 7 - \log 3$$

83.

$$m = 0,$$
$$(3, 2) = (x_1, y_1)$$
$$y - y_1 = m(x - x_1)$$
$$y - 2 = 0(x - 3)$$
$$y - 2 = 0$$
$$y = 2$$

84. Radius $= \sqrt{10}$

85. The set $\{0\}$ has a single element, namely zero.

86.

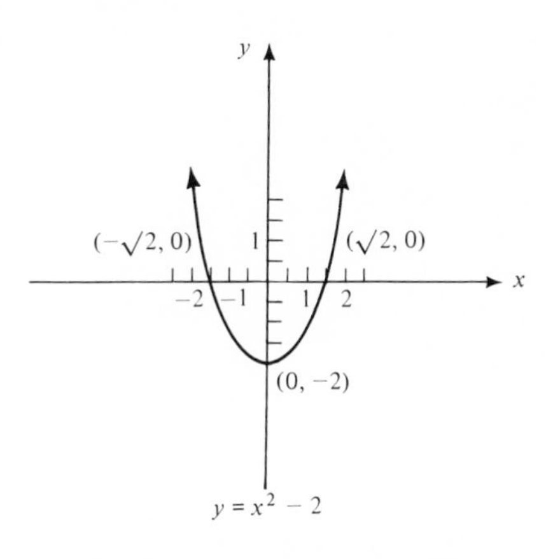

Scale: 2 squares = 1 unit

87.

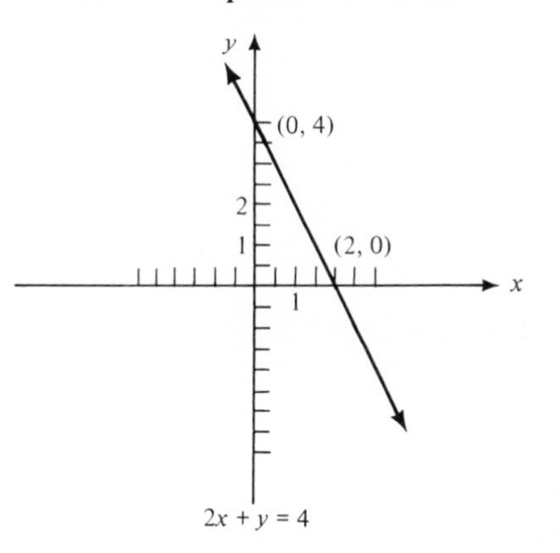

Scale: 2 squares = 1 unit

SOLUTIONS TO THE COMPREHENSIVE REVIEW EXERCISES, CONTD.

88.

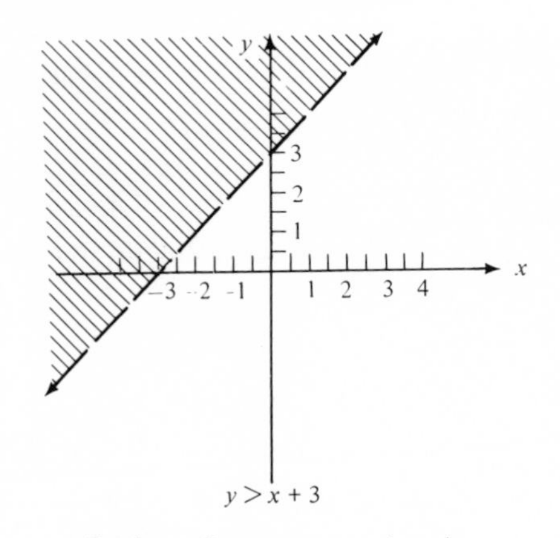

Scale: 2 squares = 1 unit

89.

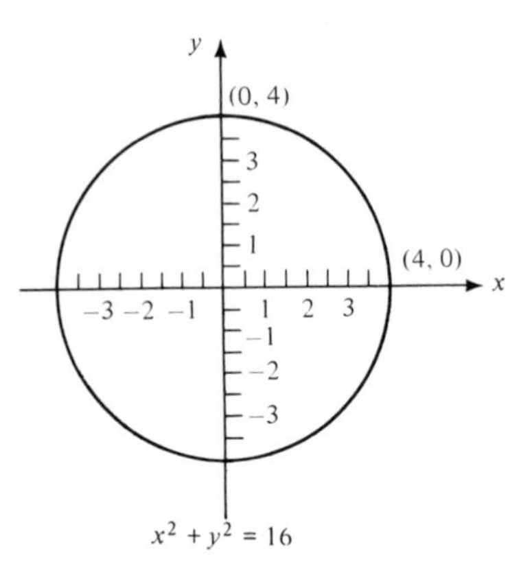

Scale: 2 squares = 1 unit

90.

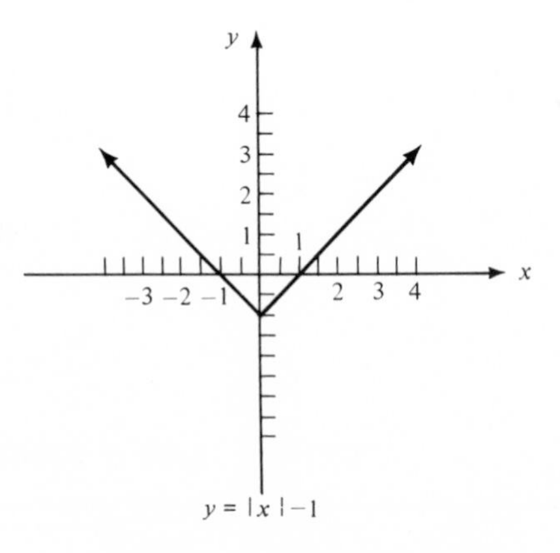

Scale: 2 squares = 1 unit

SOLUTIONS TO THE COMPREHENSIVE REVIEW EXERCISES, CONTD.

91.

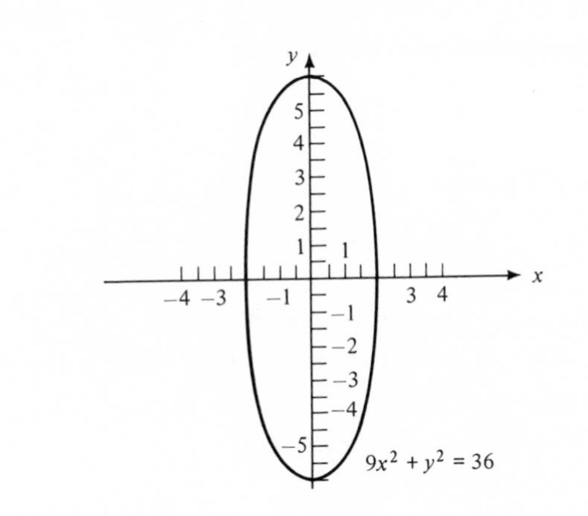

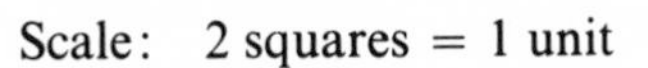

92.

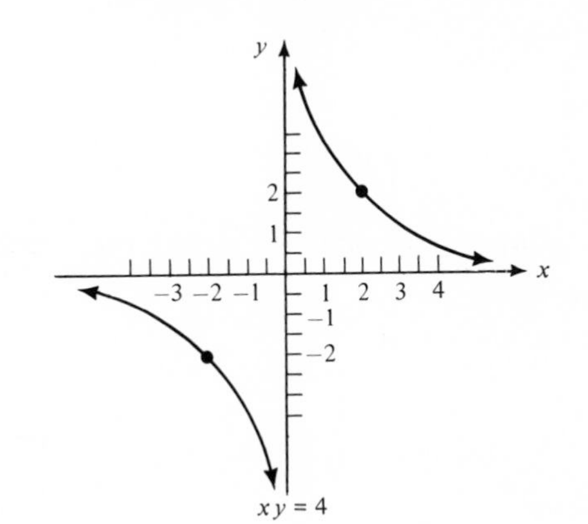

93.

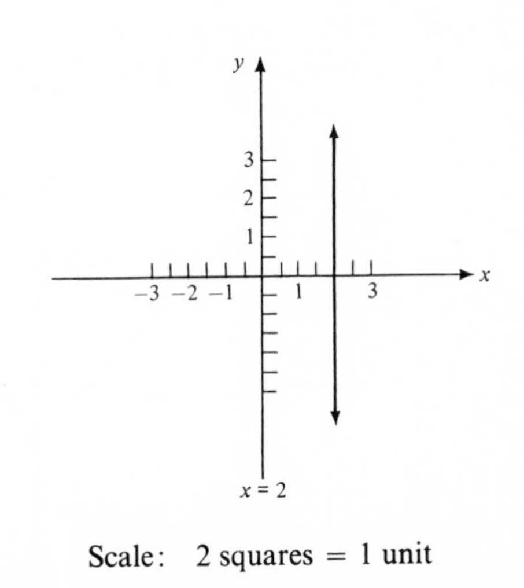

Scale: 2 squares = 1 unit

Properties and Formulas

Distributive Property of the Reals

If $a, b, c \in R$, then $a(b + c) = ab + ac$

Fundamental Principal of Fractions

If $a, b, c \in R$ and $b \neq 0, c \neq 0$, then $\frac{a}{b} = \frac{a \cdot c}{b \cdot c}$

Quadratic Formula

If $ax^2 + bx + c = 0, a, b, c \in R, a \neq 0$ then $x = \frac{-b \pm \sqrt{b^2 - 4ac}}{2a}$

Slope Intercept Form of a Straight Line

$y = mx + b$ where m is the slope and b is the y coordinate of the y intercept.

Zero Factor Law

If $a \in R$, then $a \cdot 0 = 0$
If $a \in C$, then $a \cdot 0 = 0$

Definitions

Rational Number

$$\left\{\frac{a}{b} \mid a, b \in I, b \neq 0\right\}$$

Relation

A set of ordered pairs.

Function

A set of ordered pairs (x, y) such that for each x there is one and only one y.

Absolute Value

For every $x \in R$,

$$|x| = \begin{cases} x \text{ if } x > 0 \\ 0 \text{ if } x = 0 \\ -x \text{ if } x < 0 \end{cases}$$

Logarithm

$\log_b N = x$ if and only if
$b^x = N, \quad b > 0, \quad b \neq 1, \quad N > 0$

Powers and Roots

Number	*Square*	*Square Root*	*Number*	*Square*	*Square Root*
1	1	1	26	676	5.099
2	4	1.414	27	729	5.196
3	9	1.732	28	784	5.292
4	16	2	29	841	5.385
5	25	2.236	30	900	5.477
6	36	2.449	31	961	5.568
7	49	2.646	32	1,024	5.657
8	64	2.828	33	1,089	5.745
9	81	3	34	1,156	5.831
10	100	3.162	35	1,225	5.916
11	121	3.317	36	1,296	6
12	144	3.464	37	1,369	6.083
13	169	3.606	38	1,444	6.164
14	196	3.742	39	1,521	6.245
15	225	3.873	40	1,600	6.325
16	256	4	41	1,681	6.403
17	289	4.123	42	1,764	6.481
18	324	4.243	43	1,849	6.557
19	361	4.359	44	1,936	6.633
20	400	4.472	45	2,025	6.708
21	441	4.583	46	2,116	6.782
22	484	4.690	47	2,209	6.856
23	529	4.796	48	2,304	6.928
24	576	4.899	49	2,401	7
25	625	5	50	2,500	7.071

Logarithms of Numbers

N	*0*	*1*	*2*	*3*	*4*	*5*	*6*	*7*	*8*	*9*
10	0000	0043	0086	0128	0170	0212	0253	0294	0334	0374
11	0414	0453	0492	0531	0569	0607	0645	0682	0719	0755
12	0792	0828	0864	0899	0934	0969	1004	1038	1072	1106
13	1139	1173	1206	1239	1271	1303	1335	1367	1399	1430
14	1461	1492	1523	1553	1584	1614	1644	1673	1703	1732
15	1761	1790	1818	1847	1875	1903	1931	1959	1987	2014
16	2041	2068	2095	2122	2148	2175	2201	2227	2253	2279
17	2304	2330	2355	2380	2405	2430	2455	2480	2504	2529
18	2553	2577	2601	2625	2648	2672	2695	2718	2742	2765
19	2788	2810	2833	2856	2878	2900	2923	2945	2967	2989
20	3010	3032	3054	3075	3096	3118	3139	3160	3181	3201
21	3222	3243	3263	3284	3304	3324	3345	3365	3385	3404
22	3424	3444	3464	3483	3502	3522	3541	3560	3579	3598
23	3617	3636	3655	3674	3692	3711	3729	3747	3766	3784
24	3802	3820	3838	3856	3874	3892	3909	3927	3945	3962
25	3979	3997	4014	4031	4048	4065	4082	4099	4116	4133
26	4150	4166	4183	4200	4216	4232	4249	4265	4281	4298
27	4314	4330	4346	4362	4378	4393	4409	4425	4440	4456
28	4472	4487	4502	4518	4533	4548	4564	4579	4594	4609
29	4624	4639	4654	4669	4683	4698	4713	4728	4742	4757
30	4771	4786	4800	4814	4829	4843	4857	4871	4886	4900
31	4914	4928	4942	4955	4969	4983	4997	5011	5024	5038
32	5051	5065	5079	5092	5105	5119	5132	5145	5159	5172
33	5185	5198	5211	5224	5237	5250	5263	5276	5289	5302
34	5315	5328	5340	5353	5366	5378	5391	5403	5416	5428
35	5441	5453	5465	5478	5490	5502	5514	5527	5539	5551
36	5563	5575	5587	5599	5611	5623	5635	5647	5658	5670
37	5682	5694	5705	5717	5729	5740	5752	5763	5775	5786
38	5798	5809	5821	5832	5843	5855	5866	5877	5888	5899
39	5911	5922	5933	5944	5955	5966	5977	5988	5999	6010
40	6021	6031	6042	6053	6064	6075	6085	6096	6107	6117
41	6128	6138	6149	6160	6170	6180	6191	6201	6212	6222
42	6232	6243	6253	6263	6274	6284	6294	6304	6314	6325
43	6335	6345	6355	6365	6375	6385	6395	6405	6415	6425
44	6435	6444	6454	6464	6474	6484	6493	6503	6513	6522
45	6532	6542	6551	6561	6571	6580	6590	6599	6609	6618
46	6628	6637	6646	6656	6665	6675	6684	6693	6702	6712
47	6721	6730	6739	6749	6758	6767	6776	6785	6794	6803
48	6812	6821	6830	6839	6848	6857	6866	6875	6884	6893
49	6902	6911	6920	6928	6937	6946	6955	6964	6972	6981
50	6990	6998	7007	7016	7024	7033	7042	7050	7059	7067
51	7076	7084	7093	7101	7110	7118	7126	7135	7143	7152
52	7160	7168	7177	7185	7193	7202	7210	7218	7226	7235
53	7243	7251	7259	7267	7275	7284	7292	7300	7308	7316
54	7324	7332	7340	7348	7356	7364	7372	7380	7388	7396

N	0	1	2	3	4	5	6	7	8	9
55	7404	7412	7419	7427	7435	7443	7451	7459	7466	7474
56	7482	7490	7497	7505	7513	7520	7528	7536	7543	7551
57	7559	7566	7574	7582	7589	7597	7604	7612	7619	7627
58	7634	7642	7649	7657	7664	7672	7679	7686	7694	7701
59	7709	7716	7723	7731	7738	7745	7752	7760	7767	7774
60	7782	7789	7796	7803	7810	7818	7825	7832	7839	7846
61	7853	7860	7868	7875	7882	7889	7896	7903	7910	7917
62	7924	7931	7938	7945	7952	7959	7966	7973	7980	7987
63	7993	8000	8007	8014	8021	8028	8035	8041	8048	8055
64	8062	8069	8075	8082	8089	8096	8102	8109	8116	8122
65	8129	8136	8142	8149	8156	8162	8169	8176	8182	8189
66	8195	8202	8209	8215	8222	8228	8235	8241	8248	8254
67	8261	8267	8274	8280	8287	8293	8299	8306	8312	8319
68	8325	8331	8338	8344	8351	8357	8363	8370	8376	8382
69	8388	8395	8401	8407	8414	8420	8426	8432	8439	8445
70	8451	8457	8463	8470	8476	8482	8488	8494	8500	8506
71	8513	8519	8525	8531	8537	8543	8549	8555	8561	8567
72	8573	8579	8585	8591	8597	8603	8609	8615	8621	8627
73	8633	8639	8645	8651	8657	8663	8669	8675	8681	8686
74	8692	8698	8704	8710	8716	8722	8727	8733	8739	8745
75	8751	8756	8762	8768	8774	8779	8785	8791	8797	8802
76	8808	8814	8820	8825	8831	8837	8842	8848	8854	8859
77	8865	8871	8876	8882	8887	8893	8899	8904	8910	8915
78	8921	8927	8932	8938	8943	8949	8954	8960	8965	8971
79	8976	8982	8987	8993	8998	9004	9009	9015	9020	9025
80	9031	9036	9042	9047	9053	9058	9063	9069	9074	9079
81	9085	9090	9096	9101	9106	9112	9117	9122	9128	9133
82	9138	9143	9149	9154	9159	9165	9170	9175	9180	9186
83	9191	9196	9201	9206	9212	9217	9222	9227	9232	9238
84	9243	9248	9253	9258	9263	9269	9274	9279	9284	9289
85	9294	9299	9304	9309	9315	9320	9325	9330	9335	9340
86	9345	9350	9355	9360	9365	9370	9375	9380	9385	9390
87	9395	9400	9405	9410	9415	9420	9425	9430	9435	9440
88	9445	9450	9455	9460	9465	9469	9474	9479	9484	9489
89	9494	9499	9504	9509	9513	9518	9523	9528	9533	9538
90	9542	9547	9552	9557	9562	9566	9571	9576	9581	9586
91	9590	9595	9600	9605	9609	9614	9619	9624	9628	9633
92	9638	9643	9647	9652	9657	9661	9666	9671	9675	9680
93	9685	9689	9694	9699	9703	9708	9713	9717	9722	9727
94	9731	9736	9741	9745	9750	9754	9759	9763	9768	9773
95	9777	9782	9786	9791	9795	9800	9805	9809	9814	9818
96	9823	9827	9832	9836	9841	9845	9850	9854	9859	9863
97	9868	9872	9877	9881	9886	9890	9894	9899	9903	9908
98	9912	9917	9921	9926	9930	9934	9939	9943	9948	9952
99	9956	9961	9965	9969	9974	9978	9983	9987	9991	9996

Index

Abscissa, 135
Absolute value:
  definition of, 17
  equations involving, 107, 254
  inequalities involving, 121, 254
Addition:
  law of equality, 29
  of complex numbers, 325
  of fractions, 61
  of radical expressions, 90
Additive inverse, 28
Algebraic expression, 41
Antilogarithm, 232
Arithmetic progression:
  definition of, 312
  *n*th term of an, 312
  sum of n terms of an, 312
Associative law:
  of addition, 28
  of multiplication, 28
Axiom(s):
  of equality, 27
  meaning of an, 27
Axis, 135

Base:
  of a logarithm, 219
  of a power, 39
Binary operation, 27
Binomial:
  definition of a, 41
  expansion of a, 301
  products, 47

Cancellation law:
  for addition, 29
  for multiplication, 29
Characteristics of a logarithm, 229
Circle equation of, 179
Closure for addition and multiplication, 28
Coefficient, 41
Combination, 347
Common logarithm, 227
Commutative law:
  of addition, 28
  of multiplication, 28
Completing the square, 119, 120
Complex fraction, 59
Complex numbers:
  addition of, 327
  conjugate of, 328
  definition of, 326
  equality of, 326
  imaginary part, 326
  multiplication of, 327
  real part, 326
Component of an ordered pair, 133
Conditional equation, 104
Conic section, 186
Conjugate, 92, 328
Consistent system, 271
Constant, 2
Constant function, 148
Cramer's method, 375
Cross product, 133
Cube root, 75

Degree of a polynomial, 42
Denominator:
  least common, 61
  rationalizing a, 91
Dependent systems, 271
Dependent variable, 136
Determinant(s):
  definition of a, 274, 275
  in Cramer's method, 275
Direct variation, 192
Directed distance, 149
Discriminant, 123
Disjoint sets, 7
Distance:
  between two points, 156
  directed, 149
Distributive law, 28, 40, 46

Division:
  of polynomials, 62
  of real numbers, 38
Domain:
  of a logarithmic function, 218
  of a relation, 135

e, 227
Element of a set, 1
Ellipse, 181
Empty set, 2
Equality:
  addition law of, 29
  axioms for, 27
  of complex numbers, 326
  of fractions, 54
  multiplication law of, 29
  of sets, 2
Equation(s):
  absolute value in, 107, 254
  conditional, 104
  consistent, 271
  dependent, 271
  equivalent, 104
  exponential, 247
  identities, 104
  inconsistent, 271
  logarithmic, 248
  quadratic, 115
  radical, 249
  root of an, 104
  solution of an, 104
  solution set of an, 103
  solving, for specified symbols, 108
Equivalent equations, 104
Equivalent inequalities, 253
Even integers, 7
Event, 354
Expansion of a binomial, 301
Expansion of a determinant, 274
Exponent(s):
  irrational, 209
  laws of, 45, 46, 71
  natural number, 39
  negative, 73
  positive rational, 77
  zero, 73
Exponential equation, 247
Exponential function, 213
Extraneous solution, 106

Factor(s):
  definition of, 39
  prime, 4
Factorial notation, 305
Factoring:
  by grouping, 50
  differences of cubes, 48
  differences of squares, 48
  polynomials, 48
  solving equations by, 116
  sums of cubes, 48
Field, 28
Finite sequence, 311
Finite set, 8
Fraction(s):
  addition of, 61
  complex, 59
  definition of a, 54
  division of, 59
  equality of, 54
  fundamental principle of, 55
  lowest terms of, 56
  multiplication of, 58
  signs of, 56
Function(s):
  constant, 148
  definition of a, 137
  exponential, 213
  inverse of a, 201
  linear, 146
  logarithmic, 218
  quadratic, 164
  sequence, 311
  zeros of a, 140
Fundamental principle of fractions, 55

Geometric progression:
  definition of a, 315
  infinite, 317
  *n*th term of a, 315
  ratio of a, 315
  sum of *n* terms of a, 316
Greater than, 14

Hyperbola, 183

i, definition of, 326
Identity, 104
Identity element:
  for addition, 28
  for multiplication, 28
Inconsistent system, 271
Independent variable, 136
Index of a radical, 85
Inequalities:
  with absolute value, 254
  equivalent, 253
  linear, 288
  quadratic, 257
  solutions of, 253
  solution sets of, 253
Infinite geometric progressions, 317
Infinite sequence, 311
Infinite series, 311
Infinite set, 8
Integers, set of, 4
Intercept of a graph, 147
Interpolation, 233
Intersection of sets, 7

Inverse:
  for addition, 28
  for multiplication, 28
  function, 200
  relation, 199
Inverse variation, 193
Irrational numbers:
  as exponents, 209
  set of, 4

Joint variation, 194

Least common denominator, 61
Less than, 13
Like terms, 41
Linear equation(s):
  definition of, 105
  point-slope form for, 152
  slope-intercept form for, 153
Linear function:
  definition of a, 146
  as direct variation, 192
  intercepts of the graph of a, 147
  slope of the graph of a, 149
Linear inequality, 289
Linear interpolation, 233
Logarithm(s):
  base 10, 227
  characteristic of a, 229
  common, 227
  computations using, 241
  equations, 248
  mantissa of a, 229
  reading tables of, 232
Logarithmic equations, 248
Logarithmic function, 218

Mantissa, 229
Maximum of a quadratic function, 168
Member, of a set, 1
Minimum of a quadratic function, 168
Monomial, definition of a, 41
Multiplication:
  law of equality, 29
  of complex numbers, 327
  of fractions, 58
  of radicals, 87
  of real numbers, 37
Multiplicative inverse, 28
Multiplication Principle, 338

Natural numbers:
  prime, 4
  set of, 4
Negative number exponent, 73
Null set, 2
Number(s):
  absolute value, 17
  complex, 326
  graph of a, 12
  integers, 4
  irrational, 4
  natural, 4
  ordered pairs of, 133
  prime, 4
  rational, 4
  real, 4
Numerical coefficient, 41

Odd integers, 7
One to one correspondence, 8
Open sentence, 103
Order of operations, 39
Ordered pair:
  components of an, 133
  meaning of, 133
Ordinate, meaning of, 135

Parabola, 165
Pascal's triangle, 304
Permutations, 338
Polynomial(s):
  constant, 42
  definition of a, 42
  degree of a, 42
Postulate, 27
Prime factor, 4
Prime number, 4
Probability, 356
Progression(s):
  arithmetic, 313
  geometric, 315
Proper subset, 3
Proportion, 192
Pythagorean theorem, 155

Quadrant, 136
Quadratic equation(s):
  definition of, 115
  discriminant of a, 123
  formula, 121
  nature of roots of, 123
  solution of:
    by completing the square, 120
    by extraction of roots, 118
    by factoring, 116
    by formula, 121
Quadratic function:
  definition of a, 164
  graph of a, 165
Quadratic inequalities, 257
Quotient(s):
  definition of a, 38
  of fractions, 54

Radical(s):
  addition of, 90
  changing form of, 93
  definition of, 85
  division of, 87
  equations, 249
  index of a, 85

multiplication of, 87
simplest form for, 87
Radicand, 85
Range, 135
Ratio, of a geometric progression, 315
Rational numbers, 4
Rationalizing denominators, 91
Real numbers:
number line, 12
set of, 4
Reciprocal, 28
Reflexive law, 27
Relation:
domain of a, 135
inverse of a, 199
range of a, 135
as set of ordered pairs, 133
Rise of a line, 149
Root(s) (see also Solution(s)):
cube, 76, 78, 86
of equations, 104
extraneous, 106
nth, 76, 77, 86
of numbers, 75, 86
square, 75, 77, 85
Run of a line, 149

Sample point, 356
Sample space, 356
Scientific notation, 79, 228
Sequence:
definition of a, 311
finite, 311
function, 311
infinite, 311
terms in a, 311
Series:
definition of a, 311
infinite, 311
Set(s):
disjoint, 7
elements of a, 1
empty, 2
equal, 2
finite, 8
infinite, 8
of integers, 4
intersection of, 7
of irrational numbers, 4
members of, 1
of natural numbers, 4
notation, 1, 4
null, 2
of rational numbers, 4
of real numbers, 4
solution, 103
subset of, 3
union of, 7
Sigma notation, 312
Slope of a line, 149
Slope-intercept form, 153
Solution set:
of an equation, 103
of a system, 270
Square root, 75, 85
Subset, 3
Substitution:
axiom, 27
solution by, 287
Subtraction of real numbers, 31
Sums:
of an arithmetic progression, 313
of complex numbers, 327
of a finite geometric progression, 316
of an infinite geometric progression, 317
Summation notation, 310
Symmetric law, 27
System(s):
of equations, 270
solution set of a, 270

Term(s):
of an expression, 41
like, 41
of a sequence, 311
Theorem, definition of, 29
Transitive law, 27
Tree diagram, 337
Trichotomy axiom, 252
Trinomial(s):
definition of, 41
perfect square, 119

Union of sets, 7

Variable(s):
definition of, 2
dependent, 136
independent, 136
Variation:
direct, 192
inverse, 193
joint, 194
Vertex of, a parabola, 166

Zero:
exponent, 73
of a function, 140
multiplication by, 30